PHYSICAL GEOLOGY

Second Edition

Robert J. Foster

San Jose State University

CHARLES E. MERRILL PUBLISHING COMPANY
A Bell & Howell Company • Columbus, Ohio

Published by
CHARLES E. MERRILL PUBLISHING COMPANY
A Bell & Howell Company • Columbus, Ohio

IBSN: 0–675–08776–7

Library of Congress Catalog Card Number: 74–77506

Cover photograph courtesy of H. Armstrong Roberts

3 4 5 6 7 8 9 — 80 79 78 77 76

Preface

This revision was undertaken in an effort to make this book more interesting and more useful to the student. Most of the features of the first edition have been retained.

Many chapters have been updated or rewritten. Many of these changes were dictated by the revolution in geologic thinking brought about by recent advances in the theories of sea-floor spreading and continental drift.

To make the book more interesting and more timely, geological aspects of environment have been added in every chapter. This will help the student see better the many ways that geology affects our lives. If today's students understand the environment, they may be able to make our earth a better place on which to live. Brief historical summaries have been added to many chapters to help the general education student see the development of science as part of the overall intellectual development of man and to help the geology major better understand the development of geologic concepts. Hopefully, the addition of environment and history will help to humanize geology. A summary of principles has been added at the end of each chapter to help students to focus their study.

As in the first edition, this revision begins with minerals and rocks so that they can be introduced in the early laboratory periods if the course has a required laboratory. Most laboratory courses begin with these topics, and this commonly forces the instructor to use the chapters in the text in a different sequence from what the author intended. In some courses this forces an undue hardship on the student. In many laboratory courses the emphasis is on identification and naming. To these should be added the interpretation of rocks as stressed in this text.

The Supplementary Reading listed at the end of each chapter stresses *Scientific American* articles, especially those available separately, and books available in paperback editions. Such readily available, inexpensive materials may encourage more outside reading. The end-of-chapter questions are mainly simple, factual questions that will help the student to review and test his retention. Some, however, are thought-provoking questions that may help him to see beyond the text. In a few cases some points not specifically covered in the text are covered by the questions.

Many new illustrations, especially photographs, have been added. This increase in the number of examples will help to make the principles clearer and the book more interesting. I am grateful for the generous response of many people to my requests for the photographs.

Contents

PHYSICAL GEOLOGY

Second Edition

Planet 3– An Introduction

1

Why Study the Earth?

Geology is the study of the earth. That we live on the earth is reason enough to study it. The more that we know about our planet, especially its environment and resources, the better we can understand, use, and appreciate it. To man the earth is the most important body in the universe.

In the broader view, however, the importance of the earth shrinks. At least three other stars, and possibly as many as eight, are thought to have planets. Even with our largest telescopes we cannot see these other planets, but we infer their existence from the wavering motions of the stars. The earth is a medium-sized planet orbiting a medium-sized star. Thus the earth appears to be an average planet. However, at least in the solar system, the earth is unique in having abundant water, exemplified by the oceans, and an atmosphere that can support life. The earth's surface temperature, controlled by its distance from the sun, makes these features possible, and these features, in turn, make life possible on the earth. The development and history of life are important aspects of historical geology. The space program has also revealed that the earth is unique among the planets so far studied in having a magnetic field. The earth's magnetic field, for reasons that will be discussed later, is believed to be caused by a liquid iron core that may also store the energy which causes the formation of such surface features as mountain ranges. These features, which, as far as we know, are unique to the earth, are central to the processes of erosion and deformation of the earth's surface that are the main aspects of physical geology. Thus geology is in large part the study of the consequences of the earth's unique features.

Geology has contributed a great deal to civilization both intellectually and economically. Among the great concepts gained from geological studies are an understanding of the great age of the earth, and the development of an absolute time scale. Geology differs from most other sciences in that it is concerned with absolute time. Time appears in the equations of physics and chemistry, but these sciences are generally concerned with rates of change, and the time is relative, not absolute. Geologic time extends back almost 5000 million years to when the earth formed. Thus geology is concerned with immense lengths of time when measured against human experience. It is difficult to comprehend the lengths of time involved in geologic processes, but this must be done to appreciate geology fully. (See Fig. 1-1.)

The fundamental principle that underlies most of geology is simply that the present processes occurring on the earth have occurred throughout geologic time. Thus ancient rocks can be interpreted in terms of

A group photo showing members of Hayden Survey taken at Red Buttes, Wyoming, August 24, 1870. In about a ten-year period, the Hayden Survey explored most of the Rocky Mountains geologically. The early geologic surveys of the west resulted in many concepts that helped the development of the science of geology. This scene contrasts sharply with today's space program, which is one of today's geologic frontiers. Photo by W. H. Jackson, U. S. Geological and Geographical Survey of the Territories.

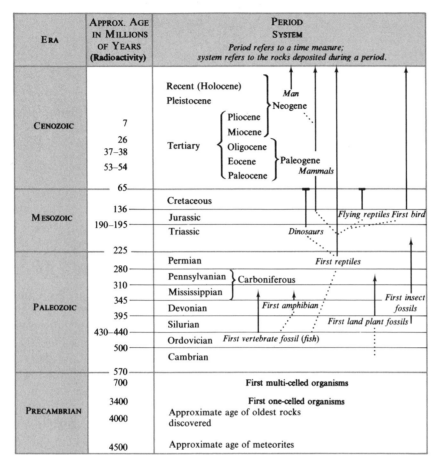

ERA	APPROX. AGE IN MILLIONS OF YEARS (Radioactivity)	PERIOD SYSTEM *Period refers to a time measure; system refers to the rocks deposited during a period.*
CENOZOIC	7 26 37–38 53–54 65	Recent (Holocene) Pleistocene } Neogene *Man* Pliocene Miocene } Tertiary Oligocene Eocene } Paleogene Paleocene *Mammals*
MESOZOIC	136 190–195 225	Cretaceous Jurassic *Flying reptiles First bird* Triassic *Dinosaurs*
PALEOZOIC	280 310 345 395 430–440 500 570	Permian *First reptiles* Pennsylvanian } Carboniferous Mississippian } *First insect fossils* Devonian *First amphibian* Silurian *First land plant fossils* Ordovician *First vertebrate fossil (fish)* Cambrian
PRECAMBRIAN	700 3400 4000 4500	**First multi-celled organisms** **First one-celled organisms** Approximate age of oldest rocks discovered Approximate age of meteorites

Figure 1-1. The geologic time scale. Shown to the right above is a very simplified diagram showing the development of life. Not included on the diagram are many types of invertebrate fossils such as clams, brachiopods, corals, sponges, snails, etc., which first appeared in the Cambrian or Ordovician and have continued to the present. This figure is discussed in Chapter 17.

present processes.

Another important point learned from geology is that constant change, both biological and physical, has been and is occurring on the earth.

The origin and development of life is part of geology. It is closely related to the history and development of the earth's surface and thus cannot be separated from the physical history of the earth. Geology shows that, in the broadest sense, all life is related. Biologists share this concern with life, but much of the evidence is geologic in nature. Thus geology and biology overlap in part.

The economic contributions of geology to civilization show that in many ways, too, geology is a very practical science. Geologic knowledge is used to locate and to exploit our mineral resources. Except for water and soil, all mineral resources, such as sand and gravel, petroleum, coal, and metals, are non-renewable. Once mined they are gone, and new deposits must be found. Geologists have discovered the deposits of metal and energy-producing minerals on which our civilization is based.

We take these things for granted now; but a hundred years ago, when the West was opening up and the industrial revolution was occurring, these mineral deposits were being discovered at a rapid rate and geologists were the most influential scientists of the day. At this time, too, the principles of geology were being formulated.

Today the study of mineral resources is of utmost importance, but geologists are also concerned with other economic problems such as urbanization. The development of large cities has resulted in the building of large structures, such as tall buildings and dams. Geology helps in designing foundations for these structures. Examples of both large and small structures that have failed through neglect of simple geologic principles, easily understood by elementary students, are common. (See Fig. 1-2.) Dams fail because they are built near active faults or on porous foundations. During their first rainy season, new freeways are washed out or blocked by landslides. Homes built on hillsides are destroyed by landslides and mudflows. Geologists have also recognized the need for earthquake-resistant structures in some areas and have helped in their design. These and many other geological aspects of our environment will be described.

Scope of Geology

Geology is concerned both with the processes operating in and on the earth, and with the history of the earth including the history of life. In the broadest sense, geology includes the study of the continents, the oceans, the atmosphere, and the earth's magnetic and radiation fields.

Figure 1-2. Landsliding at Point Firmin, California. Geologic studies can prevent losses such as this. Spence air photo.

Clearly, this scope is too broad for any one scientist, so geologists generally, but not exclusively, limit themselves to the solid earth that can be studied directly. Other specialties have developed to study the other aspects of the earth. Geophysicists study the deep parts of the earth and its fields, mainly by indirect methods; oceanographers study the hydrosphere; and meterologists study the atmosphere.

Even with this restriction, geology is a very broad field. Most geologists specialize in one or more facets of geology, much as engineers specialize in various fields of physical science such as electronics or construction. However, geology is even broader than engineering because it encompasses both physical and biological science.

Mention of a few of the specialties in geology will illustrate. Those who study minerals and rocks need specialized training in chemistry and physics, as does the geochemist who is concerned with chemical processes in the earth. Those who study fossils must be trained in biology of plants and animals, both vertebrate and invertebrate. Those who study deformed rocks must know mechanics. Ground water and petroleum geologists must be familiar with hydrodynamics. A complete listing would be very long, but these examples will illustrate the point. All of these specialties overlap somewhat, and, no matter what his specialty, a geologist must be familiar with all facets of geology.

Discovering the Earth's Place in the Universe

The science of geology developed around the end of the eighteenth century, but long before then, the earth's place in the universe was explored. In many early civilizations, the priests kept track of the movements of the sun, moon, and planets relative to the stars. This was of practical value, as well as religious, because it led to the development of calendars and prediction of such events as eclipses. The Greeks went further and developed a system to explain the motions of the heavens. This was the first big step in finding the earth's place in the universe.

The work of the Greeks was summarized in 140 A.D. by Ptolemy. In the Ptolemaic system, the earth was at the center of the universe and all of the other objects in the solar system—the sun, moon, and planets—revolved around the earth. Although not all of the earlier Greeks would have agreed with this system, the great advantage of Ptolemy's work was that it could be used to predict future movements of the sky. Thus, it was of great use in astrology and navigation. Columbus and the early explorers used the Ptolemaic system.

The first important challenge to the idea that the earth and man's place was at the center of the universe came at the Renaissance about 1400 years after Ptolemy. Putting the sun at the center and having the planets revolve around the sun made a much simpler system in the sense that fewer motions are necessary. This system was published posthumously in 1543 by Nicholas Copernicus. Moving man away from the center created a controversy that raged for a hundred years because proof of the earth's daily rotation and yearly revolution came much later. The main advantage of the Copernican system over the Ptolemaic system was that it was simpler. Both had many inaccuracies in prediction. In the early part of the seventeenth century, Johannes Kepler, using data collected by Tycho Brahe, was able to describe the motions of planets and showed that Copernicus was basically correct.

Galileo Galilei, an Italian, was a contemporary of Brahe and Kepler. He worked in many fields, but only parts of his studies of astronomy and mechanics will be mentioned. In 1609, Galileo built one of the first telescopes and used it to observe the heavens. He saw things that no man had ever seen before, and many of his observations could not be explained by the Ptolemaic system. He announced his discoveries in two books but was later forced to recant by the Inquisition. As a result, he had to spend the last years of his life in virtual house arrest. It was during this period that he carried out many of his experiments in mechanics, experiments which later helped Newton.

Isaac Newton, an English physicist, using the work of Kepler and Galileo, discovered the laws of mechanics and of gravity. He worked out these laws about 1664 and published them in 1686. It was then possible to understand why the planets orbit the sun. This discovery profoundly influenced eighteenth century thought, and it was widely believed that almost everything could be explained by Newton's laws. It was not until the early part of the twentieth century that Einstein showed that Newton's laws do not apply in all situations.

Thus by Newton's time, the earth's place in the sun's family was established. The distance to the sun, 93 million miles, was measured in 1672. But where is the sun in the universe? New instruments, especially big telescopes, had to be built before this question could be answered and man's place in the universe known, if indeed it can ever be fully known.

The first step out of the solar system was to measure the distance to some nearby stars. This was not done until 1838 when Bessel measured the very small change in angle to a nearby star, as the earth moves from one side of its orbit to the other. This geometric method can only be used to an angle of 0.008 second of arc which corresponds to a distance of 400 light-years. Beyond this distance the angle is too small to measure accurately. The closest star is about 4 light-years away.

It was not until the early twentieth century that greater distances could be measured. During much of the nineteenth century, spectroscopic analysis was developed and applied to starlight. In 1911, Ejner Hertzsprung in Holland and H. N. Russell in the United States noted that there is a relationship between spectral type and absolute brightness of a star. The absolute brightness can be calculated from the apparent or measured brightness for those stars whose distance is known. It then became possible to calculate the distance from the apparent brightness of any star whose spectral type is known, because from the spectral type, one knows the absolute brightness.

The size of the known universe was growing rapidly, but for the most part, the distances measured were to stars in our galaxy, the Milky Way. (All of the stars seen with the naked eye are within our galaxy, although some more distant objects such as the Andromeda galaxy can be seen without a telescope.) During the nineteenth century, astronomers noted many objects whose light was soft and fuzzy. By 1850, several spiral galaxies had been noted, but their nature was in doubt. In 1908, some stars were resolved in some of these objects, suggesting that they might be vast groups of stars like the Milky Way. It was not until 1925, when the new 100-inch telescope at Mount Wilson became available, that Edwin Hubble was able to report that these objects were composed of many stars. Objects the size of the Milky Way, so distant that they appear to be stars except in the largest telescopes, reveal a universe whose size is beyond comprehension. The diameter of the Milky Way itself is 80,000 light-years.

In 1912, Henrietta Leavitt made a discovery that enabled measurement of the distance to a galaxy. She observed that there is a relationship between the brightness and the period of a type of variable star called a Cepheid. Using this, as well as other methods, the size of the known universe quickly expanded to over a 1000 million light-years. The total number of galaxies may be as great as 100,000 million.

Perhaps the most astonishing thing revealed was Hubble's discovery in 1929 that all of these galaxies are moving away from us — that is, the universe is expanding. The further away a galaxy, the faster it is receding from us. This motion is measured from the *red shift,* a movement of the lines in a galaxy's spectrum toward the red. This change in frequency is similar to the change in pitch of a train whistle as the train moves away from the listener.

Thus, in about 500 years, man's perception of the earth has changed from its being the center of the universe to its being a small planet orbiting an average star in one of perhaps 100,000 million galaxies. (See Fig. 1-3.) Research continues in astronomy, and new, strange objects (qua-

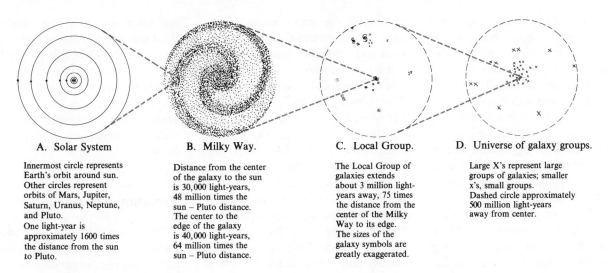

A. Solar System	B. Milky Way.	C. Local Group.	D. Universe of galaxy groups.
Innermost circle represents Earth's orbit around sun. Other circles represent orbits of Mars, Jupiter, Saturn, Uranus, Neptune, and Pluto. One light-year is approximately 1600 times the distance from the sun to Pluto.	Distance from the center of the galaxy to the sun is 30,000 light-years, 48 million times the sun – Pluto distance. The center to the edge of the galaxy is 40,000 light-years, 64 million times the sun – Pluto distance.	The Local Group of galaxies extends about 3 million light-years away, 75 times the distance from the center of the Milky Way to its edge. The sizes of the galaxy symbols are greatly exaggerated.	Large X's represent large groups of galaxies; smaller x's, small groups. Dashed circle approximately 500 million light-years away from center.

Figure 1-3. The earth's place in the universe.

sistellar objects—QSO's) have been discovered. Some of these QSO's emit radiowaves, and some have very large red shifts, suggesting very high velocities and perhaps great distances. These objects, or others yet to be discovered, may again change our ideas on man's and the earth's place in the universe.

Exploring the Earth

Concurrent with finding his place in the universe, man also explored his own planet. The Crusades, which lasted from 1095 to 1291, marked the end of the Dark Ages. In the Crusades, the Europeans discovered that the people of the near east were in many ways more advanced than the Europeans. Marco Polo also described much more advanced peoples in the far east. He returned to Venice about 1300 after a trip, mainly overland, to China and India and wrote a book describing the things he had seen. Although he attracted much interest, he was not widely believed because few people thought that the

Orient could be more advanced than Europe. At this time very little was known of the world outside of Europe, although missionaries had ventured into Africa and Asia. Marco Polo's journey and the Crusades were the first step in awakening Europeans to the world around them.

Some attempts at systematic exploration began during the first half of the fifteenth century. Portuguese captains under the direction of their ruler Henry the Navigator (1394–1460) sailed further and further along the African coast. The age of exploration, however, began with the fall of Constantinople to the Turks in 1453. This severed the trade route to Asia and left Europe with no source of silks or spices. They could do without the silk but they had to have spices. The price of spices rose sharply, adding a commercial incentive to the existing religious and political reasons for finding a route to Asia.

Spices were very important at that time because in the winter Europeans had almost nothing to eat except pickled meat. They had no fodder

crops, so each fall most of the cattle were slaughtered and pickled. Without spices this was a very dreary diet.

The great voyages of exploration began in 1492 with Columbus. He sailed west, hoping to reach the Indies. This was the first time anyone had ventured into the Atlantic far away from the coast. One reason for this reluctance to leave the sight of land was the lack of knowledge of navigation. Columbus used a plumb bob and a quadrant to sight stars and could find his latitude from sighting the North Star, but with his equipment even this was difficult in a wave-tossed ship. He had only an hourglass to measure time, so he was unable to find his longitude with any accuracy. He thought that he had reached the Indies because he believed that the earth was much smaller than it is.

After Columbus came the other great explorers. In 1497 Da Gama sailed around Africa to India. Balboa discovered the Pacific in 1513 and was the first of the Spanish Conquistadors to explore the Americas. In 1519 Magellan sailed around

South America to India. There were many other voyages, but interestingly, it was almost one hundred years later that Australia was discovered by the Dutch. The last of the great voyages was that of Captain Cook who in 1768 was sent to find the mythical southern continent. It was not until the eighteenth century that the chronometer and other navigational methods and instruments became available.

Exploration of the continents lagged somewhat behind that of the oceans. Exploration of the Americas was started almost as soon as they were discovered, but Africa was explored much later. As late as 1725, guidebooks to the Alps described dragons who reportedly lived there. The opening of the Alps is especially interesting because many important geological discoveries were made in those mountains. Until well into the last half of the eighteenth century, very few of the Alps had been climbed although some climbing began around 1500. In 1769 Rousseau wrote *La Nouvelle Héloïse,* a very popular and influential book which glorified nature and encouraged exploration. In 1786 Mont Blanc, the highest peak in the Alps, was climbed, an important milestone. By 1850 about one hundred mountains had been climbed. Between 1854 and 1865 almost all of the peaks in the Alps were conquered.

Although Africa is close to Europe, exploration of its interior by Europeans did not really begin until the late eighteenth century. North Africa is defended by great deserts, and much of the rest is shielded by tropical jungles. Much of the exploration was done by the British, with some by Germans and other nationalities. The objective of most expeditions was to determine the course

of one of the major rivers. These expeditions generally lasted longer than one year, and their stories are gripping examples of privation and heroic personal accomplishment. The culmination was the search for the source of the Nile. Expeditions by Burton and Speke, 1857–1859, Speke and Grant, 1860–1863, and others had narrowed the problem.

David Livingstone, a Scottish doctor and missionary, had spent many years in Africa and was well known throughout the world. No word from him had reached the outside world for five years when, in 1871, the *New York Herald* sent Henry Stanley to look for him. Stanley returned in 1872 after spending several months exploring with Livingstone. Stanley led other expeditions in the 1870's and 80's and was responsible for exploring much of central Africa. He also helped in the colonization and opening up of Africa by building roads.

Thus the physical geography of Africa was determined after the time of the American Civil War. This was

the same time that the American West was being explored and opened up. It was in 1869 that Major John Wesley Powell went down the Colorado River in small boats to find its course. (See Fig. 1-4.) Such geographical exploration must precede or accompany geologic study of an area.

Beginnings of Geology

At first, men were only interested in rocks and minerals that were useful. Later interest expanded to ornamental materials and much later to attempting to understand the origins of the earth and the materials of which it is made.

The first observers of whom we have any record were the early Greeks. They drew logical conclusions from their observations, but many of their discoveries were lost during the Middle Ages. One of the first of these observers was Herodotus (500 B.C.), who discussed the Nile River and saw that

A.   B.

Figure 1-4. A. Major Powell's famous armchair boat. This photo was taken in 1872 on his second trip down the Colorado River. B. Photo of the same place taken in 1968. Note the small amount of erosion in nearly one hundred years. Photo A. by J. K. Hillers, Photo B. by H. G. Stephens, U. S. Geological Survey.

the Nile deposited silt during the floods. Thus he recognized some of the slow, continuous processes that modify the surface of the Earth; but like his contemporaries, he believed that most of the surface features were caused by rapid, violent actions. This is not unreasonable because he lived in an area of many earthquakes, which were then thought to be caused by winds inside the earth.

Aristotle (384–322 B.C.) was the most important of the Greeks. He was Plato's student. He recognized river-deposited material and realized that fossil sea shells from rock strata were similar to those found along the beach, indicating the fossils were once living animals. From this he concluded that the position of land and sea had changed, and so he was the first to recognize the great length of time necessary for these changes. Some of his ideas that persisted through the Middle Ages were that the earth is a sphere because that is the most perfect shape, that the orbits of planets are circular for the same reason, and that a heavy object falls faster than an lighter one.

Theophrastus (374–287 B.C.) was Aristotle's pupil, and he wrote the first book on mineralogy, called *Concerning Stones*. This was a very important book and was the basis of most mineralogies through the Middle Ages and even after. Unfortunately he also suggested that fossils might not have been once living animals and started a controversy that lasted through the Middle Ages.

After the fall of Rome (500 A.D.) the Dark Ages began, and little more is heard about science until the Renaissance. During this time the work of the earlier philosophers and observers was preserved in two ways— by the Church and by the Arabs. The Arab Avicenna or Ibn Sina

(980–1037) wrote *On the Conglutination of Stones*. This work was in part original and in part probably copied from Theophrastus. He also translated much of Aristotle's work. He noted that mountains could result from either of two processes—either from uplift by earthquakes or from erosion. In recognizing that mountains can be uneroded remnants, he was very advanced. The Dark Ages ended about 1100 with the establishment of the Feudal System and the First Crusade.

The Dark Ages were followed by the Middle Ages which lasted until the Renaissance. During the Middle Ages the concept of the Ptolemaic universe with the Earth at the center of the universe was firmly entrenched. The belief that the position of the sun and moon and planets had a great effect on the lives of men was widely held. This idea has persisted to modern times as what we now call astrology. Also most believed that everything in nature had some special meaning or use and certain minerals were believed to have special powers such as curing or protecting from some sickness. It was also believed that either fire or the sun's rays at the center of the earth produced metals. The alchemists believed, following Aristotle, that base metals could be transformed into precious metals. Ores that contained several metals, such as the common association of lead and silver with a little gold, were thought to be the proof of this. Stones were believed to grow in the soil, and even today, anyone who farms or gardens will know how this idea started. The source of water was thought to be underground as the Bible seems to say this.

A rebirth of learning did occur around the middle of the thirteenth

century. About this time the great universities of Europe were formed. Probably an important factor in this rebirth was the final failure of the Crusades, which shook the foundation of European thought. Translation of Arabic books into Latin, which at that time was the universal language of scholars, helped bring about the resurgence of learning. Typical of this period were Dante, St. Francis of Assisi and Marco Polo. The most influential was probably St. Thomas Aquinas (1225–1274), who attempted to weave Aristotle's theories and the Bible into a single system covering all phases of life. As a result of this work, future people were to regard Aristotle's writings as having almost the same stature and authority as the Bible. With the invention of the printing press about 1450, the Renaissance began.

With the Renaissance, men began to observe again as the early Greeks had many years before. The medieval scholars, following Aristotle's methods, had used only logic and philosophy as their tools. Man now began to rediscover the world around him. This was the age of exploration which culminated with Columbus and Magellan.

Geologic Methods

Geology is based mainly on observations and seeks to determine the history of the earth by explaining these observations logically, using other sciences such as physics, chemistry, and biology. Only a small part of geology can be approached experimentally. For example, although the important use of fossils to date or establish contemporaneity of rock strata is based on the simple, basic principle that life has changed during the history of the earth, this principle could not be established experimentally; it was the result of careful observations and analyses over a long period of time by many people of varied backgrounds.

Geologic problems are many, diverse, and complex; almost all must be approached indirectly, and in some cases, different approaches to the same problem lead to conflicting theories. It is generally difficult to test a theory rigorously for several reasons. The scale of most problems prevents laboratory study; that is, one cannot bring a volcano into the laboratory, although some facets of volcanoes can be studied indoors. It is also difficult to simulate geologic time in an experiment. All of this means that geology lacks exactness and that our ideas change as new data become available. This is not a basic weakness of geology as a science, but means only that much more remains to be discovered; this is a measure of the challenge of geology.

Reasoning ability and a broad background in all branches of science are the main tools of the geologist. The geologist uses the method of multiple working hypotheses to test his theories and to attempt to arrive at the best-reasoned theory. This thought process requires as many hypotheses as possible and the ability to devise ways to test each one. Not always is it possible to arrive at a unique solution—but this is the goal. In the sense that the geologist uses observation, attention to details, and reasoning, his methods are similar to those of fictional detectives.

The most important method used by the geologist is to plot on maps the locations of the rock types exposed at the earth's surface. The rocks are plotted according to their

Figure 1-5. Block diagram showing a geologic map with cross-sections on the sides. A geologic map shows the distribution of rock types on the surface. From such maps, the history of an area is interpreted.

type and age on most maps. (See Fig. 1-5.) From such maps it is possible to interpret the history of the area. The early geologists had to make their own maps, and work in a remote area was very difficult in many cases. Now, excellent maps produced from aerial photographs are available for most areas. Much geologic mapping is done directly on

either black and white or color aerial photographs, which have proved to be unexcelled for accurate location of rock units. In addition, the outcrop pattern of the rock units generally shows well on air photos. (See Fig. 1-6.) Radar images that show the surface beneath thick forest cover now extend the use of aerial photographs into such areas. (See Fig. 1-7.) Helicopters and jeeps have largely, but not entirely, replaced pack animals and back packing. (See Fig. 1-8.) However, in spite of all of these advances that have accelerated geologic mapping, much of the earth is not mapped geologically or at best has been mapped only in reconnaissance fashion.

Recent advances have extended classical geologic mapping by providing new things to map. As an example, it is possible to obtain, with some complex instrumentation, aerial views of areas in infrared or heat

Figure 1-6. Vertical aerial photograph of folded rocks in Wyoming. Erosion has etched out the differences in resistance of the various rock layers. Such a photo is essentially a geologic map of the area. Photo from U.S. Geological Survey.

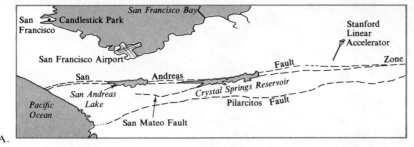

A.

Figure 1-7. Radar images. A. San Francisco Peninsula. Unlike conventional aerial photos, radar images can be obtained in cloudy weather or even at night. The radar penetrates the vegetation and reveals the actual surface. The bottom (west) part of this area has thick redwood forests. From U.S. Geological Survey in cooperation with NASA and Westinghouse Electric Co.

wavelengths rather than visible light. These are used in the study of active volcanic areas and ground water studies as well as other types of studies. (See Fig. 1-9.) Other examples of the types of data now mapped include the earth's magnetic field, radioactivity, heat flow, seismic properties, and many more. Mapping this new data is giving geologists new insights in understanding the earth.

The other more obvious method used extensively by the geologist is the detailed study of rocks themselves. (See Figs. 1-10 and 1-11.) This can be done at several levels, and the first chapters of this book are largely concerned with these topics. The first methods applied were optical microscopy and chemical analysis, and these are both still in wide use. Newer methods such as the electron microscope; X-ray analysis; optical, infrared, and X-ray spectros-

copy; and the electron microprobe now permit very detailed studies of rocks.

Recent Trends

Recent discoveries have expanded the interest of the geologist far beyond his classical realm, the rocks of the earth's surface. In addition, geology is becoming more quantitative, although it is still qualitative when compared to a science such as physics. The early geologists soon understood in a general way the surface processes such as erosion. A geologic time scale based on fossils was also soon developed, and this was used to decipher the history of the continents. In the early part of the twentieth century, radioactive dating enabled the geologist to work with the very old rocks, and the problem

of the origin and early history of the earth began to take form. Since World War II oceanography has developed rapidly, and the features of the ocean bottoms have been mapped. (See Figs. 1-12 and 1-13) These oceanographic discoveries, together with the study of magnetism in ancient rocks, resulted in a new theory explaining the structures of the earth's surface. One aspect of this theory is that the continents are not fixed on the surface, but have moved during geologic time. (See Fig. 1-14.)

Continental drift had been suggested in the past; but in the light of these recent discoveries it is now widely accepted and the change in geological thinking has been nothing short of revolutionary. This new theory has brought many of the concepts of geology into a single unified theory. Chapter 16 presents this theory and will hopefully also serve to summarize much of what we know about the earth.

At present, the space programs are giving us information about other parts of the solar system. (See Fig. 1-15.) Most astronomers believe that all of the solar system formed at the same time, so this new information gives much help to the geologist. This has given rise to a new science of "geology" of planets, termed astrogeology, or perhaps better, planetology. The origin of the solar system is, of course, a classic problem of astronomy, but is also part of geology. These advances, plus the sharpening by geophysicists of the tools that probe the depths of the earth, have greatly expanded the viewpoint of the geologist. At the present time geology, astronomy, oceanography, geophysics, and meteorology are coming closer together; and it is this

B.

Figure 1-7. Radar images. (cont.) B. Appalachian Mountains in western Virginia. The structures of these folded rocks are well shown. Courtesy of Raytheon Company, Autometric Facility.

Figure 1-8. Mapping on the northwest coast of Alaska. The helicopter is used to transport geologists to this remote area and for movement about this trackless region. Photo by Standard Oil Co. of California.

combination of sciences that must be used together to study the whole earth.

This Book

This book is about the solid rock earth, especially its history and the processes that formed its surface features. In this sense it covers classical geology, but in order to understand these features, it will be necessary to consider evidence from a wide range of fields. Thus, the viewpoint will be wide, and many of the topics would not have appeared in a survey of geology only a few years ago. This broadness, however, makes modern geology a more valuable course for the non-science major.

We begin our study by considering the crust—the outer few miles—

Figure 1-9. Infrared image and aerial photograph of the same area. A. Infrared image shows warm areas in light tones and cool areas in dark tones. The curving light line is a blacktop road, and the straight lines are fences separating fields with different kinds of plant cover. The light pattern near the bottom is landslides, and the dark areas are poorly drained areas between the landslides.

Figure 1-9. (cont.) B. Aerial photograph shows landslides at the bottom and hills near the top. The landslides moved from the bottom toward the top. From an unpublished report to NASA by U.S. Geological Survey, by R. E. Wallace and R. M. Moxham with J. Vedder.

Figure 1-10. A geologist and some of his tools. The gravity meter measures the earth's gravity. The plane table, alidade, and stadia rod are used in surveying. Cores from wells, if available, reveal the rocks at depth. The geophones are used in seismic studies of the buried rocks. The Brunton compass and clinometer is used to measure the attitude of layered rocks. Photo by Standard Oil Co. of California.

Figure 1-11. This geologist is examining a sample of a sedimentary rock that has abundant fossil fragments. The hand lens is used in the field and is commonly followed by microscopic, X-ray, or many other more detailed methods in the laboratory. Photo by Standard Oil Co. of California.

of the earth. We study, first, the minerals and rocks of the crust, because they tell us much about the composition of the crust and the processes that take place within the crust as well as on its surface. Then we consider erosion—the dominant surface process. Next, we consider the interactions of the processes in the crust and in the deeper parts of the earth and learn how the present surface features of the earth formed. This leads us to a model of the entire earth. Finally, the methods of dating and deciphering the earth's history are described.

Figure 1-12. U.S. Geological Survey research vessel *Don J. Miller* at Prince William Sound, Alaska. Photo by W. R. Hansen, U.S. Geological Survey.

Figure 1-13. A two-man submersible research vessel. Note the extended sampling arm. Photo courtesy of Noel P. James, University of Miami.

Figure 1-14. Movement of continents. A. Late Paleozoic time, 250 million years ago. B. End of Jurassic Period, 136 million years ago. C. End of Cretaceous Period, 65 million years ago, is shown dotted and present positions are shown by the solid lines. Based on R. S. Dietz and J. C. Holden, 1970.

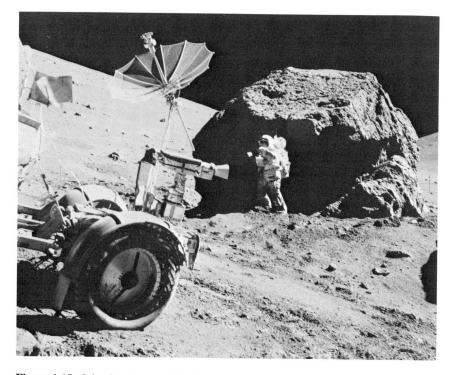

Figure 1-15. Scientist-astronaut Harrison Schmitt beside a huge boulder on the moon. Photo from NASA.

Supplementary Reading

Adams, F. D., *The Birth and Development of the Geological Sciences.* New York: Dover Publications, Inc., 1954, 506 pp. (paperback). Originally published in 1938.

Albritton, C. C. (ed)., *The Fabric of Geology.* Reading, Mass.: Addison-Wesley Publishing Co., Inc., 1963, 372 pp.

American Geological Institute, *Geology: Science and Profession.* Washington, D.C.: American Geological Institute, 1965, 28 pp.

Chamberlin, T. C., "The Method of Multiple Working Hypotheses," *Journal of Geology,* 1897, Vol. 5, pp. 837-848.

Geikie, Archibald, *The Founders of Geology.* New York: Dover Publications, Inc., 1962, 486 pp. (paperback). Originally published in 1905 by Macmillan and Co.

Gillispie, C. C., *Genesis and Geology.* New York: Harper and Row, 1951, 306 pp. (paperback).

Mather, K. F., and S. L. Mason, *Source Book in Geology.* New York: McGraw-Hill, 1939, 702 pp.

Roy, C. J., *The Sphere of the Geological Scientist.* Washington, D.C.: American Geological Institute, 1962, 28 pp.

Atoms,
Elements, and
Minerals

2

Introduction

Mineralogy, which is now only a branch of geology, was probably the parent of geology. The first geological books were mineralogies, and they bear very little resemblance to modern mineralogy books. They made no distinction between minerals and fossils and included myths and folklore associated with the minerals.

Mineralogy came out of the Middle Ages with the writings of Georg Bauer (1494–1555) who, following the custom of the time, wrote in Latin under the Latinized name of Agricola. Like many of the early people in geology, he was a physician, as this was the only scientific training available. He was a Saxon German and lived in Joachimstal, near the famous silver mines. His hobby was the study of minerals and fossils, and he wrote several books on these subjects. His most important book was *De Re Metallica* which was published in 1556. This was the first book on mining and metallurgy. It was very popular and was one of the first books translated into local languages that could be read by the common people. This book was published at the right time, about 100 years after the invention of printing, enabling it to be widely circulated; and it was on a practical subject that was well received. It is interesting to note that this took place during the age of such men as Da Vinci, Copernicus, and Luther.

Another of Agricola's books, *De Natura Fossilum,* was on minerals and fossils. It was published in 1546 and classified minerals on the basis of physical properties.

Nicolas Steno (1638–1687) was about one hundred years ahead of his time in understanding geology. He will be introduced here for his work in mineralogy, and we will come back to him later when stratigraphy is considered. He was born a Dane, Niels Steensen, but he also

Calcite crystals from Iron County, Missouri. Photograph by and courtesy of Nathaniel Ludlum, Department of Mineralogy, The Ohio State University.

Latinized his name because he wrote in Latin. In 1668 he published a book in which he recognized the true nature of fossils. He also noted from his study of quartz crystals that the angles between the faces are always the same. This was the beginning of the science of crystallography, one of the off-shoots of mineralogy.

Abraham Gottlob Werner (1749–1817) did much for mineralogy but probably held geology back for 100 years. He studied at Freiberg from 1769–71 and then went on to the University of Leipzig where he wrote a mineralogy book, *Von der äusserlichen Kennzeichen der Fossilien,* in 1774. On the basis of this book, he was invited back to Freiberg the same year and became a Professor at the age of 25. He was the chief proponent of the so-called mixed system of mineralogy as the most practical at the time. He knew that this was not the best system and strived for the ideal system based on the true composition. He constantly improved his system during the 40 years that he stayed at the Academy at Freiberg. In his last revision in 1817, he recognized 317 minerals in classes such as earthy minerals, saline minerals, combustible minerals, and metallic minerals.

Werner also founded the Neptunist theory of the origin of the Earth. This latter work was not at all in the same class with his mineralogy. This theory, which will be considered in Chapter 3, was based on his idea that all rocks were precipitates from a universal ocean. Although Werner wrote very little, even when his theory of the earth was under attack, he was extremely popular, and his fame, based in part on his mineralogy, was spread far and wide by his students.

During Werner's lifetime, further advances were made in crystallography. Abbé René-Just Haüy, (1743 – 1822) a French botanist, became interested in mineralogy after hearing a lecture at the Royal Gardens in Paris. The consistency of form of plants and minerals intrigued him. After accidentally dropping a calcite crystal, he noted that no matter what the shape of the original crystal, the cleavage fragments all have the same (rhombohedral) shape. From this he deduced that all minerals were made of tiny building blocks, thus beginning our modern ideas of unit cells. He also developed the six crystal systems.

Mineralogy developed very rapidly in the nineteenth century. Crystallography advanced with the development of instruments to measure accurately the angles between crystal faces. The most important advance, however, was the development of chemical analysis so that the composition of minerals could be determined.

The new mineralogy based on chemistry came from the great Swedish chemist Johann Jacob Berzellius (1779 – 1848). He discovered selenium and thorium, and was the first to isolate calcium, barium, strontium, silicon, and zirconium. He also devised the system of writing chemical formulas and equations that we still use. In 1814 he published in Stockholm *A new system of mineralogy.* In this book he recognized silicates, bringing order out of chaos.

With these chemical and crystallographic methods available, mineralogists studied many minerals and much new data were produced. This, in turn, led to the publication of books that described and classified minerals. Two outstanding books of this era that are still in use (in much more modern editions) are a German text by Groth published in Germany and Dana's *System of Mineralogy* published in the United States. James Dwight Dana (1813 – 1895) was a professor at Yale and will be discussed again in structural geology, a field in which he made important advances.

In the twentieth century, the discovery of X rays enabled probing the internal structure of minerals. This was first done in 1912 by Max von Laue and his students W. Friedrich and P. Knipping. Their discovery proved that the atoms in a crystal have an orderly arrangement. Very soon after this, Sir William Bragg and his son, Laurence, used X rays to determine the internal structure of halite (rock salt), and a new era in mineralogy began. Since then X-ray methods have been refined, and the structures of many minerals have been determined.

Abundance of Elements

On casual examination, the surface of the earth appears to be quite complex. Many different-appearing rocks and soils are to be seen. The complexity is much more apparent than real, and close examination reveals a very simple composition. The chemical composition of the crust of the earth will be considered first.

Over 100 elements are known, and, of these, 80 are stable. The crust of the earth is composed almost

Table 2-1. The average composition of continental crustal rocks. After Brian Mason, 1966, and Konrad Krauskopf, 1968. (A few very rare elements and shortlived radioactive elements are omitted.)

Symbol	Name	Weight (per cent)	Symbol	Name	Weight (per cent)	Symbol	Name	Weight (per cent)
O	Oxygen	46.4	Nd	Neodymium	0.0028	Mo	Molybdenum	0.00015
Si	Silicon	28.15	La	Lanthanum	0.0025	Ho	Holmium	0.00015
Al	Aluminum	8.23	Co	Cobalt	0.0025	Eu	Europium	0.00012
Fe	Iron	5.63	Sc	Scandium	0.0022	Tb	Terbium	0.00011
Ca	Calcium	4.15	N	Nitrogen	0.0020	Lu	Lutetium	0.00008
Na	Sodium	2.36	Li	Lithium	0.0020	I	Iodine	0.00005
Mg	Magnesium	2.33	Nb	Niobium	0.0020	Tl	Thallium	0.000045
K	Potassium	2.09	Ga	Gallium	0.0015	Tm	Thulium	0.000025
Ti	Titanium	0.57	Pb	Lead	0.00125	Sb	Antimony	0.00002
P	Phosphorus	0.14	B	Boron	0.0010	Cd	Cadmium	0.00002
Mn	Manganese	0.105	Th	Thorium	0.00096	Bi	Bismuth	0.000017
F	Fluorine	0.095	Sm	Samarium	0.00073	In	Indium	0.00001
Ba	Barium	0.0625	Gd	Gadolinium	0.00073	Hg	Mercury	0.000008
Sr	Strontium	0.0425	Pr	Praseodymium	0.00065	Ag	Silver	0.000007
S	Sulphur	0.0375	Dy	Dysprosium	0.00052	Se	Selenium	0.000005
C	Carbon	0.026	Yb	Ytterbium	0.0003	A(r)	Argon	0.000004
H	Hydrogen	0.020	Hf	Hafnium	0.0003	Pd	Palladium	0.000001
Zr	Zirconium	0.0165	Cs	Cesium	0.0003	Pt	Platinum	0.000001
V	Vanadium	0.0135	Er	Erbium	0.00028	Te	Tellurium	0.000001
Cl	Chlorine	0.013	Be	Beryllium	0.00028	Ru	Ruthenium	0.000001
Cr	Chrome	0.010	U	Uranium	0.00027	Rh	Rhodium	0.0000005
Rb	Rubidium	0.009	Br	Bromine	0.00025	Os	Osmium	0.0000005
Ni	Nickel	0.0075	Ta	Tantalum	0.0002	Au	Gold	0.0000004
Zn	Zinc	0.0070	Sn	Tin	0.0002	He	Helium	0.0000003
Ce	Cerium	0.0067	As	Arsenic	0.00018	Re	Rhenium	0.0000001
Cu	Copper	0.0055	Ge	Germanium	0.00015	Ir	Iridium	0.0000001
Y	Yttrium	0.0033	W	Tungsten	0.00015			

entirely of only eight elements, and all of the other elements make up less than one per cent, as shown in Table 2-1. Thus, in spite of its heterogeneous aspect, the crust of the earth has a very simple composition. Why the crust is composed of these elements will be discussed in later chapters. Notice that many of our common, cheap metals, such as chrome, nickel, zinc, copper, lead, and tin, are really rather rare and form only about 0.04 per cent of the crust. How these elements are concentrated into mineable deposits will be discussed in Chapters 3 and 6.

Before discussing the important implications of these data, we should look at their origin. All abundance data are calculated from estimates of the amounts of the different rock types and their chemical analyses. Obviously, the main problem is estimating the amounts of the rock types; however, geologic maps showing the distribution of rock types have been made for many parts of the earth's surface. In spite of this, such estimates are difficult, and different workers have made different estimates. The amounts shown in the tables are subject to some revision, but they show the general features of the composition of the crust. Estimates of the composition of the whole earth or of the solar system are quite different.

Definitions

So far we have discussed *elements,* which are materials that cannot be subdivided by ordinary chemical methods. The smallest unit of an element is the *atom.* The atom can be broken down into smaller particles such as electrons, neutrons, protons,

Table 2-2. Main elements in the continental crust.

ELEMENT	WEIGHT PER CENT	ATOM PER CENT	VOLUME PER CENT
Oxygen	46.40	62.17	94.05
Silicon	28.15	21.51	.88
Aluminum	8.23	6.54	.48
Iron	5.63	2.16	.48
Calcium	4.15	2.22	1.19
Sodium	2.36	2.20	1.11
Magnesium	2.33	2.05	.32
Potassium	2.09	1.15	1.49
Totals	99.34	100.00	100.00

etc., but except in the case of radio-active elements, this requires the large amounts of energy available in atom-smashing machines such as the cyclotron.

In nature, elements are combined to form *minerals,* which can be defined as

1. naturally occurring, crystal-line,
2. inorganic substances with
3. a definite small range in chemical composition and physical properties.

Note that all three conditions must be met by a mineral.

Minerals are *crystalline* sub-stances; that is, they have an orderly internal structure (arrangement of atoms), as opposed to such things as glasses which are super-cooled liq-uids. A *crystal* is a solid form bound-ed by smooth planes which give an outward manifestation of the orderly internal structure. (See Fig. 2-18.) Note the distinction between crystal and crystalline. Although all minerals are crystalline, they do not neces-sarily occur as geometric crystals.

The minerals group themselves na-turally to form *rocks,* which can be defined as

1. a natural aggregate of one or more minerals, *or*
2. any essential and appreci-able part of the solid por-tion of the earth (or any other part of the solar system).

Note that this definition does not re-quire all rocks to be composed entirely of minerals. Some non-crystalline material is present in many quickly chilled volcanic rocks.

Oxygen and Silicon, the Building Blocks of the Rock-forming Minerals

Because the rocks of the crust are composed predominantly of eight elements, we would suspect at first glance that these eight elements, which must form most minerals, would occur in many different com-binations. However, most rocks are composed of only a few combinations of elements—or minerals. The rea-son for this happy situation, which means that a beginner needs to learn only a few minerals to understand most rocks, can be seen in the volume data in Table 2-2. Rocks are com-posed of over 90 per cent oxygen, by volume, because oxygen atoms (ions) are relatively large. Thus, the number of possible combinations is controlled by the ways in which oxy-gen atoms can be arranged to build minerals. This is especially true be-cause relatively few other atoms are readily available to form minerals. Therefore, although about 2,000 minerals have been discovered, only 20 are common, and fewer than 10 form well over 90 per cent of all rocks.

Thus the key to understanding the structure of minerals lies in how the atoms (more correctly called *ions,* because they have gained or lost one or more electrons) in a mineral are arranged, i.e., how the other atoms are fitted among the oxygen atoms. Clearly, the size of the atoms in question, compared to the size of the oxygen atom, will determine the structure of any mineral. This re-lationship is shown in two dimen-

Small atom fits in space between 3 large atoms

Larger atom fits in space between 4 large atoms

Figure 2-1. The relative sizes of atoms determine how they can fit together. Shown here in two dimensions.

Figure 2-2. Three views of an SiO₄ tetrahedron. The large spheres are oxygen, and the small spheres are silicon.

sions in Fig. 2-1. The number of oxygen ions that surround an element can be calculated from simple geometric relationships.

As can be predicted from the abundance data, most of the rock-forming minerals are composed largely of oxygen and silicon, gen-

erally with aluminum and at least one more of the "big eight." The sizes of oxygen and silicon are such that the small silicon ion fits inside four oxygen ions. This unit, which is the building block of the silicate minerals, is called the *SiO₄ tetrahedron,* because the four oxygen ions can be considered to be situated on the four corners of a tetrahedron with the silicon ion at the center. (See Fig. 2-2.)

The structures of the silicate minerals are determined by the way these SiO₄ tetrahedra are arranged. Two of these tetrahedra, if joined at a corner, can share a single oxygen ion. This forms a very strong bond. With this type of linkage, through sharing of oxygen, the structure of the silicate minerals is formed. The possible structures are shown in Fig. 2-3.

Atomic Structure

The sizes of ions determine the internal geometry of minerals and other crystalline substances; however, we must learn some features of atomic structure to understand how minerals are held together. This study will also provide an understanding of natural radioactivity, which is an important source of energy in the earth and will be discussed in later chapters.

Atoms are composed of smaller particles, and although physicists have discovered many such particles, only a few need be discussed here. Most of the mass of an atom is in its nucleus. The nucleus is composed of protons and neutrons (except in the case of the hydrogen nucleus that is composed of only a single proton). Protons are heavy and have a positive electrical charge. The nucleus is surrounded by a cloud of the much

	FORMULA OF SILICON-OXYGEN UNIT	NUMBER OF OXYGEN SHARED	EXAMPLE
1. Single tetrahedron	(SiO_4)	0	Olivine
2. Double tetrahedron	(Si_2O_7)	1	Epidote
3. Ring	(Si_6O_{18})	2	Tourmaline
4. Chains	(SiO_3)	2	Pyroxene (augite)
5. Double chains	(Si_4O_{11})	2 & 3	Amphibole (hornblende)
6. Sheets	(Si_2O_5)	3	Micas (muscovite, biotite)
7. Three-dimensional networks	(SiO_2)	4	Quartz, feldspars

Figure 2-3. Structures of silicate minerals.

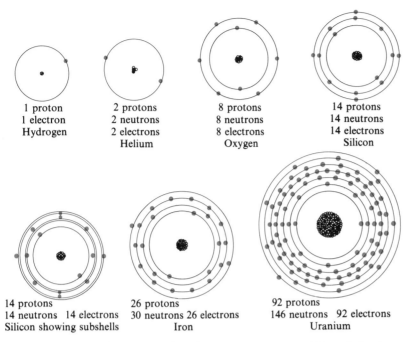

1 proton
1 electron
Hydrogen

2 protons
2 neutrons
2 electrons
Helium

8 protons
8 neutrons
8 electrons
Oxygen

14 protons
14 neutrons
14 electrons
Silicon

14 protons
14 neutrons   14 electrons
Silicon showing subshells

26 protons
30 neutrons 26 electrons
Iron

92 protons
146 neutrons   92 electrons
Uranium

Figure 2-4. Diagrammatic representation of some common atoms. Although the electron shells are shown in only two dimensions here, the electrons actually move in spherical shells. The electron shells are complicated by subshells shown only for silicon. The nuclei contain almost all of the mass of the atoms, but the atomic radii are between 20,000 and 200,000 times the radii of the nuclei.

lighter, negatively charged electrons. (See Fig. 2-4.) An atom is mostly empty space because the diameter of the outer electron orbits, which is the size of the atom and averages nearly 0.00000002 centimeter, is between 20,000 and 200,000 times the diameter of the nucleus. The number of electrons equals the number of protons; hence, an atom is electrically neutral. The number of protons and electrons determines the chemical properties and so determines the identity of an element. The number of protons (or electrons) is called the *atomic number* of an element. In addition, an atom may have neutrons in its nucleus. Neutrons have a mass slightly greater than that of a proton plus an electron, and are electrically neutral. The number of neutrons in the atom of an element may vary, giving rise to *isotopes*.

Changing the number of neutrons in the atom of an element does not change the chemical properties, but it does change the mass of the atom. The *atomic mass number* is the number of neutrons and protons in the nucleus. Most elements have naturally occurring isotopes, as shown in Table 2-3.

Bonding

The bonding of atoms to form minerals is done in several ways, most of which involve the electrons surrounding the nucleus. The electrons are not randomly distributed around the nucleus, but move in distinct orbits. Each orbit or shell contains a fixed number of electrons, as shown diagrammatically in two dimensions in Fig. 2-4. Most of the electron shells have subshells as indicated in

Fig. 2-4 and Table 2-4. Up to seven shells are possible to accommodate the number of electrons necessary to balance the number of protons in the nucleus. As the number of protons in the nucleus increases, the electron shells are filled. The first shell contains 2 electrons when filled, and the second, 8. The third shell can contain up to 18 electrons; the fourth, 32; the fifth, 32; the sixth, 9; and the seventh, 2. The electrons are distributed among the outer shells of any element in such a manner that the outer shell always contains eight or fewer electrons. Larger numbers can be accommodated only in interior shells.

The main principle involved in mineral bonding is to achieve a stable configuration with the outer shell filled. This can be accomplished in different ways. One possibility is for the atom to gain or lose electrons. An example of this is an atom with one electron in its outer shell losing this electron to another atom with seven electrons in its outer shell. The resulting *ions*, as atoms that have gained or lost electrons are called, have opposite charges and so are mutually attracted. (See Fig. 2-5.) This is called *ionic bonding*, and many minerals have ionic bonds.

Another way that the outer shell of an atom can be filled is by two or more atoms sharing electrons. In this way the outer shells of the atoms are filled even though the total number of electrons is not enough to satisfy both atoms individually. This is called *covalent bonding* and is shown in Fig. 2-6. Diamond is one of the few minerals with covalent bonding, and this accounts for the hardness of the diamond. Bonding of the silicate minerals is midway between ionic and covalent.

A third common type of bonding is the *metallic bond*. In a metal, the

Table 2-3. Examples of naturally occurring isotopes.

ELEMENT	SYMBOL*	ATOMIC NUMBER (= NUMBER OF PROTONS)	NUCLEUS		ATOMIC MASS NUMBER (= NUMBER OF PROTONS AND NEUTRONS)	ELECTRONS (= NUMBER OF PROTONS)	REMARKS
			Protons	Neutrons			
Hydrogen	$_1H^1$	1	1	0	1	1	Common form
	$_1H^2$	1	1	1	2	1	Deuterium
Helium	$_2He^4$	2	2	2	4	2	Common form
	$_2He^3$	2	2	1	3	2	
Oxygen	$_8O^{16}$	8	8	8	16	8	Common form
	$_8O^{18}$	8	8	10	18	8	
Carbon	$_6C^{12}$	6	6	6	12	6	Common form
	$_6C^{14}$	6	6	8	14	6	
Potassium	$_{19}K^{39}$	19	19	20	39	19	Common form
	$_{19}K^{40}$	19	19	21	40	19	
Lead	$_{82}Pb^{206}$	82	82	124	206	82	
	$_{82}Pb^{207}$	82	82	125	207	82	
	$_{82}Pb^{208}$	82	82	126	208	82	
Uranium	$_{92}U^{235}$	92	92	143	235	92	
	$_{92}U^{238}$	92	92	146	238	92	

*The subscript is the atomic number and the superscript is the atomic mass number. Sometimes only the superscript is used because only the mass number can vary for a given element.

atoms are closely packed so that the electrons are not owned by any particular nucleus. This allows the electrons to move rather freely through the metal and so accounts for the good thermal and electrical conductivity of metals. In minerals, metallic bonds are found in some sulfides and in native elements, such as gold.

The size of atoms or ions is very important in determining the structure of minerals. Ions that are within about 10 per cent of the same size may mutually substitute in many minerals. Some examples of ion pairs that can mutually substitute are silicon and aluminum, sodium and calcium, iron and magnesium, and gold and silver. See Table 2-4 and Fig. 2-7. The type of bond also affects possible substitution, and although potassium and gold are nearly the same size, they do not substitute for each other. Potassium

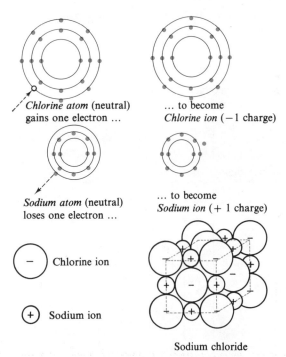

Chlorine atom (neutral) gains one electron ...

... to become Chlorine ion (−1 charge)

Sodium atom (neutral) loses one electron ...

... to become Sodium ion (+ 1 charge)

− Chlorine ion

+ Sodium ion

Sodium chloride

Figure 2-5. Ionic bonding. Ions are formed by the completion of an electron ring by the gain or loss of electrons. The resulting ions are held together (bonded) by the attraction of unlike electric charges.

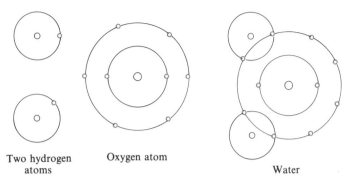

Two hydrogen atoms

Oxygen atom

Water

Figure 2-6. Covalent bonding, shown diagrammatically, results when electrons are shared by more than one atom.

forms ionic bonds, and gold forms covalent bonds with oxygen.

Our knowledge of the internal structure of minerals comes from X-ray studies. X rays have wavelengths in the same size range as atoms, and so are deflected by the atoms in a crystal. This deflection occurs because the atoms in a crystal are packed in an orderly way. If the packing were not orderly, the X rays would be deflected in a random way. Instead, the regular pattern of the deflected X rays can be used to determine the internal structure of a mineral. (See Fig. 2-8.)

The material in the rest of this section is not directly related to mineralogy—the main topic of this chapter. The geological application of the rest of this section appears in later chapters. It is included here because it is directly related to atomic structure.

Natural Radioactivity

Not all nuclei are stable. Some nuclei are radioactive and disintegrate

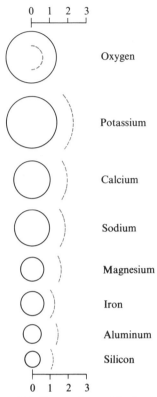

Oxygen

Potassium

Calcium

Sodium

Magnesium

Iron

Aluminum

Silicon

Figure 2-7. Comparison of ionic sizes of some common elements. Their atomic sizes are indicated by the dashed lines. The scale is angstroms (1 angstrom is one ten-millionth of a millimeter).

Table 2-4. Atomic and ionic radii of some elements. Note the general relationship between gain or loss of electrons and the difference between the atomic and ionic radii.

ELEMENT	ATOMIC NUMBER	NUMBER OF ELECTRONS PER SHELL AND SUBSHELL						ION FORMED	NUMBER OF ELECTRONS GAINED OR LOST TO FORM ION	RADII (ANGSTROMS)	
		1st	2nd	3rd	4th	5th	6th			Ionic	Atomic
Oxygen	8	2	2·4					O^{-2}	2 gained	1.40	0.60
Sodium	11	2	2·6	1				Na^+	1 lost	0.97	1.86
Magnesium	12	2	2·6	2				Mg^{+2}	2 lost	0.66	1.60
Aluminum	13	2	2·6	2·1				Al^{+3}	3 lost	0.51	1.43
Silicon	14	2	2·6	2·2				Si^{+4}	4 lost	0.42	1.17
Sulfur	16	2	2·6	2·4				S^{-2}	2 gained	1.85	1.04
Chlorine	17	2	2·6	2·5				Cl^-	1 gained	1.81	1.07
Potassium	19	2	2·6	2·6	1			K^+	1 lost	1.33	2.31
Calcium	20	2	2·6	2·6	2			Ca^{+2}	2 lost	0.99	1.96
Titanium	22	2	2·6	2·6·2	2			Ti^{+3}	3 lost	0.76	1.46
								Ti^{+4}	4 lost	0.68	1.46
Iron	26	2	2·6	2·6·6	2			Fe^{+2}	2 lost	0.74	1.24
								Fe^{+3}	3 lost	0.64	1.24
Silver	47	2	2·6	2·6·10	2·6·10	1		Ag^+	1 lost	1.26	1.44
Gold	79	2	2·6	2·6·10	2·6·10	2·6·9	1	Au^+	1 lost	1.37	1.44

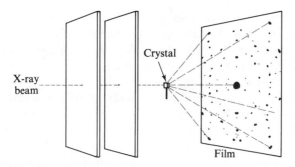

Figure 2-8. A beam of X rays is deflected by the atoms in a crystal into a regular pattern, thus revealing the internal structure of the crystal.

spontaneously, releasing energy in the process. Understanding why and how this takes place was one of the great advances in physics during the first half of the twentieth century. In natural radioactivity, the unstable nucleus emits several types of high energy particles and also releases energy in the form of electromagnetic waves similar to light energy.

The more important particles released in natural radioactivity are called alpha and beta particles. Alpha particles are helium nuclei (ions) and so consist of two protons and two neutrons. They are therefore positively charged. Beta particles are electrons moving at velocities near the speed of light and are negatively charged, as well as much smaller and less massive than alpha particles. The amount of energy released by emission of a particle depends on the mass and the velocity of the particle.

The other type of radiation in natural radioactivity is gamma radiation. Gamma rays are energy in the form of electromagnetic waves. Visible light, radio waves, and X rays are also energy in the form of electromagnetic waves and differ only in their frequency or wavelength. Gamma rays differ from X rays only in frequency; gamma rays have a somewhat higher frequency.

In radioactive decay, an atom is changed to an atom of another element. This means that the number of atoms of a radioactive element decreases with the passage of time. If the rate of disintegration is known, then by measuring the amounts of the parent and daughter elements, the age of crystallization of the mineral containing the parent element can be found. This is the principle of radioactive dating, and works only because the rate of disintegration is not affected by temperature, pressure, etc.

The rate of radioactive decay is expressed in terms of the half-life. In one half-life, one-half of the radioactive atoms present will decay to the daughter element. In the next half-life, one-half of the remaining radioactive atoms will decay, and so on. (See Fig. 2-9.)

The products of radioactive decay can be understood rather easily from the foregoing descriptions of nuclei and radiation particles. Every element has a certain equal number of protons and electrons, and a variable number of neutrons that determines the isotope of that element. If the element decays by alpha emission, it loses the two protons and two neutrons that make up the alpha particle. Thus, the atomic number is reduced by two, and the atomic mass is reduced by four as shown in the following equation:

$$_{62}Sm^{147} \rightarrow {}_{60}Nd^{143} + {}_2\alpha^4$$

Alpha particles can be seen on a luminous dial watch. In a dark room the flash of light that is produced each time an alpha particle hits the zinc sulfide on the dial of the watch can be seen with a magnifying lens.

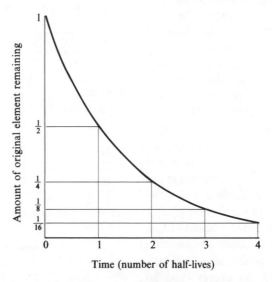

Figure 2-9. Rate of radioactive decay. During each half-life, one-half of the remaining amount of the radioactive element decays to its daughter element.

In beta decay an electron is lost. This can be thought of as occurring because a neutron separates into a proton and an electron; the electron is emitted at high velocity, and the proton is retained by the nucleus. Thus the atomic number is increased by one, and the atomic mass remains the same. The additional proton in the nucleus is balanced by gaining an electron for the outer shell. Free electrons are generally available and can be captured by the atom. In equation form

$$_{37}\text{Rb}^{87} \rightarrow {}_{38}\text{Sr}^{87} + \beta^-$$

The third geologically important type of radioactivity occurs by electron capture. In this case, the nucleus takes an electron from one of the inner shells. This electron unites with a proton to form a neutron. Thus the atomic number is reduced by one, and the mass remains the same, as in

$$_{19}\text{K}^{40} + e^- \rightarrow {}_{18}\text{A}^{40}$$

The only changes in rate of radioactive decay occur in decay by electron capture, in which the rate changes slightly as the abundance of nearby electrons changes.

In most radioactivity, gamma rays are also emitted. Emission of a particle may leave the nucleus in an excited state, and emission of gamma rays can allow the atom to settle down. In any case, mass must be conserved in a radioactive event, and this can explain why gamma rays are emitted. Note in Table 2-5 that the mass of a neutron is slightly greater than the combined masses of a proton and an electron. The question then, using beta decay as an example, is: Where did the excess mass go

when the neutron changed into a proton and an electron? The answer is that the mass was changed into energy in the form of gamma rays. The equivalence of mass and energy was postulated by Einstein, and experiments have so far shown this to be correct. His famous equation is

$$E = mc^2$$

or, in words, energy equals mass times the speed of light squared. However, in beta decay the energy of the gamma rays is less than the amount of mass lost. This led Fermi to postulate the emission of a small neutral particle he called the neutrino. Many years later the existence of neutrinos was proved. The neutrino is only one of a few dozen tiny particles found or postulated by atomic scientists. How they are formed and whether they exist in the nucleus as particles or in the form of energy is not known. With this in mind, it is clear that the descriptions above of how radioactive decay takes place are only convenient fictions to account for the main results.

Radioactivity in Geology

The naturally radioactive isotopes are shown in Table 2-6 and Fig. 2-10. Note that there are three radioactive series that go through many steps before a stable isotope is formed, as well as a number of isotopes that decay to a stable product in a single step. Also some isotopes, such as potassium 40, decay by two processes to two different daughter products. In addition to the three series shown in the figures is the neptunium series; however, the half-life of this series is much shorter than that of the others, so only very small amounts of some of the members of this series are found in nature. Only a few of the natural radioactive isotopes are of geologic importance. Some of these are useful in age de-

Table 2-5. Mass and charge of some particles and atoms. Masses are in atomic mass units (amu) based on oxygen atom = 16.0000 amu (old unit).* Note that the mass of an electron plus a proton is less than the mass of a neutron.

	CHARGE	MASS		CHARGE	MASS
Electron	−1	0.000549	$_8\text{O}^{16}$	0	16.000000 (standard)
Proton	+1	1.007593	$_8\text{O}^{18}$	0	18.004855
Neutron	0	1.008982	$_{82}\text{Pb}^{204}$	0	204.0361
α-particle	+2	4.002675	$_{82}\text{Pb}^{206}$	0	206.0386
$_1\text{H}^1$	0	1.008142	$_{82}\text{Pb}^{207}$	0	207.0403
$_1\text{H}^2$	0	2.014735	$_{82}\text{Pb}^{208}$	0	208.0414
$_1\text{H}^3$	0	3.016997	$_{90}\text{Th}^{232}$	0	232.11034
$_2\text{He}^3$	0	3.016977	$_{92}\text{U}^{235}$	0	235.11704
$_2\text{He}^4$	0	4.003873	$_{92}\text{U}^{238}$	0	238.12493

*1 amu = 1/16 of the weight of an oxygen atom. A molecular weight of oxygen is 16 grams and there are 6.02×10^{23} atoms in a molecular weight (Avogadro's number). Therefore

$$1 \text{ amu} = 1/16 \times \frac{16 \text{ grams}}{6.02 \times 10^{23}} = 1.66 \times 10^{-24} \text{ grams}$$

10^{23} means 1 followed by 23 zeros
10^{-24} means 1/1 followed by 24 zeros

Table 2-6. Summary of naturally occurring radioactive isotopes. A few isotopes with very long half-lives and of very small abundances are not included.

A. SINGLE-STEP DECAY

Isotope	Decay	Half-life (in years unless noted)	Per Cent of Total Amount of Element in Crust
Neutron	$^1n \longrightarrow {}^1H^+ + \beta^-$	12 minutes	—
Tritium ($_1H^3$)	$_1H^3 \longrightarrow He^3 + \beta^-$	12.26	0.0001—0.00001
Carbon 14	$_6C^{14} \longrightarrow {}_7N^{14} + \beta^-$	5660 ± 30	(in biologic material)
Potassium 40	$_{19}K^{40}$ $\xrightarrow{89\%} {}_{20}Ca^{40} + \beta^-$; $\xrightarrow{11\%} {}_{18}A^{40}$ (electron capture)	1.3×10^9	0.0118
Rubidium 87	$_{37}Rb^{87} \longrightarrow {}_{38}Sr^{87} + \beta^-$	4.7×10^{10}	27.85
Indium 115	$_{49}In^{115} \longrightarrow {}_{50}Sn^{115} + \beta^-$	6×10^{14}	95.72
Lanthanum 138	$_{57}La^{138}$ $\xrightarrow{85\%} {}_{58}Ce^{138} + \beta^-$; $\xrightarrow{15\%} {}_{56}Ba^{138}$ (electron capture)	1.1×10^{11}	0.089
Samarium 147	$_{62}Sm^{147} \longrightarrow {}_{60}Nd^{143} + {}_2\alpha^4$	1.06×10^{11}	15.1
Lutetium 176	$_{71}Lu^{176}$ $\xrightarrow{33\%} {}_{72}Hf^{176} + \beta^-$; $\xrightarrow{67\%} {}_{70}Yb^{176}$ (electron capture)	2.1×10^{10}	2.59
Rhenium 187	$_{75}Re^{187} \longrightarrow {}_{76}Os^{187} + \beta^-$	4×10^{10}	62.93

B. MULTIPLE-STEP DECAY

All isotopes with atomic number greater than 83 are radioactive, and all of these that occur naturally are members of the three radioactive families listed below. Some isotopes with atomic number less than 83 also occur in these families. The complete decay of these families is shown in Fig. 2-10.

Isotope	Decay	Half-life (in years unless noted)	Per Cent of Total Amount of Element in Crust
Uranium 235	$_{92}U^{235} \longrightarrow {}_{82}Pb^{207} + 7_2\alpha^4 + 4\beta$	7.1×10^8	0.72
Uranium 238	$_{92}U^{238} \longrightarrow {}_{82}Pb^{206} + 8_2\alpha^4 + 6\beta$	4.5×10^9	99.28
Thorium 232	$_{90}Th^{232} \longrightarrow {}_{82}Pb^{208} + 6_2\alpha^4 + 4\beta$	1.4×10^{10}	100

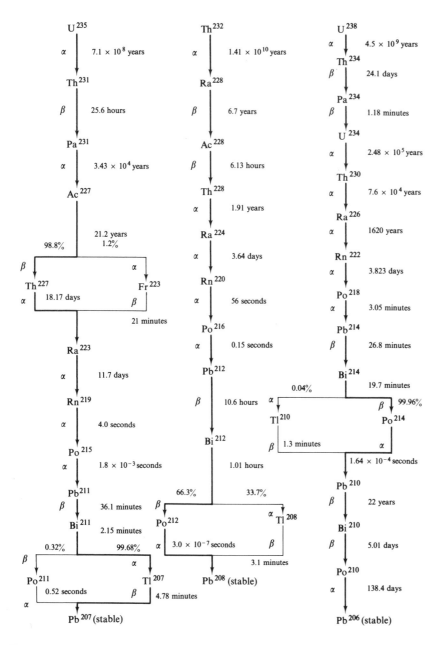

Figure 2-10. Decay schemes for uranium 235, uranium 238, and thorium 232. Type of decay, branching ratios, and half-lives are indicated. This figure is an expansion of Table 2-6B.

containing the parent element. Those most important as a source of radioactive heating can be determined from consideration of abundance data and amount of energy released, both of which are shown in the tables. They are uranium 235, uranium 238, thorium 232, and potassium 40.

Calculation of the amount of energy released by radioactivity is simple. One needs only to know the masses of the parent and daughter elements and the masses of the particles released. (See Table 2-5.) These data and Einstein's famous equation, $E = mc^2$ (energy equals loss of mass multiplied by the speed of light squared), tell the amount of energy released by the decay of one atom of the parent isotope. The rate of decay (half-life) is then used to determine how often an atom of the parent decays. The data for the main heat-producing isotopes are shown in Table 2-7. Notice that because all of these isotopes have been decaying throughout geologic time, more of them were present in the early history of the earth and so produced more heat at that time. Note also that potassium 40 and uranium 235 both have much shorter half-lives than uranium 238 and thorium 232 and so are assumed to have been much more abundant at that time and thus produced much more heat during the early stages of the earth.

In the earth the energy released by radioactivity goes mainly into heat. Each kind of radiation behaves differently. The heavy, slow-moving alpha particles are stopped by a sheet of paper, but their impact is very destructive. They impart some of their kinetic energy to electrons and, being heavy, may disrupt whole atoms. Beta particles are much

termination, and others are sources of radioactive heating of the earth. Those that have proved most useful in dating are carbon 14, potassium 40, rubidium 87, uranium 235, uranium 238, and thorium 232. The requirements for dating are a reasonable rate of decay (half-life), the retention of daughter isotopes, and the existence of common minerals

Table 2-7. Radioactive heat production in rocks. At the time the earth formed 4500 million years ago, the shorter-lived radioactive elements were more abundant and so produced more heat than at present. The units used are 10^{-14} calories per second per gram of rock. From J. W. Winchester, "Terrestrial heat flow, radioactivity, and the chemical composition of the earth's interior," *Journal of Geological Education,* Vol. 14, No. 5, December, 1966, pp. 200-204.

RADIOACTIVE ISOTOPE	GRANITIC ROCKS		BASALT	
	Present	4500 million years ago	Present	4500 million years ago
Uranium 238	10.60	22.2	1.34	2.68
Uranium 235	0.46	37.5	0.06	4.74
Thorium 232	11.47	14.4	1.67	2.10
Potassium 40	3.30	40.1	0.73	8.89
Total	25.83	114.2	3.80	18.41

lighter and faster. They are about one hundred times more penetrating than alpha particles, but do about one hundred times less damage. Because they are too light to affect nuclei, they lose their kinetic energy by interaction with electrons. Gamma rays move with the velocity of light and have no mass. They are about one hundred times more penetrating than beta particles but do much less damage. Like beta rays, gamma rays only disrupt electrons. Heat is molecular motion, so the main effect of radioactivity is to heat the sur-roundings by imparting motion to electrons and atoms. Other effects are damage to a crystal lattice and increase of potential energy by moving electrons from one shell to another. These less important features are discussed below.

It should also be clear that only gamma rays have high energy and can penetrate very far, so most radiation detectors work mainly on gamma rays. Alpha rays are absorbed by about four inches of air, and beta rays by about a yard of air, but gamma rays can travel over a half-mile in air. Less than an inch of rock will absorb alpha and beta radiation, but gamma rays can penetrate over a foot of rock. These points should be taken into account in prospecting for radioactive mineral deposits. Figure 2-11 shows one means of detecting radioactivity.

The crystal structure of minerals containing alpha-emitting elements can be damaged or destroyed by the alpha particles, especially if the mineral has weak bonding. In other cases alpha radiation produces discoloration, and this can cause gray (smoky) quartz, violet fluorite, and blue or yellow halite. Beta particles can also cause discoloration in some minerals. Another common example of alpha damage is the halos or rings that commonly surround tiny crystals of radioactive minerals. These small crystals generally occur as inclusions in later-formed minerals, and dark halos are common in biotite. Uranium and thorium minerals emit alpha particles with various amounts of energy, and because the alpha particles do most of their damage near the end of their movement,

A.

B.

Figure 2-11. The mineral uraninite. A. An ordinary photograph. B. An autoradiograph made by placing photographic film next to the mineral in a darkroom. The radiation from the uraninite exposes the film even though no visible light reaches the film. This can be done without a darkroom by wrapping the film in opaque paper and laying the flat mineral on the film. Exposure times are generally a few weeks. Photo from Ward's Natural Science Establishment, Inc. Rochester, N.Y.

traveling a distance determined by the energy of each particle, they form a series of concentric halos. (See Fig. 2-12.) The amount of damage done to a mineral (the thickness and darkness of the halo) can be used to calculate the age of the mineral, but so far such dating methods have met with only limited success. Uranium 238 also decays by spontaneous fission at a very much slower rate than it does by alpha decay. The high-energy fission particles released in this way do much damage to the enclosing crystal. Under high magnification, the paths of such particles can be seen in some minerals. The path is a small tube, not unlike a bullet hole. (See Fig. 2-13.) By counting the tubes, the number of fission disintegrations that have occurred can be determined. The number of remaining uranium atoms, the source of the particles, can be determined by instrumental analysis, and thus the age of the mineral can be calculated.

Radiation can increase the energy of an atom by moving its electrons into different orbital shells and by trapping within the atom other electrons from beta particles or from other atoms. Remember that each

Figure 2-12. Radiation halo. Such halos are commonly seen in biotite crystals where they surround tiny crystals of radioactive minerals such as uraninite. Typical size is less than one-tenth of a millimeter.

Figure 2-13. Fossil fission tracks. The tubes are caused by high-energy radioactive particles escaping from the mineral at the center. The tubes are much like bullet holes. The longer tubes are about one one-hundredth of a millimeter in length.

electron belongs in a certain shell. This energy can be released by freeing the electrons so that they can escape or move back to the original shell. One way to do this is to excite the atom by heating it. The energy is released in the form of light. This phenomenon can be seen easily by heating a mineral such as fluorite in a frying pan in a dark room. The complete explanation of this phenomenon would require much deeper knowledge of atomic structure. Because the amount of energy stored depends on the length of time a mineral has been exposed to radiation, the age of some minerals can be determined by measuring the amount of stored energy. So far this has not proved to be an accurate method of dating minerals.

Neutrons—Cosmic Rays

To complete our understanding of natural radiation, it is necessary to consider another particle. This particle is the neutron, and it is particularly interesting because of its central role in atomic and nuclear

bombs. The main source of natural neutrons is cosmic rays, and only a few of the possible types of neutron reactions will be considered here.

Primary cosmic rays are nuclei (atoms that have lost their electrons) moving at nearly the speed of light. They are mainly hydrogen and helium, but most other elements are also present in small quantities. That they come from outer space is shown by their increase with height. The number does not change appreciably from night to day, so they do not come from the sun. The atoms are believed to come from exploding stars (supernova), and are accelerated to very high energy levels by the associated magnetic fields.

The primary cosmic rays collide with the atoms of gas that form the earth's atmosphere, producing secondary cosmic rays. These secondary cosmic rays are particles knocked out of the atoms; they consist of neutrons, protons, and other particles. Neutrons outside a nucleus are not stable, but decay, with a half-life of 12.8 minutes, to a proton and a beta particle. Each primary cosmic ray can produce several secondary cosmic rays, and each of the secondaries can produce more cosmic rays, and so on. Each successive generation has less energy; however, primary cosmic rays and certain of the early-formed particles have great penetrating power. Neutrons have great penetrating power because they have no charge, and thus are not deflected by protons or electrons.

Some of the less energetic nuclei of the primary cosmic rays form the Van Allen radiation belts around the earth. Because these nuclei have electric charges, they are deflected by the earth's magnetic field. The interaction is such that some of the less energetic charged particles move in a spiral course around lines of force of the earth's magnetic field. Following such a path, these particles move from pole to pole and back. (See Fig. 2-14.) The more energetic particles are able to penetrate the magnetic field; these are the primary cosmic rays discussed above.

In addition to cosmic rays, there are also particles from the sun that affect the earth. These particles are electrons and protons, and though they are constantly expelled from the sun, at times many more are expelled. These times of intense solar activity correspond with intense sun spot activity and solar flares. Because these charged particles disrupt the earth's magnetic field, these events are called magnetic storms. Such changes, of course, affect the earth's radiation belts, as well as radio communications. These particles also follow the lines of force of the earth's magnetic field and so fall on the earth near the magnetic poles. It is the interaction between these particles and the atoms of oxygen and nitrogen in the upper atmosphere that causes the aurora, or northern and southern lights. The processes may be likened to the effect of an electrical current on neon gas that produces the red light in a neon sign.

Of the cosmic rays, only the neutron is important geologically. A neutron colliding with a nitrogen atom knocks a proton out, converting the nitrogen to carbon 14:

$$_7N^{14} + n^1 \rightarrow {}_6C^{14} + {}_1H^1$$

Carbon 14 is radioactive and decays by beta emission to nitrogen 14:

$$_6C^{14} \rightarrow {}_7N^{14} + \beta^-$$

These reactions are utilized in the carbon 14 method of dating.

Cosmic rays also make possible the dating of meteorites. When cosmic rays hit iron atoms, helium 3 is produced. Thus the amount of helium 3 reveals how long the iron meteorite has undergone cosmic-ray bombardment. This effect was first noticed when attempts were made to date meteorites by measuring the amount of helium produced by alpha decay. More helium than expected was found, and analysis of the helium revealed both helium 4 from alpha activity and helium 3 from cosmic-ray activity. It was also discovered that the helium 3 concentration was greatest near the surface of a meteorite and became less dense in the interior, as would be expected from cosmic-ray activity. Similar effects were noted with other isotopes formed by cosmic-ray bombardment.

Thus ordinary radioactive dating can tell the time of crystallization of a meteorite, and cosmic-ray dating tells when the body broke up to form a meteorite.

The principle involved in these cosmic-ray transformations is used commonly by nuclear physicists. By bombarding atoms with high-speed particles, alpha particles and electrons, as well as neutrons, they have succeeded in making many isotopes. Radioactive isotopes of almost every element have been made artificially.

The energy used in geologic processes ultimately comes from the atom. It is neutrons that cause the splitting or fission of atoms. The large, unstable uranium 235 atom releases neutrons when it undergoes fission, making a continuous, or chain, reaction possible. Natural uranium consists of only 0.7 per cent uranium 235. The 99.3 per cent uranium 238 absorbs neutrons, preventing natural chain reactions. Thus to produce an atomic bomb, uranium 235 must be concentrated in a large enough mass (the critical mass) that the neutrons cannot escape.

The opposite of fission is fusion, the making of a more massive atom from lighter atoms. This is the process that releases energy in the sun and other stars. An example is

$$_1H^1 + {}_1H^1 \rightarrow {}_1H^2$$
$$_1H^2 + {}_1H^1 \rightarrow {}_2He^3$$
$$_2He^3 + {}_2He^3 \rightarrow {}_2He^4 + 2_1H^1$$

or, in summary, the net reaction is

$$4_1H^1 \rightarrow {}_2He^4 + 2e^-$$

These reactions take place at very high temperatures of the order of several thousand million degrees.

Figure 2-14. The earth's magnetic field and radiation belts.

Such temperatures occur in uranium atomic bombs. The hydrogen bomb, which is a fusion device, is made by adding deuterium ($_1H^2$) and tritium ($_1H^3$) to a uranium bomb, producing what is called a thermonuclear device.

Identification of Minerals— Physical Properties

With this knowledge of atoms and their structure, minerals, the natural materials formed by atoms, can be considered.

The definition of mineral suggests that minerals can be identified in several ways. The fact that minerals have reasonably fixed chemical compositions suggests that ordinary methods of chemical analysis can be used, and, historically, this approach was important in the development of both mineralogy and chemistry. The drawbacks to its current use are that a laboratory, as well as much skill and knowledge, is needed, and the relatively recent discovery that certain minerals having identical chemical compositions have very different internal (crystal) structures. In this latter case, the polymorphs (many forms) generally have very different characteristics, such as graphite and diamond, which are both composed of pure carbon. Proof that certain minerals were indeed polymorphs had to wait until the use of X rays to probe crystal structure was perfected. Another problem in the use of chemical analysis is that analysis of the all-important silicate minerals is an extremely difficult task, even in the best-equipped laboratories. Fortunately, this problem can be bypassed by using the easily determined physical properties to identify minerals, and this can be supplemented where necessary with spot chemical tests.

The physical properties include such things as color, luster, hardness, weight (specific gravity), crystal form, fluorescence, taste, solubility, cleavage, magnetism, radioactivity, and many more. The use of the more important physical properties follows.

Hardness

The hardness of a mineral is its resistance to scratching. As might be guessed, it is a difficult property to measure exactly because the amount of force, the shape of the scratches, and the relation of the surface tested to the crystal structure will all affect the measured hardness. In spite of this, the gross hardness is easily measured and is of great help in identification. Because surface alteration may change the apparent hardness, fresh surfaces only must be tested.

The hardness scale (Mohs) follows. (The steps are not of equal value but are arbitrarily defined. The steps are reasonably equally spaced except that the step between corundum and diamond is very large.)

1. Talc (Softest mineral)
2. Gypsum
3. Calcite
4. Fluorite
5. Apatite
6. Orthoclase
7. Quartz
8. Topaz
9. Corundum
10. Diamond (Hardest mineral)

A useful hardness scale is:

2.
—Fingernail
3.
—Copper cent (Use a bright shiny one, or you will only test the hardness of the tarnish.)
4.
5.
—Knife blade or glass plate
6. —File
7. —Quartz
8.

To determine the hardness of a mineral, you must find the softest mineral on the hardness scale that will scratch it. With a little practice, only a knife blade and your fingernail will be necessary to estimate the hardness. Note that two substances of the same hardness can scratch each other. A pitfall that can trap the neophyte is accepting any mark

Two cleavages meeting at right angle

Two cleavages meeting at oblique angle

Perfect cleavage in one direction – Note two cleavage surfaces

Poor cleavage in one direction – Note step effect

Figure 2-15. Cleavage.

A.

B.

C.

D.

Figure 2-16. Minerals showing cleavage. A. Mica crystal with one cleavage. B. Galena, cubic cleavage. C. Calcite. Note the double refraction of the string behind the crystal. D. Serpentine asbestos with fibrous cleavage. Photos from Ward's Natural Science Establishment, Inc., Rochester, N.Y.

on the mineral as a scratch, because just as soft chalk leaves a mark on a hard blackboard, a soft mineral may leave a mark on a hard one. You can check this by trying to remove the mark with your moistened finger, by feeling the mark with your fingernail (even a tiny depression can be detected), and by looking very closely at the mark with a lens. The final test should always be to reverse the test and see whether the mineral can scratch the test mineral.

Cleavage and Fracture

Cleavage and fracture describe the way a mineral breaks. Any irregular break is termed a fracture. Many terms have been coined to describe fracture, but most, such as even, fibrous, splintery or hackly, are self-explanatory. A common type of fracture, called *conchoidal,* is the hollow, rounded type of break that occurs in glass. (See Fig. 3-33.)

A mineral has a cleavage if it has a direction of weakness that, when broken, produces a smooth plane that reflects light (Figs. 2-15 and 2-16). The common minerals may have up to six such directions. Note that a cleavage is a direction; therefore, one cleavage direction will, in general, produce two cleavage surfaces. Cleavages can be recognized by their smoothness and by their tendency to form pairs or steps, as suggested in the sketches. A series of steps is more common in minerals with poor cleavage, that is, minerals in which the cleavage direction is only slightly weaker than any other direction in the mineral. Even in this case, each of the cleavage surfaces will be smooth and parallel and, so, will reflect light as a unit.

The planes of weakness that are the cleavage directions in a mineral are caused by the atomic structure of the mineral. Cleavages form along directions of weak bonding as shown in Fig. 2-17.

Crystal Form

All crystalline substances crystallize in one of six crystal systems (Figs. 2-18, 2-19, and 2-20). If the mineral grows in unrestricted space, it develops the external shape of its crystal form; if it cannot grow its external shape, its crystalline nature can be determined only under the microscope or by X-ray analysis. Many external forms are possible in each of the systems; however, the system can be determined by the symmetry of the crystal. Crystallography is a fascinating subject, and more information can be found in the mineralogy references listed at the end of this chapter.

Figure 2-17. Structure of graphite. The bonds between the carbon sheets are weak, causing the cleavage in graphite. Graphite and diamond are both composed of carbon only; the differences between them are caused by the internal structure.

Figure 2-18. Examples of some of the more common crystal forms. They are shown and named for reference only.

Color

Strictly speaking, color means the color of a fresh, unaltered surface, although in some cases the tarnished or weathered color may help in identification. For some minerals color is diagnostic, but many, such as quartz, may have almost any color, due to slight impurities.

Streak

Streak is the color of the powdered mineral. To see the streak, the mineral is rubbed on a piece of unglazed

A.

B.

C.

D.

Figure 2-19. Mineral specimens showing crystal forms. Parts A., Quartz, and B., Galena, from Ward's Natural Science Establishment, Inc., Rochester, N.Y.; Parts C., Orthoclase (potassium feldspar), and D., Pyrite, are from the Smithsonian Institution.

porcelain called a streak plate. The color of the streak in many minerals is a more constant property than the color. The streak may be a very different color from the mineral color, and this, too, is of help in identification. Care should be taken, especially when working with small disseminated crystals, that only the mineral, and not the matrix, is rubbed on the streak plate. Many powdered minerals are white and are said to have no streak. Minerals with a hardness greater than the streak plate will, of course, have no streak.

Luster

Luster is the way a mineral reflects light. The two main types of luster are metallic and nonmetallic. The distinction is difficult to describe in words; either it looks like metal, or it does not. A few fall in between and are sometimes called sub-metallic. Dozens of terms have been proposed to describe all of the types of luster. A simplified outline of the most commonly used terms follows. Most terms are self-explanatory.

Metallic
  Bright
  Dull
(Sub-metallic) best to avoid if
  possible.
Nonmetallic
  Adamantine (brilliant luster,
    like a diamond)
  Vitreous (glassy)
  Greasy
  Resinous

A.

B.

C.

Figure 2-20. Minerals whose crystalline nature is not readily apparent. In most of the specimens shown, the individual crystals are too small to be seen with the naked eye. A. Agate. B. Hematite. C. Native copper. Part A. is from Ward's Natural Science Establishment, Inc., Rochester, N.Y., Parts B. and C. are from the Smithsonian Institution.

Waxy
Pearly
Silky
Dull or earthy

Specific Gravity

Specific gravity is a measure of the relative weight of a substance. It is the ratio of the mass of a substance to the mass of an equal volume of water. It is measured by weighing the substance in air and in water,

$$\text{Specific gravity} = \frac{\text{weight in air}}{\text{weight in air} - \text{weight in water}}$$

This is a difficult measurement to make with most specimens. Estimations of specific gravity can be made fairly accurately by one who is experienced in handling minerals. It is of help in identification, especially with the very heavy minerals.

The *density* of a substance is its mass per unit volume; when ex-

pressed in grams per cubic centimeter, density is numerically equal to specific gravity.

Taste and Solubility

Taste and solubility can be determined by touching the specimen with the tongue. A soluble mineral will have the feel of a lump of sugar, and an insoluble mineral may feel like glass.

Reaction with dilute hydrochloric

acid is another solubility test. Calcite will react, producing effervescence. Dolomite (see Tables A-1 and A-2 in Appendix A) must be powdered to increase the surface area before effervescence occurs. Dolomite will also react with hot dilute acid or with strong acid, both of which also increase the rate of chemical reaction.

Other physical properties may require special equipment to test. Examples are radioactivity (Geiger counter), magnetism (magnet), fluorescence (ultraviolet light).

Rock-forming Minerals

The main rock-forming minerals are silicates, minerals in which elements such as iron, magnesium, sodium, calcium, and potassium are combined with aluminum, silicon, and oxygen. Their identification will be mentioned here and summarized in Tables A-1 and A-2 in Appendix A. Their structures are shown in Fig. 2-3.

Feldspars

Feldspars are the most abundant minerals. (See Fig. 2-21.) They are composed of sodium, calcium, and potassium, combined with silicon, aluminum, and oxygen. There are two main types, and they are complicated by mixing.

Plagioclase feldspar is an example of a continuous mineral series. The high-temperature plagioclase is *calcic,* $CaAl_2Si_2O_8$; at low temperature *sodic plagioclase,* $NaAlSi_3O_8$, forms. A continuous series of intermediate plagioclase containing both sodium

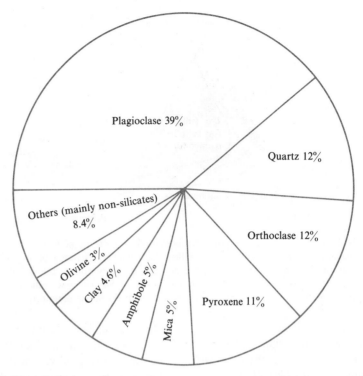

Figure 2-21. Relative abundances of minerals in the earth's crust. Data from Ronov and Yaroshevshy, 1969.

and calcium can form, depending on the temperature and composition of the melt. This is possible because sodium and calcium ions are about the same size. Aluminum and silicon, which are also nearly the same size, also mutually substitute. The microscope must be used to determine the relative amounts of sodium and calcium in plagioclase.

Potassium feldspar, $KAlSi_3O_8$, is the third, more complicated type of feldspar. A number of varieties have been recognized, but their identification requires microscopic or X-ray techniques. The variety that generally forms in granitic rocks is commonly, although generally incorrectly, called *orthoclase;* and the variety in volcanic rocks, which crystallize at higher temperatures than granitic

rocks, is called *sanidine.*

Limited amounts of sodium can be accommodated in potassium feldspar in spite of the differences in ionic sizes. This limited mixing is further complicated in that the temperature of formation further controls the amount of mixing. (See Fig. 2-22.)

At high temperatures, such as in volcanic rocks, all mixtures of sodium and potassium are possible and can be preserved by rapid chilling. Slow cooling will allow the potassium feldspar and the sodic plagioclase to unmix. In some rocks the unmixing is complete and two distinct crystals form; but in many, the two feldspar crystals are intergrown and form *perthite.*

Feldspars are identified by their

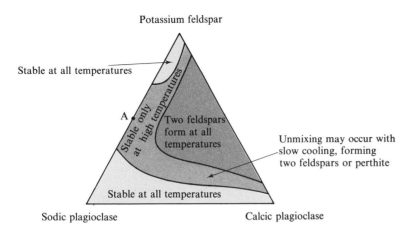

Figure 2-22. This triangular composition diagram shows the possible mixtures of the three feldspars and their temperature stability. This diagram is based on experimental data. Compositions on the left side of the diagram are common in the granitic rocks discussed in the next chapter. A feldspar composed of 50 per cent sodic plagioclase and 50 per cent potassium feldspar would fall on point A on the diagram.

Figure 2-23. Plagioclase striations appear on only one of the two cleavage directions.

light color, hardness (6), and two cleavages at nearly right angles. Striations on one cleavage direction will distinguish plagioclase (Fig. 2-23). Perthite will appear as mottled, irregular veining of different color (Fig. 2-24).

Olivine

Olivine, $(Mg,Fe)_2SiO_4$, is the simplest of the ferromagnesian or dark-colored minerals. The parentheses in the formula mean that magnesium and iron may be present in varying amounts because they are nearly the same size. Magnesium olivine forms at higher temperatures than the iron-rich types. Olivine generally forms small, rounded crystals that, because of alteration, may give deceptively low hardness tests. It can be recognized by its distinctive olive-green color.

Pyroxene

Pyroxene is the family name of a very large and complex group of minerals. Here we will consider only

augite, $(Ca,Na)(Mg,Fe^{II},Fe^{III},Al)$-$(Si,Al)_2O_6$, the main rock-forming mineral of the family. The possible range in composition is suggested by its complex formula. It is generally recognized by its dark color, hardness (5–6), and two cleavages that meet at nearly right angles.

Amphibole

Amphibole is a family name of another very large and complex group of minerals. Here we will consider only *hornblende,* $Ca_2Na(Mg,Fe^{II})_4$-$(Al,Fe^{III},Ti)(Al,Si)_8O_{22}(O,OH)_2$, the main rock-former in the group. Its composition is even more complex than that of augite. It is black, has a hardness of 5–6, and has two cleav-

ages that meet at oblique angles, distinguishing it from augite.

Mica

The micas are easily recognized by their colors and perfect cleavage. The common micas and their formulas are

Biotite—black mica,
$K(Mg,Fe)_3(AlSi_3O_{10})(OH)_2$

Muscovite—white mica,
$KAl_2(AlSi_3O_{10})(OH)_2$

Quartz

Quartz, SiO_2, is generally clear or white but may have any color. It is recognized by its hardness (7) and lack of cleavage.

As in many other minerals, the color of quartz can be caused by very minor amounts of impurities. How this can occur is shown in the following diagrams of crystal defects (Fig. 2-25). In other cases color is caused by radiation as described earlier.

Substitutional impurity   Interstitial impurity   Vacancy

Figure 2-25. Some types of crystal defects shown diagrammatically.

Figure 2-24. Cleavage surface of perthite, showing intergrowth of two types of feldspar. Photo from Smithsonian Institution.

Summary

The crust of the earth is composed almost entirely of 8 elements: oxygen, silicon, aluminum, iron, calcium, sodium, magnesium, and potassium.

Elements are materials that cannot be subdivided by ordinary chemical methods. A little over 100 are known.

Atoms are the smallest units of an element. They are composed of electrons, neutrons and protons.

Minerals are
  1. naturally occurring, crystalline
  2. inorganic substances with
  3. a definite small range in chemical and physical properties.

Crystalline substances have orderly internal structures.

A *crystal* is a solid form bounded by smooth planes which give an outward manifestation of the orderly internal structure.

A *rock* is
  1. a natural aggregate of one or more minerals *or*
  2. any essential and appreciable part of the solid portion of the earth (or any other part of the solar system).

There are about 2000 minerals, only about 20 are common, and fewer than 10 form well over 90 per cent of all rocks.

Ions are atoms that have gained or lost electrons and so have an electrical charge.

Silicon and oxygen, the two most abundant elements, form the SiO_4 *tetrahedron* which is the basic building block of the rock-forming minerals.

The shape of the SiO_4 tetrahedron is determined by the relative sizes of oxygen and silicon ions, and consists of one small silicon ion surrounded by four large oxygen ions.

SiO_4 tetrahedra are linked by sharing oxygen ions, and the resulting arrangements determine the structures of silicate minerals.

The small number of rock-forming minerals is determined by the limited number of ways that SiO_4 tetrahedra can join together and how the other abundant ions can fit into these structures.

The *nucleus* of an atom contains most of the mass of an atom and is composed of protons and neutrons.

Protons are heavy and have a positive electrical charge.

Neutrons have slightly more mass than protons and are electrically neutral.

The nucleus is surrounded by a cloud of light, electrically negative *electrons*.

An atom is mostly empty space.

The number of electrons equals the number of protons, so an atom is electrically neutral.

The number of protons determines the identity of an element and is called the *atomic number* of an element.

The number of protons and neutrons is the *atomic mass number*.

The number of neutrons in the atom of an element may vary, giving rise to *isotopes*.

Ionic bonding is due to the attraction of unlike electrically charged ions.

Covalent bonding results when atoms complete their outer electron shells by sharing electrons.

In *metallic bonding,* the atoms are closely packed and the electrons are free to move throughout the metal.

Ions of similar size can substitute for each other in certain minerals.

In radioactive decay, an atom changes into an atom of another element. In this process, alpha, beta and gamma radiation may be emitted.

Alpha particles are helium nuclei (ions) and so consist of two protons and two neutrons.

Beta particles are high-velocity electrons.

Gamma rays are electromagnetic waves somewhat higher in frequency than X rays.

Half-life is the time required for one half of the remaining radioactive atoms to decay. This rate is unaffected by temperature, pressure, etc., and so radioactivity can be used to determine age.

The energy released in radioactivity is due to conversion of mass to energy and can be calculated from the equation

$$E = mc^2$$

The main radioactive elements used in geologic dating are carbon 14, potassium 40, rubidium 87, uranium 235, uranium 238, and thorium 232.

The main radioactive sources of energy in the earth are uranium 235, uranium 238, thorium 232, and potassium 40.

Cosmic rays are nuclei (atoms that have lost their electrons), mainly hydrogen and helium, moving at nearly the speed of light. They come from outer space. They collide with atoms in the atmosphere and knock out neutrons, protons, and other particles.

The *Van Allen radiation belts* are caused by the interaction of some of the less energetic cosmic rays with the earth's magnetic field.

Neutrons from cosmic rays colliding with nitrogen in the atmosphere produce carbon 14.

Minerals are best identified by their physical properties, such as cleavage, fracture, crystal form, hardness, color, streak, luster, specific gravity, taste, solubility, magnetism, and fluorescence.

Cleavage is a smooth break caused by planes of weakness in a mineral.

Conchoidal fracture is the hollow, rounded type of break that occurs in glass.

Hardness is the resistance of a mineral to scratching.

Streak is the color of the powdered mineral.

Luster is the way a mineral reflects light. The main types are *metallic* and *non-metallic*.

Specific gravity is the ratio of the mass of a substance to the mass of an equal volume of water.

The main rock-forming minerals are:

> *Feldspars*
> > plagioclase
> > > calcic plagioclase $CaAl_2Si_2O_8$
> > > sodic plagioclase $NaAlSi_3O_8$
> > potassium feldspar (orthoclase, sanidine) $KAlSi_3O_8$.

All mixtures of calcic and sodic plagioclase are possible. Limited mixing of potassium feldspar with plagioclase is possible at high temperatures, but only a few of these mixtures are stable at ordinary temperatures. The unmixed feldspar produced by slow cooling of these mixes is called *perthite*.

> *Olivine* $(Mg,Fe)_2SiO_4$
> *Pyroxene*
> > augite $(Ca,Na)(Mg,Fe^{II},Fe^{III},Al)(Si,Al)_2O_6$
> *Amphibole*
> > hornblende $Ca_2Na(Mg,Fe^{II})_4(Al,Fe^{III},Ti)(Al,Si)_8O_{22}(O,OH)_2$
> *Mica*
> > biotite (black mica) $K(Mg,Fe)_3(AlSi_3O_{10})(OH)_2$
> > muscovite (white mica) $KAl_2(Al,Si_3O_{10})(OH)_2$
> *Quartz*, SiO_2

Questions

2-1. The four most abundant elements (weight per cent) in order are _____, _____, _____, and _____.

2-2. Define mineral.

2-3. Distinguish between crystal and crystalline.

2-4. Why are there only eight important rock-forming minerals?

2-5. Can you relate the perfect cleavage in mica to its internal structure?

2-6. How are minerals distinguished from rocks?

2-7. Distinguish between silicon and silicate.

2-8. Why are the physical properties of graphite and diamond so different?

2-9. What factors determine whether one element can substitute for another in a mineral?

2-10. Why are silicon-oxygen tetrahedra the building blocks of most minerals?

2-11. Describe the units of which atoms are made.

2-12. What are ions, and what is ionic bonding?

2-13. What is an isotope?

2-14. Which are the geologically more important radioactive elements?

2-15. Define half-life. Why was radioactive heating more important in the early history of the earth than it is now?

2-16. How is radioactivity used to determine the age of a mineral by decay products? By crystal damage?

Supplementary Reading

General

Akasofu, Syun-Ichi, "The Aurora," *Scientific American* (December, 1965), Vol. 213, No. 6, pp. 54-62.

Ernst, W. G., *Earth Materials*. Englewood Cliffs, N.J.: Prentice-Hall, Inc., 1969, 150 pp. (paperback).

Fleischer, R. L., P. B. Price, and R. M. Walker, "Nuclear Tracks in Solids," *Scientific American* (June, 1969), Vol. 220, No. 6, pp. 30-39.

Hurley, P. M., *How Old Is the Earth?* (Science Study Series). Garden City, New York: Doubleday & Co., Inc., 1959, 153 pp. (paperback).

O'Brien, B. J., "Radiation Belts," *Scientific American* (May, 1963), Vol. 208, No. 5, pp. 84-96.

Rankama, Kalervo, *Isotope Geology*. London: Pergamon Press, Inc., 1954, 477 pp.

Reynolds, J. H., "The Age of the Elements in the Solar System," *Scientific American* (November, 1960), Vol. 203, No. 5, pp. 171-182. Reprint 253, W. H. Freeman and Co., San Francisco.

Minerals

Berry, L. G., and Brian Mason, *Mineralogy*. San Francisco: W. H. Freeman and Co., 1959, 612 pp.

Hurlbut, C. S., Jr., *Dana's Manual of Mineralogy*. New York: John Wiley & Sons, Inc., 1959, 609 pp.

Hurlbut, C. S., Jr., and H. E. Wenden, *Changing Science of Mineralogy*. Boston: D. C. Heath and Co., 1964, 128 pp. (paperback).

Pough, F. H., *A Field Guide to Rocks and Minerals*. Boston: Houghton Mifflin Co., 1955, 349 pp.

Rapp, George, Jr., "Color of Minerals," *ESCP Pamphlet Series,* PS-6, Boston: Houghton Mifflin Co., 1971, 30 pp.

Igneous Rocks and Volcanoes

3

Introduction

The study of igneous rocks developed from the study of mineralogy. It began with a controversy started by Werner who, as noted earlier, contributed much to mineralogy. Werner was a proponent of Neptunism, the theory that rocks were precipitated in layers from a universal ocean. Thus he taught that all rocks are what would now be termed sedimentary rocks. He thought that granite, metamorphic rocks, and volcanic rocks were all precipitates from this universal ocean. The active volcanoes, such as those in Italy, were explained as caused by the burning of coal beds. The eventual downfall of Neptunism came from the proof that basalt, which we now know is the most common volcanic rock, is a volcanic rock and not a chemical precipitate. One reason basalt was thought to be crystallized from a solution was that it commonly forms six-sided prisms that resemble crystals such as those of quartz. The basalt prisms are the result of contraction during cooling (see Fig. 3-4).

Abraham Gottlob Werner (1749–1817) began teaching mineralogy at Freiberg in 1774. He was an extraordinary teacher, and, as a result, students came to him from all over Europe. It was his students who spread his theories far and wide because he published almost nothing. His aversion to writing was so great that in his later years he would not even write letters. His great success in mineralogy was because he was an organizer, and he organized what was known about minerals into a useful classification. His interest in organization was such that he was said by some to buy books so that he could arrange them in his library. Neptunism, in a sense, is an attempt to organize rocks into a simple, easily understood system. Probably this simplicity was the great appeal of Neptunism, for no further study was necessary. A rock's place in the onion-skin layers of the earth could be determined by merely identifying and naming it.

The Neptunists had a number of problems, such as what happened to the water in the universal sea. They had no answer to this question. Another problem was that enough was known about chemistry at the time to know that the volume of rocks assumed to have precipitated was much greater than could have been in solution in the sea. No one seems to have been troubled by this.

The main problem, however, was the origin of basalt. Unknown to the antagonists of the Neptunism controversy, the volcanic origin of basalt had been demonstrated in 1774, which was before Werner had taken up Neptunism. The story begins with Jean Étienne Guettard (1715–1786) who, after visiting Auvergne, an area in central France, recognized the ancient volcanoes there. (See Fig. 3-1.) He published this observation in 1752. In 1770, he published the idea that basalt is not lava; by this he meant that the rocks he had seen

A volcano and lava flows. Pic Tousside, Tibesti Mountains, Chad, Africa. Apollo 9 photo from NASA.

45

at Auvergne were different from the similar rocks with no remaining vestige of volcanoes that were called basalt. Ancient basalt is commonly found with sedimentary rocks, and erosion has removed the more obvious indications of its volcanic origin, such as volcanic cones. Thus, in a sense, Guettard was on both sides of the basalt and Neptunism questions. Nicholas Desmarest (1725–1815) was another Frenchman who also studied the Auvergne volcanoes. His studies began in 1763 and were published in 1774. He clearly showed that the lava of Auvergne is basalt. This should have stopped the controversy before it started, but these studies went unnoticed for years.

The opponents of the Neptunists were called Plutonists because they believed that igneous rocks originated deep in the earth, the home of Pluto, the god of the underground. Plutonism was just one aspect of Uniformitarianism, the idea that present processes occurring on earth can be used to interpret ancient rocks. This idea will be expanded in later chapters. Uniformitarianism was the idea of James Hutton (1726–1797), a Scotsman. He first stated it in 1788 and then published a book *Theory of the Earth, with Proofs and Illustrations* in 1795. The book was written in a difficult style and so was not widely read. After Hutton's death in 1902, his friend Playfair wrote a readable book *Illustrations of the Huttonian Theory*.

The defeat of Neptune did not occur because of the work of the Plutonists as much as it did from the studies of Neptunists. Two of Werner's former students visited Auvergne. The first was Leopold von

Buch (1774–1853), a German, who was one of the most influential geologists of the time and certainly the most traveled. He visited central France in 1802 and recognized the volcanic nature of the basalt there, but he did not concede that Werner's theory was completely wrong. From that point on, he gradually abandoned more and more of the Neptunist ideas as his travels brought more

evidence before him. The other visitor to Auvergne was a Frenchman, Jean Françoise D'Aubuisson de Voissins (1769–1819). He abandoned Neptunism in 1804 after this visit. The journey was suggested by the referees of a paper embracing Neptunism that he had published the year before. This was a dramatic change. Thus Neptunism was challenged while Werner was still alive,

A.

B.

Figure 3-1. The volcanoes of Auvergne, France. A. Chaine des Puys. B. Puy de Pariou in foreground and Puy de Dôme in Chaine des Puys. Photos courtesy Maurice Krafft, Équipe Vulcain.

and soon after his death Plutonism was widely accepted.

One reason that this controversy went on so long was the limited means available to study rocks. The hand lens was the only tool for close examination. It worked reasonably well on rocks with large grains such as granite, but it could give little insight as to the nature of fine-grained rocks such as basalt. Chemistry was developing at this time so some analyses were available to aid the study of fine-grained rocks.

The microscope was available, but its adoption by geologists was slow. Microscopic study is now one of the most important ways that rocks are studied. The technique used is to slice chips of rocks to the thinness of paper so that they are nearly transparent, and to examine these thin sections, as they are called, by passing light through them. The first thin sections were made by William Nicol who also invented the Nicol prism, a device that polarizes light. These thin sections were of petrified wood and were described in 1831. No further use was made of thin sections for about 25 years until Henry Clifton Sorby came to the University of Edinburgh and saw Nicol's thin sections. He clearly saw their value in the microscopic study of rocks. Interestingly, his work was not quickly accepted because many geologists said in effect that mountains cannot be studied under the microscope. In 1862, Sorby chanced to meet the German Ferdinand Zirkel (1838–1912) while on an excursion in Germany. Zirkel's adoption of Sorby's microscopic study of rocks put the Germans in the forefront of this phase of geology, and lead to the general acceptance of microscopic study of rocks.

The only other common way that natural rocks are studied is by chemical analysis. This method is applied mainly to rocks too fine-grained for microscopic study. Chemical studies and classification of rocks probably reached a peak near the turn of the present century. They may again become important because modern methods of instrumental analysis may make chemical analysis of rocks as cheap and easy as microscopic study.

Experimental study of igneous rocks has contributed much to our understanding. Such studies began early in the eighteenth century but

Figure 3-2. The middle of this diagram, showing the relationships among igneous, sedimentary, and metamorphic rocks, is what is generally considered the rock cycle. The upper and lower parts of the figure show how material is added to and subtracted from the rock cycle. The external processes of the upper half of the diagram are discussed in Chapters 7 through 13 of this book, and the internal processes in the lower half are discussed later.

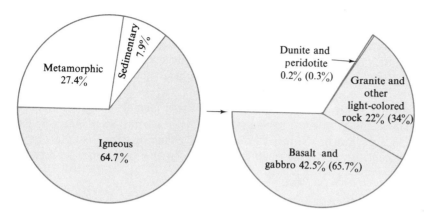

Figure 3-3. Relative abundance of igneous rocks in the earth's crust. For example, basalt and gabbro form 42.5 per cent of all rocks and 65.7 per cent of the igneous rocks. The igneous rock types are described later in this chapter. Data from Ronov and Yaroshevsky, 1969.

did not become important until much later. Fouqué and Michel-Lévy did significant work in France, starting in the 1880's. Perhaps the most important work was that of N. L. Bowen (1887–1956) who worked mainly at the Geophysical Laboratory of the Carnegie Institution in Washington, D. C. His work on basalt melts will be described later in this chapter. Much of this work was done at atmospheric pressure and so may not duplicate conditions within the earth where most rocks crystallize. After World War II, furnaces were developed that could melt rocks under pressure so the later studies of Bowen and his coworkers came much closer to duplicating natural conditions.

Rock Cycle

The rocks of the crust are classified into three types according to their origin. Two of these types are formed by processes deep in the earth and, therefore, tell us something about conditions within the crust. They are

1. *Igneous rocks,* which solidify from a melt or *magma;* and
2. *Metamorphic rocks,* which are rocks that have been changed — generally by high temperature and pressure within the crust.

The third type, which records the conditions at the surface is

3. *Sedimentary rocks,* which are deposited in layers near the earth's surface by water, wind, and ice.

Much of geology is concerned with the interactions among the forces that produce these three rock types. The relationships are quite involved, but can be illustrated by the rock cycle (Fig. 3-2). The relative abundance of igneous rocks is shown in Fig. 3-3.

Igneous rocks are formed by the crystallization of magma. A magma is a natural, hot melt composed of a mutual solution of rock-forming materials (mainly silicates) and some volatiles (mainly steam) that are held in solution by pressure. Magmas may

contain some solid material, but are mobile. Magma probably originates near the bottom of the crust, but its origin is an important problem.

Volcanoes

A great deal can be learned about igneous processes from study of volcanoes. They are, after all, the only direct evidence we have for the existence of magma in the crust.

The active volcanoes today are of several different types, and the differences among them seem to depend on the composition of their magmas, more particularly on the behavior of the volatiles in the magmas. The main volatile in magma is water, which escapes from volcanoes in the form of steam. Carbon dioxide is a common volcanic gas, but the sulfur gases (hydrogen sulfide and the oxides of sulfur), because of their strong odors, are the gases most easily noted near volcanoes. In addition, minor amounts of other gases, such as carbon monoxide, hydrochloric and hydrofluoric acid, ammonia, hydrogen, and hydrocarbons, are also released. Ordinary air also escapes from some volcanoes, especially those with porous rocks. Volcanic emanations are believed to have played a major role in the formation of the oceans and the atmosphere. (See Table 3-1.) Recent eruptions at Hawaii are estimated to release one to two per cent gas during the early stages of eruption and about one-half per cent during later stages. Laboratory studies suggest that, at depth, magmas may contain up to five to eight per cent dissolved water.

A simple analogy will illustrate the

behavior of the water and other gases in volcanic magma. Most magmas are believed to contain at least a few per cent water in solution. The water is dissolved in the magma just as carbon dioxide gas is dissolved in a bottle of soda pop. Most magma is able to dissolve this water because the magma is under considerable pressure due to the weight of the overlying rocks. The gas is dissolved in the soda pop as a result of the pressure inside the bottle. When the bottle is opened, the pressure is reduced and the gas begins to bubble out of the liquid. In the same way, when magma is erupted to the surface, the confining pressure is reduced and the gas, in this case mainly steam, is released.

If the magma has a low viscosity, the gas merely escapes or forms bubbles in the resulting rock. The tops of lava flows can commonly be distinguished from the bottoms by the presence of these bubbles at the top. (See Fig. 3-4.) In some cases, the bubble holes are filled by material deposited by the fluids, or are filled at

Figure 3-4. A section through a lava flow.

Table 3-1. Composition of volcanic gas at Hawaii.

	VOLUME PER CENT
Water	70.75
Carbon dioxide	14.07
Hydrogen	0.33
Nitrogen	5.45
Argon	0.18
Sulfur dioxide	6.40
Sulfur trioxide	1.92
Sulfur	0.10
Chlorine	0.05
	99.65%

After Shepherd, 1921 in Macdonald, G. A., "Physical Properties of Erupting Hawaiian Magmas," *Geological Society of America Bulletin* (August, 1963), Vol. 74, No. 8, pp. 1071-1078.

a later time. The fillings are commonly silica materials such as opal or chalcedony, but many other minerals also occur here. Large filled holes of this kind are actively sought by rock collectors who cut and polish them.

If a fluid magma is erupted very suddenly, it may form a froth of tiny bubbles which quickly solidifies to form the light, glassy rock *pumice*. As a result of the many bubble holes, pumice is light enough to float. The formation of pumice can be likened to the opening of a bottle of warm soda pop or champagne, which results in the production of a froth. A rock with coarser bubbles is called *scoria*.

More violent events occur when more viscous magmas are erupted to the surface and the gas attempts to escape. In this case, the plastic rock is shattered by the pressure of the expanding gas. The gas may expand so rapidly that the rock explodes, blowing the fragments high in the air. The fragments produced in this way are called *volcanic ash* if sandsize or finer, and *volcanic blocks* if larger. The term ash implies burning, but, of course, these rocks do not result from burning. The term was first applied when volcanoes were thought to result from subterranean burning and so is a historical mistake. The rocks formed by this process are called *pyroclastic rocks* (fire-broken rocks). Pyroclastic rocks are named according to the size of the fragments without reference to composition.

Figure 3-5. An outcrop of volcanic breccia in the Thomas Range, Utah. This rock is composed of coarse, angular fragments of volcanic rocks in a fine-grained matrix. The dark fragments are obsidian, and the light are pumice. Photo by M. H. Staatz, U.S. Geological Survey.

Consolidated ash is called *tuff,* and a pyroclastic rock composed of mainly large fragments is called *volcanic breccia.* (See Fig. 3-5.) The main problem with these rocks is distinguishing them from normal sedimentary rocks formed by erosion of a volcanic area. The latter rocks are called volcanic sandstone and volcanic conglomerate. In general, field study is needed to distinguish between these two types of accumulation.

Volcanic bombs are formed by pyroclastic material that is still molten when thrown from the volcano. Again, the name is not really appropriate. Bombs are recognized by their shapes. A liquid assumes a streamlined shape while in flight. Most volcanic bombs are somewhat viscous when thrown from a volcano and solidify during flight. Most spin during flight and so have an elongate shape. (See Fig. 3-6.) Some are still fluid when they hit the ground, and these flatten out.

Volcanic eruptions range from quiet to violent, as will be evident from the following descriptions. The examples used are currently active volcanoes and those that are recent enough that erosion has not greatly modified them. A volcano that has erupted in historical time is generally considered an active volcano; but only at one or two closely studied, well-instrumented volcanoes is it possible to predict eruptions, and even at these, warning comes just before eruption, and the place of eruption is known only generally. Thus, designating active volcanoes is difficult.

The rocks covering the surface of much of the western United States are volcanic, and many of these are tens of millions of years old. This may seem old but is relatively young by geologic standards. Many of the original features of these rocks have been removed by erosion, but they are readily recognizable as volcanic. Less easily recognized are buried volcanic terranes. Such rocks are exposed by uplift and erosion in all parts of the earth, and the age of these rocks may be hundreds of millions of years. Thus study of present volcanoes will help in interpretation of ancient rocks — an important principle in geology.

Present-day volcanoes occur in

A.   B.

Figure 3-6. Volcanic bombs. A. An unusual volcanic bomb 13 feet long. Photo taken in 1902 in southern Idaho by I. C. Russell, U.S. Geological Survey. B. Typical volcanic bombs from Hawaii. Sizes range from under an inch to several feet. Many other shapes occur.

relatively few areas, such as the rim of the Pacific Ocean, the Mediterranean Sea, and the Mid-Atlantic Ridge. These are all geologically active areas and will be discussed in Chapter 15. (See Fig. 3-7.)

It will be convenient to use rock names in the following paragraphs. Dark colored volcanic rock is called basalt; light colored, felsite; and intermediate between the two is andesite. The naming of igneous rocks will be described later in the chapter.

The quietest eruptions are lava flows that come from dikes or dike swarms. The lava is basalt and apparently is very fluid. These quiet flows can produce great thicknesses and cover huge areas. An example is the Columbia River Basalt of southeast Washington and northeast Oregon that covers an area of over 30,000 square miles and is over 5,000 feet thick at places. Thus this unit comprises at least 30,000 cubic miles of basalt. The source of this basalt field was four widely separated dike swarms. The very fluid basalt must have flowed like water because individual flows can be followed for over one hundred miles. (See Fig. 3-8.) Similar basalt fields are found at several places on the earth, such as the Deccan Plateau of India.

At other places similar basalt erupts from central vents, and large *shield volcanoes* form. They are called shield volcanoes because the very fluid magma builds gentle slopes that resemble shields. (See Fig. 3-9.) The Hawaiian Islands are examples of this type of quiet eruption whose most violent displays are fountains of incandescent lava thrown up from lava pools in the craters. (See Figs. 3-10 and 3-11.) Despite the gentle slopes, the Hawaiian Islands rise a total of 28,000 feet from the floor of the Pacific to 13,784 feet above sea level.

The fluid basalts discussed so far are easily recognized by their ropy surfaces. (See Fig. 3-12.) However, even in Hawaii, not all flows are of this type. Some flows are more viscous; their surfaces harden during flowage and are broken into blocks by the traction of the fluid lava a few inches below the crust. These are the blocky lava flows. (See Fig. 3-13.) The terms *blocky* and *ropy*, although somewhat descriptive, are

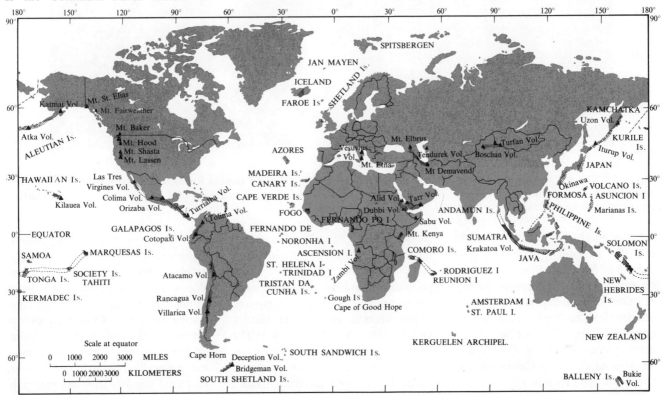

Figure 3-7. Volcanic areas of the world. Comparison with Fig. 16-20 will show the relationship with global structures. From G. A. Waring, U.S. Geological Survey, *Professional Paper* 492, 1965.

Figure 3-8. Oblique aerial view looking east near the confluence of the Snake and Imnaha Rivers on the Oregon-Idaho border. The bedded rocks on the valley sides are basalt flows of the Columbia River Basalt. The lower parts of the valleys are cut in metamorphic rocks that underlie the basalt. Photo from U.S. Geological Survey.

not too good; and many geologists use the Hawaiian words *aa* and *pahoehoe* for blocky and ropy lava respectively. At some places the top of a flow may solidify, and the melted interior flows on, forming a lava tube or tunnel. Examples can be seen at many places, including the Modoc lavas in northeastern California and Craters of the Moon in Idaho.

Cinder cones probably result if the lava is more viscous than that in blocky flows. Pyroclastic material thrown out of the vent builds up a cone around the vent. A central depression around the vent remains at the summit. The cone is composed of tuff, breccia, ash, cinders, blocks, and bombs. (See Fig. 3-14.) The slope is generally near 30 degrees, which is the steepest slope that can be made of loose material. Cinder cones are very common features and, because the loose material is easily eroded, are generally young features. They are rarely more than a few hundred feet high because these slopes are so unstable; such loose slopes are commonly tiring and frustrating to climb. Cinder cones are commonly made of basalt or andesite.

The very high volcanic mountains are generally *composite volcanoes,* so called because they are generally made of a combination of pyroclastic material and lava. Such volcanoes

Figure 3-9. Maps of the island of Hawaii. A. Shows cinder cones, pit craters, rift fissures, and the margins of the five component volcanoes. B. Shows the historic lava flows thru 1950 (color). C. Cross-section of Mauna Loa, a shield volcano, without vertical exaggeration. Parts A. and B. are from C. K. Wentworth and G. A. Macdonald, U.S. Geological Survey, *Bulletin* 994, 1953. Part C. is from K. O. Emery and others, U.S. Geological Survey, *Professional Paper* 260A, 1954.

are generally made of andesite (intermediate composition) and combine explosive eruptions with outpouring of lava. Mt. Shasta and Mt. Rainier (Fig. 3-15) are examples. The lavas cover the pyroclastic material and form the skeleton that enables the mountain to grow large. The lava is more viscous here, and the vents are blocked from time to time by congealing lava. These plugs cause the gas pressure to build up until the pressure exceeds the strength of the plug, and a greater or lesser explosion results. Under these conditions new vents may form on the sides of the volcano. Obviously, as the magma becomes

more viscous, the explosive eruptions become more numerous and/or more destructive. Crater Lake, Oregon, (Fig. 3-16) was formed by the explosive destruction of the top of a composite volcano similar to Mt.

Figure 3-10. Lava fountains 300 feet high on Kilauea, September 23, 1961. Note the trees burning around the active margin of the flow fed by the fountains. Photo by D. H. Richter, U.S. Geological Survey.

A.

B.

Figure 3-11. A. Oblique aerial view of the summit of Kilauea. The craters are believed to be collapse features. The inner crater, Halemaumau, has been the site of eruptions with lava fountaining and, at times, has been partially filled by lava forming a lava lake. This view is toward the north and shows the profile of 13,784 foot Mauna Kea. Photo from U.S. Geological Survey. B. Looking southwest into Halemaumau at the collapse features formed during February 1960. An eruptive phase followed. The deep pit is about 750 feet below the rim. Photo by D. H. Richter, U.S. Geological Survey.

Rainier or Mt. Shasta. Interestingly, the volume of material blown out of Crater Lake is much smaller than the volume of the missing top of the mountain, suggesting that after explosive discharge of the magmatic gases the magma subsided. Most craters and calderas, as large craters are called, are formed by subsidence.

Krakatoa, between Java and

A.

Figure 3-12. Ropy or pahoehoe lava flow near McCartys, New Mexico. The downstream end of the flow is to the right. Photo by H. R. Cornwall, U. S. Geological Survey.

B.

Figure 3-13. Blocky or aa lava. A. Blocky lava flow and cinder cone. B. Steep front of a flow of viscous, blocky volcanic glass (obsidian). Photos from U.S. Forest Service.

Figure 3-14. Aerial view of Paricutin, a large cinder cone that began in a cornfield in 1943 and was active for about nine years. The lava flow spread from vents beneath the cone. Photo from Chicago Natural History Museum.

fluid because of the gas and can move very rapidly down even a gentle slope. When it comes to rest, the hot glass fragments weld together, forming a *welded tuff.* Eruptions of this type are called glowing clouds *(nuées ardentes)* or glowing ava- lanches. It was an eruption of this kind from Vesuvius that buried Pompeii in 79 A.D.

The 1902 eruption of Mt. Pelée was witnessed by a number of people. They reported that the hot cloud traveled at about 100 miles per hour. It killed all but about four of the perhaps 30,000 inhabitants of St. Pierre. The survivors were a prisoner in a dungeon, two people who were near the edge of the cloud, and one who was spared although all around him died. In the harbor all but two ships were sunk. Most of the wit- nesses were from those ships. They reported a sudden darkness and hot dust mixed with water which caused a short intense heat and a feeling of suffocation. Many of these people were badly burned.

Many welded tuffs closely re- semble obsidian (volcanic glass) or

Sumatra, has a somewhat similar history. It was a volcano 2,600 feet above sea level that formed within a ring of islands marking the rim of an older volcano that lost its summit, probably in much the same way as Crater Lake formed. In 1883, the island of Krakatoa erupted ex- plosively, destroying the island. The ash from this great explosion was the cause of beautiful sunsets all over the world long after the eruption.

Lavas ranging in composition from andesite to felsite tend to be viscous, and they cause explosive eruptions. Plugs tend to block the vents, resulting in violent explosions of various kinds. One type has just been described. In 1902, Mt. Pelée on Martinique Island was in eruption, and a large spine formed in the vent, plugging it. This spine was pushed up about 1,500 feet, forming quite a spectacle. Finally, the pressure of the magmatic gases was enough to enable the gas-charged magma to break through. The resulting erup- tion was one of the most destructive

ever witnessed, and led to the recog- nition of a very important type of pyroclastic rock. (Recall the earlier description of the formation of pumice.) At Mt. Pelée, the magmatic gas expanded so violently that in- stead of forming a frothy pumice, the expanding bubbles completely shattered the pumice, producing a mixture of hot gas and small glass fragments. Such a mixture is quite

Figure 3-15. Mt. Rainier, a massive composite volcano in the Cascade Mountains, Washington. Photo from Washington State Department of Commerce and Economic Development.

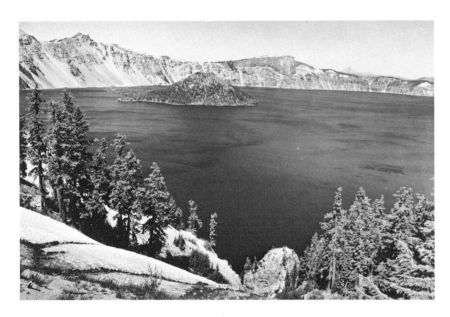

Figure 3-16. Crater Lake, Oregon, fills a depression formed partly by explosive eruptions and mainly by subsidence of the magma. Wizard Island is a small volcanic cone with a crater at its summit. Photo from Oregon State Highway Department.

Underwater volcanic eruptions are common, and lavas sometimes flow into water bodies. This causes rapid chilling of the lava, and causes the lava to subdivide into ellipsoidal bodies. The term *pillow lava* is applied to the resulting rocks. (See Fig. 3-18.)

Erosion may remove the fragmental rocks from a volcano, leaving the lava that congealed in the central conduit as shown in Fig. 3-19.

Volcanic eruptions cannot yet be predicted; however, recent studies have detected the movement of magma at a few very closely studied, well-intrumented volcanoes. The movement is shown by many thousands of shallow, very weak earthquakes. The location of these earthquakes may ultimately reveal the magma plumbing system. It also appears that the magma moves upward to shallow magma chambers just before eruption. The latter hy-

felsite, and only recently has the abundance of welded tuffs been recognized. Their abundance on the surface of the continents may be as great as that of basalt. All welded tuffs are not associated with dome volcanoes. Much of Yellowstone National Park is covered by welded tuffs, and the geysers there probably get their heat from the cooling lava.

Lava plugs or domes, such as the spine at Mt. Pelée, are believed to form during the waning phase of some composite volcanoes. The only recently active volcano in the United States, except those in Alaska and Hawaii, is Mt. Lassen and is of this type. The plug or dome is composed of very viscous lava, and its extrusion is probably very similar to the squeezing of toothpaste from a tube. At Mt. Lassen the plug dome where exposed has steep, smooth cliffs with vertical grooves that developed when the dome was squeezed up. The eruptions of 1914 – 1915 were small; the most violent events were mudflows caused by melting snow. (See Fig. 3-17.)

Figure 3-17. Mount Lassen in eruption. This is the only volcano that has erupted in recent times in conterminous United States. Photo by B. F. Loomis, from National Park Service.

Figure 3-18. Pillow lava at Depoe Bay, Oregon.

pothesis is suggested by the swelling of a volcano just before eruption. This is detected by very sensitive tilt meters. After an eruption the magma apparently returns to the depths. This is indicated by the tilt meters and, at Hawaii, by the draining of the lava lakes that form in the craters there. (See Fig. 3-20.)

Mt. Rainier, Washington, is a volcano with some steam activity in its crater. It is being monitored near the summit by sensors that transmit their data via satellite. An eruption of Mt. Rainier would endanger nearby cities.

Environmental Hazards of Volcanoes

The destruction caused by volcanoes is of two kinds — the direct destruction by material issuing from the volcano (see Fig. 3-21) and the indirect damage caused by mudflows

and floods. The former has already been described in the discussion of explosive volcanoes, such as Mt. Pelée and Krakatoa. There is little defense against such rapid destruction except evacuation at the first indication of activity by the volcanoes. The first eruption may be of the glowing cloud (nuée ardent) type and may occur anywhere on the vol-

cano. In such cases there may be no escape. In the quieter eruption of basaltic volcanoes, such as those in the Hawaiian Islands, the fluid basalt moves at a few miles an hour and generally there is enough time to evacuate. This may seem of little value if your home and fields are inundated by either lava or air-borne ash, but at least you are alive and able to start over.

Indirect damage from volcanoes can also be swift and without warning. Many volcanoes are high mountains and even in the tropics are glacier covered. An eruption under a glacier can cause extensive melting that results in floods. Also many volcanoes, especially those that have been inactive for long periods, have large lakes in their depressed summit areas. Crater Lake in Oregon is an example. An eruption that breeches the lake, or an eruption into the lake, can cause the draining of the lake and so cause floods. The eruption of a submarine or shallow-island volcano such as Krakatoa can cause flooding by seismic sea waves.

Probably the largest areas inundated by volcanoes are those swept by mudflows. Although some mudflows are concurrent with the erup-

Figure 3-19. Shiprock, New Mexico, is probably part of an ancient volcano. Dikes radiating from it can be seen on the left skyline and the right foreground. Photo from New Mexico Department of Development.

Figure 3-20. Map of Kilauea volcano, Hawaii, showing principal features of the 1959 eruption of Kilauea Iki. Small-scale inset map shows tilt pattern on the summit of Kilauea prior to the eruption. From J. P. Eaton and D. H. Richter, *Geotimes*, Vol. 5, 1960.

tion, most occur at the next rainy season. Although the latter mudflows can be predicted, many lives are lost because many people refuse to leave their homes and fields. Mudflows are probably the most important way that pyroclastic volcanic rocks are distributed from volcanoes and so are a normal geologic process. A pyroclastic eruption can deposit ash and coarser material on the slopes of the volcano and on other slopes downwind. This loose material is not stable and at the next rain becomes mud that can flow even on very

gentle slopes. Rain commonly accompanies volcanic eruption, and, although we don't know why, it may be related to the large amounts of steam that are part of most eruptions. Thus mudflows can accompany eruptions. Pyroclastic eruptions commonly kill vegetation by covering low plants, and the falling debris strips the leaves from trees. Without vegetation, the runoff is increased, and the possibility of mudflows is greater.

Mount Rainier was reported to have erupted 14 times between 1820 and 1894, and debris from some of

these eruptions have been identified and dated. The recent history of Mt. Rainier shows both the potential hazards and gives some estimate of the frequency of such events. The main hazards at Mt. Rainier are eruptions of lava, eruptions of pyroclastics, and mud or debris flows. The youngest lava flows are probably those that formed the summit cone. Renewed activity there would melt snow, perhaps forming lakes in the two overlapping craters or causing floods or debris flows if the water moves down the mountain. Climbers have descended as deep as 450 feet in ice caves near the summit, and stones thrown from there caused splashes from deeper water. During the descent, many fumaroles and areas of hot rock were seen. Thus the main dangers of renewed lava eruptions are floods; and although the lavas themselves might cover some areas, they would move slowly enough to allow evacuation and probably would not flow far from the mountain.

Mudflows, or debris flows, which are similar to mudflows but contain much material coarser than mud, are common and are the most devastating type of activity at Mt. Rainier. Water and available rock material are needed to form mudflows. Therefore, they can result from either lava or pyroclastic eruptions, or heavy rainfall, or any other event that causes melting of glacial ice; because much rock material mantles the mountain, only water must be added. The largest postglacial mudflow, the Osceola Mudflow, occurred about 5000 years ago and had a volume of almost one-half cubic mile. It is believed to have originated by steam explosions near the summit. It travel-

Figure 3-21. In early 1973, an eruption began at Heimaey, an island that is part of Iceland. The cone in the left background is composed of pyroclastics which also partially cover the houses. Photo from Consulate General of Iceland.

ed 40 miles down valley and spread up to 70 feet thick over 65 square miles in the Puget Lowlands. (See Fig. 3-22.) Remnants of the flow show that it was at least 500 feet thick at places as it moved down the valley. The other large mudflow, the Electron Mudflow, occurred about 500 years ago. In 1947 heavy rainfall caused a debris flow at Kautz Creek that moved about 50 million cubic yards of debris. The banks of that creek expose six previous debris flows. In 1963 a rock avalanche from Little Tahoma Peak moved 4.3 miles and stopped only 2000 feet from White River Campground. It may have been caused by a small steam explosion.

The types of activity that have occurred at Mt. Rainier are listed in Table 3-2. Their effects, possible warning signs, and possible frequen-

cy, are also shown. Wise planning of land use and observation of the Cascade volcanoes can probably prevent a major disaster. The main problem is possibly complacency in view of the infrequency of major eruptions.

Crystallization

One of the most important concepts in igneous geology is the reaction series first described by Bowen, an American geochemist who studied silicate melts in the laboratory. This series shows the sequence in the crystallization of a basaltic melt, which is the most common type of volcanic rock. It also suggests how the earliest formed crystals might be separated from the melt or magma, forming rocks with a composition

different from that of the original melt. Such a process is called *differentiation,* and we will discuss its role when we consider the origin of magmas and of the various types of igneous rocks. Although Bowen originated the idea of a reaction series from studies of experimental melts, the general sequence of crystallization of most rocks was already known as a result of intensive microscopic studies. The reaction series is an oversimplification but is very useful. (See Fig. 3-23.)

The reaction series really consists of three series. The right-hand side is a *continuous series;* that is, all compositions of plagioclase from entirely calcic to entirely sodic and all compositions in between occur. The left-hand side is a *discontinuous series;* that is, the changes from one mineral to the next occur in discrete steps. Each of the minerals named in the

Figure 3-22. The extent of the Osceola and Electron Mudflows from Mt. Rainier. The lighter colored area is the Osceola Mudflow, and the lined area the Electron Mudflow. From U. S. Geological Survey *Bulletin* 1238, 1967.

Table 3-2. Types of eruptions that have occurred at Mount Rainier in post-glacial time, anticipated effects and frequency of similar eruptions in the future, and possible warning signs of an impending eruption. From U.S. Geological Survey *Bulletin* 1238, 1967.

	VOLCANIC ACTIVITY NOT NECESSARILY ASSOCIATED WITH THE RISE OF NEW MAGMA	TYPES OF VOLCANIC ACTIVITY ASSOCIATED WITH THE RISE OF NEW MAGMA			
	Steam explosion (generally on small scale)	Steam explosion (may be on small to very large scale)	Pumice eruption	Eruption of bombs and (or) block-and-ash avalanches	Eruption of lava flows
DIRECT EFFECTS	Formation of small-scale rockfalls and avalanches. Effects confined chiefly to flanks of volcano and valley floors immediately adjacent.	Formation of rockfalls and avalanches which grade down-valley into debris flows. Effects confined chiefly to flanks of volcano and valley floors. Formation of air-laid rubble deposits whose distribution would be limited to flanks of volcano and areas closely adjacent.	Fall of pumice on flanks of volcano; widest distribution beyond flanks in a downwind direction, thus probably greatest to the northeast, east, or southeast. Thickness probably will be less than a foot beyond 5-mile radius of summit. Anticipate thin ashfall over broad area without regard for topography.	Fall of hot to incandescent bombs, ash, and rock fragments. Distribution chiefly on flanks of volcano, but avalanches may extend into valleys.	Distribution probably limited to flanks of volcano. Major effects expectable only in immediate vicinity of flow.
INDIRECT EFFECTS	Possible floods and (or) debris flows caused by damming of rivers by avalanche deposits. Principal effects probably would be limited to valley floors within and closely adjacent to park.	Very large avalanches of moist altered rock on flanks of volcano grading directly into debris flows on valley floors. Debris flows may be very long, extending tens of miles beyond park boundaries.	Extensive melting of glaciers possible, caused by internal volcanic heat and by steam moving toward outside of volcano. Expected result: floods and debris flows on valley floors. Ejection of water from summit craters in early stage of eruption. Possible result: floods and debris flows on valley floors. Debris flows may result from extensive downslope movements of pumice onto valley floors.	Ejection of water from craters if flows occur at summit of volcano. Possible floods and debris flows caused by eruption of hot debris onto snow.	Extrusion of lava flow beneath or onto glacier might cause catastrophic floods.
POSSIBLE WARNING SIGNS OF IMPENDING ACTIVITY	Appearance of steam jets and clouds of water vapor, possibly accompanied by explosions and rockfalls. Abnormal melting of glaciers at hot spots; appearance of melt pits in glaciers.		Increase in frequency and magnitude of local earthquakes. Large increase in stream discharge unrelated to meteorological conditions. Appearance of clouds of water vapor associated with steam jets, possibly accompanied by small steam explosions and rockfalls. Increase in fumarolic activity and increased melting of snow at summit cone. Abnormal glacier melting at hot spots; appearance of melt pits in glaciers. Increase in temperature of fumaroles. Increase of sulfur and chlorine in fumarolic gases.		
INDICATED POSSIBLE FREQUENCY	1 in 10 to 100 years	1 in 2,000 years	1 in 2,500 years	1 in 5,000 years	1 in 10,000 years

[The flanks of the volcano include an area within 4 or 5 miles from the summit]

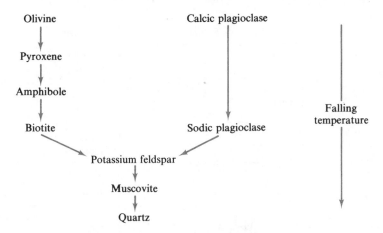

Figure 3-23. Bowen's reaction series. Shows the general sequence of mineral formation with falling temperature in a basaltic melt.

left-hand series is actually a continuous series in composition of that mineral group. The lower series is only a sequence of crystallization. The arrangement shows the relative, but not exact, sequence of crystallization; calcic plagioclase and olivine tend to crystallize at the same time, and the lower series is last to crystallize. *It should be emphasized that Bowen's series holds for only some basaltic magmas,* but, as we shall see, basalt magma is very important in igneous geology.

The interpretation of the crystallization of a basaltic magma in terms of Bowen's series is as follows. In general, the first mineral to crystallize from the melt is olivine. In most cases, at the same time, or a little later, calcic plagioclase begins to crystallize. The discontinuous series continues with falling temperature; the olivine reacts with the melt to form pyroxene, which at the same time also starts to crystallize directly from the melt. Plagioclase continues to crystallize during falling temperature; the composition, however, continuously becomes more sodic. If cooling is slow enough, these two

processes go on: the minerals on the left side react with the melt to form the next lower mineral, while that mineral is crystallizing from the melt, and the plagioclase continuously reacts with the melt to form more plagioclase that is more sodic in composition. This process ends when all the ferromagnesian minerals and plagioclase are formed. Then the potassium feldspar-muscovite-quartz series begins. This series may in part overlap the sodic part of the plagioclase series.

Another way to explain the discontinuous series is to consider what happens on heating one of the minerals. When one of the discontinuous series minerals is heated, it does not

melt at a single temperature the way ice or similar substances do. Instead, when an amphibole, for example, is heated, it begins to melt when a certain temperature, depending on its composition, is reached, and melting continues over a range of temperature. If we examine the remaining solid during this first stage of melting, by quick quenching to prevent reaction during slow cooling, we would see that the amphibole has been transformed to a pyroxene and a liquid — just the reverse of what takes place during the slow cooling of a magma. On further heating, a temperature is reached at which further melting would occur. As before, melting would take place over a range in temperature, resulting in a melt and olivine. Finally, a temperature would be reached at which all the olivine would melt.

The evidence for the reaction series is abundant. Under the microscope, for instance, pyroxene crystals with jackets of amphibole are seen, recording a discontinuous reaction that was interrupted, probably by cooling too rapidly for the reaction to be completed. (See Fig. 3-24.) In differentiated rock bodies, slow cooling and low viscosity have combined to preserve some of the main points of the reaction principle. In such

Figure 3-24. Evidence for the Bowen reaction series seen in microscopic study of igneous rocks. A. Augite crystal with remnants of olivine inside it. B. Hornblende crystal with augite remnants. In both cases the reaction was incomplete, thus preserving the remnants of the earlier mineral. Note the cleavage angles in augite and hornblende. The minerals shown here are typically about one-eighth inch in diameter, but may be much larger or smaller.

rock bodies, the early-crystallized olivine sank to the bottom and so was removed from contact with the main mass of the magma, allowing progressively lighter-colored rocks to form in the upper parts of the body. (See Fig. 3-25.)

Some believe that such differentiation explains the great variety of igneous rocks, but this is probably an oversimplification. The reaction series applies to basaltic magmas and the vast areas of basalt lava certainly

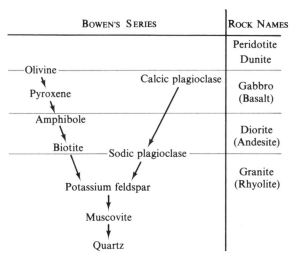

Figure 3-26. This diagram shows in a very general way how differentiation by separation of successively formed crystals could result in several rock types originating from a single basalt magma. (Volcanic rock types are in parentheses.) It also shows why minerals widely separated in the reaction series, such as quartz and olivine, rarely appear in the same rock.

indicate that basalt magmas are common in the earth's crust. But, as we will see later, rocks of granitic composition are also very abundant. The vast amount of granitic rocks and their volcanic equivalents suggests that these rocks also represent important magma types in the crust, because they are far too abundant to be derived from basalt by differentiation. Complete differentiation of basalt can produce, at the most, 15 per cent granitic-composition rocks. This points out a major problem of geology — the origin of magmas. Understanding the crystallization of basalt gives some insight into the genesis of rock types as shown in the next section.

Classification

Igneous rocks are classified on the bases of composition and texture. The classification is an attempt to show the relationships among the various rock types; hence, it is an attempt at natural or genetic classification.

Composition, in the classification of igneous rocks, follows the reaction series. Most rock names came into use long before Bowen described the reaction series. These names had been applied to the more common rocks. It is of more than passing interest that these common groups of rocks fall into the various steps in the differentiation of a basaltic magma according to Bowen's reaction series. This correlation suggests that the compositional part of igneous rock terminology may be genetic. This is shown in Fig. 3-26.

Because almost all coarse-grained rocks contain feldspar, the most common mineral family, the *type* of feldspar is the most important factor in the composition. Of secondary importance is the type of dark mineral which generally accompanies

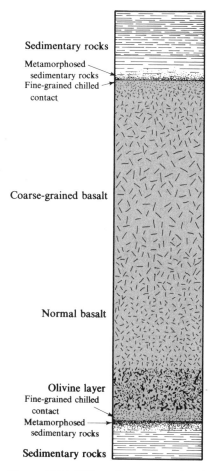

Figure 3-25. Differentiation in a basalt sill. A narrow column through a sill is shown here. The olivine layer develops by the sinking of the early-formed, heavy olivine crystals.

each type of feldspar. See classification diagram, Fig. 3-42.

In a similar way, the texture tells much about the cooling history of an igneous rock. In igneous rocks the texture refers mainly to the grain size. Rocks that cool slowly are able to grow large crystals; but quickly chilled rocks, such as volcanic rocks, are fine grained. To see the significance, we must consider the mode of occurrence of igneous rocks.

Texture and Mode of Occurrence

Igneous rocks occur in two ways, either as *intrusive* (below the surface) bodies or as *extrusive* (surface) rocks. The ultimate source of igneous magma is probably deep in the crust or in the upper part of the mantle, and the terms intrusive and extrusive refer to the place where the rock solidified. Intrusive igneous rocks can be seen only where erosion has uncovered them. They are described as *concordant* if the contacts of the intrusive body are more or less parallel to the bedding of the intruded rocks, and *discordant* if the intrusive body cuts across the older rocks. (See Fig. 3-27.)

The largest discordant bodies are called *batholiths*. These are very large features, the size of mountain ranges such as the Sierra Nevada in California, which is nearly 400 miles long by about 50 miles wide. Many other ranges, both larger and smaller, also contain batholiths. (See Fig. 3-28.) Because batholiths are large and also because they probably were emplaced at least several thousand feet below the surface, they cooled very slowly. This slow cooling per-

Figure 3-27. Intrusive igneous rock bodies. The laccolith and sills are concordant with the enclosing sedimentary beds, and the batholith and dikes are discordant. The heat from the batholith metamorphoses the surrounding rocks.

mitted large mineral grains to form; therefore, it is not surprising that batholiths are composed mainly of granitic rocks with crystals large enough to be easily seen. (See Fig. 3-29.) Smaller bodies of coarse-grained igneous rocks are termed *stocks* or *plutons.*

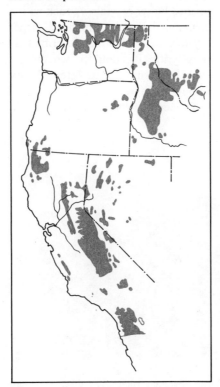

Figure 3-28. Batholiths of western United States.

Dikes are tabular, discordant intrusive bodies. They range in thickness from a few inches to several thousand feet, but generally are of the order of a few feet to tens of feet in thickness. (See Figs. 3-30 and 3-31.) They are generally much longer than they are wide, and many have been traced for miles. Most dikes occupy cracks and have straight, parallel walls. Because the intrusive rock is commonly resistant, dikes generally form ridges when exposed by erosion.

The corcordant intrusive bodies are *sills* and *laccoliths.* They are very similar and are intruded between sedimentary beds. Sills are thin and do not noticeably deform sedimentary beds. (See Fig. 3-32.) Laccoliths are thicker bodies and up-arch the overlying sediments, in some cases forming mountains.

Dikes and sills are small bodies compared to batholiths and have much more surface area for their volume; thus, these bodies cool much more rapidly and are commonly fine grained. Laccoliths are generally fine grained but, depending on their size, may be coarse grained.

The extrusive rocks result from lava flows and other types of volcanic activity. These rocks commonly cool rapidly and are fine grained, or glassy

Figure 3-29. Granite has a coarse, interlocking texture formed by crystallization from a melt. Photo from Ward's Natural Science Establishment, Inc., Rochester, N.Y.

(see Fig. 3-33) if cooled so rapidly that no crystallization occurs.

Many igneous rocks are a mixture of coarse and fine crystals, and this texture is called *porphyritic*. (See Fig. 3-34.) Such a texture, in general, records a two-step history of the rock. The large crystals form as a result of slow cooling, perhaps in a deep magma chamber. Then the magma is moved higher in the crust to form a dike, sill, or flow. The remainder of the rock cools rapidly, resulting in a fine-grained or even glassy matrix surrounding the early-formed larger crystals. A rock with a coarse matrix and some much larger crystals may also be described as porphyritic. It is doubtful whether all such rocks have a two-step history as outlined here.

The temperature of basalt lava is in the range from 1,000°C to 1,200°C, felsite lava is about 900°C, and granitic rocks are believed to be emplaced at about 820°C. Advancing basalt flows at Hawaii have been measured at temperatures as low as 745°C. The cooling time of igneous rocks is not well known. Batholiths are believed to solidify in one to a few million years. Dikes, depending on size, probably solidify in several tens of thousands of years. Basalt lava is known to have hardened to a depth of over ten feet in two years. Rocks are poor conductors of heat; hence, intrusive rocks take a very long time to cool because the heat must be conducted away by the surrounding rocks. Lava flows lose their heat to the atmosphere more rapidly.

A summary of textural terms will be of help in learning to recognize igneous rock types:

Glassy—no crystals

Fine-grained—grains can be seen in sunlight with a good hand lens

Coarse-grained—crystals easily seen

Pegmatitic—crystals over one centimeter. (This texture will be discussed below.)

Porphyritic rocks—generally are named for the texture of the groundmass, giving such terms as porphyritic basalt for a dark, fine-grained rock with some larger crystals, and porphyritic granite for a granite with some much larger crystals.

Origin and Emplacement

The origin of magma is a problem about which only some suggestions can be made. More will be said in the

Figure 3-30. Dikes cutting through horizontal sedimentary rocks. The dikes are more resistant to erosion than are the sedimentary rocks. West Spanish Peak, Colorado, is in the background. Photo by G. W. Stose, U.S. Geological Survey.

Figure 3-31. Teanaway dike swarm, Cascade Mountains, Washington. At places the dikes, shown diagrammatically here, comprise over 90 per cent of the surface, recording much dilation. The dikes were the source of the Teanaway Basalt. From R. J. Foster, *American Journal of Science,* Vol. 256, 1958.

LEGEND (from figure):

Younger than Teanaway Basalt:
- POST-ROSLYN RHYOLITE
- ROSLYN ARKOSE
- CLASTIC TEANAWAY LAVA BASALT
- DIKES, SHOWN DIAGRAMMATICALLY (FROM USGS FOLIOS 106 & 139)

Older than Teanaway Basalt:
- SWAUK ARKOSE
- PERIDOTITE & METAMORPHICS

CLE ELUM LAKE — TEANAWAY RIVER — SWAUK CR. — NORTH — ONE MILE — FAULT — FOLD AXIS

In this area the salt beds are very deep and thus are much lighter than the surrounding rocks. The salt of lesser density tends to rise and float to the top. This seems unlikely because we are used to considering rocks as strong, rigid materials. It can be easily shown, however, that rocks are not strong enough to support themselves. This topic will be pursued further in Chapter 14, but a moment's thought will show that rocks have finite strength and that the weight of overlying rocks will exceed this strength at some depth. Below this depth rocks behave as viscous fluids. In the present case, salt behaves as a viscous fluid and the overlying sedimentary rocks are very weak. To start the formation of a salt dome, some initial irregularity is probably necessary, such as a high point on the salt bed, or a weakness or break in the overlying sediments. The rising salt may remain connected to the salt bed or might, in

structure chapters, where volcanoes and other igneous bodies will be considered in relation to the other features with which they are associated. Seismic data suggest that volcanic magma forms near the base of the crust and moves upward to a shallow magma chamber before erupting to the surface. (See Fig. 3-35.)

Once the magma is generated, the next problem is: why does it rise in the crust? Several suggestions have been made. Rocks, like most substances, expand when they are melted. Thus, magma is less dense than the surrounding rocks and tends to rise, much as oil mixed with water rises and floats on the top.

Salt domes may illustrate intrusion caused by gravitational rise of lighter material. Salt domes occur in many parts of the world, and because some are associated with oil fields, they have been closely studied. Salt domes occur in sedimentary rocks; the salt originates from salt beds formed by the evaporation of sea water. These sedimentary rocks will be described more fully in Chapter 4. Salt, or halite as the mineral is called, has a specific gravity of 2.2. The specific gravity of sedimentary rocks increases with depth of burial because the weight of the overlying rocks compresses the deeper rocks. At the Gulf Coast of the United States, the near-surface sedimentary rocks have a specific gravity of 1.9, at 2,000 feet the specific gravity has increased to 2.2, and at 10,000 feet it is 2.4.

Figure 3-32. The black horizontal band is a sill. The inclined band is a cross-cutting dike. The dark stripes in the sandstone are carbonaceous sediments. Taylor Glacier region, Victoria Land, Antarctica. Photo by W. B. Hamilton, U.S. Geological Survey.

Figure 3-33. Obsidian or volcanic glass. The round, shell-like fractures are called conchoidal fractures. Photo from Ward's Natural Science Establishment, Inc., Rochester, N.Y.

some cases, become completely detached from the salt bed. The tops of many salt domes are mushroom shaped. (See Fig. 3-36.) Some salt domes may have risen up to eight miles in the Gulf Coast. The mechanics of salt dome formation are complex, and some downward movement of the surrounding sedimentary rocks may have occurred.

Magmas are less viscous than salt and, like salt, are less dense than the rocks at depth. Thus, the same mechanism, which may account for the rise of salt domes, may also account for the rise of some magmas. The internal structure of salt domes is somewhat similar to the internal structures of some intrusive igneous rocks, further suggesting a similarity. The heat of igneous rocks could weaken the surrounding rocks and thus assist the rise.

A possible source of pressure

within the magma is the expansion of the volatiles. This was discussed earlier in connection with the surface manifestations of volcanism. In the present context, the volatiles in the newly formed magma would tend to

expand as the magma rises because the confining pressure (weight of overlying rocks) is reduced. This might crack the enclosing rocks, allowing the magma to rise further.

Expanding gas bubbles might also be able to erode the enclosing rocks. This principle is used in a type of drilling and may be the mechanism by which some volcanic rocks reach the surface. Possible examples are the cylindrical, pyroclastic-filled conduits found in some volcanic areas.

The classic explanation of the emplacement of batholiths in the crust is *stoping*. The name comes from the mining term for the process of removing ore overhead. The hot magma may cause blocks of the overlying enclosing rock to be broken off and fall through the melt. Such a process requires a low-viscosity magma and would allow much contamination of the magma by reaction with the stoped blocks. Most batholiths contain many inclusions, probably of the enclosing rocks, giving some credence to this

Figure 3-34. Porphyritic texture. The light-colored feldspar crystals grew during slow cooling, and subsequent chilling probably produced the fine-grained dark matrix. Photo from Ward's Natural Science Establishment, Inc., Rochester, N.Y.

Figure 3-35. Hypothetical cross-section through Hawaii, showing possible source and accumulation zones. After G. A. Macdonald, *Science*, Vol. 133, 1961.

hypothesis. (See Figs. 3-37 and 3-38.)

Another possible way that magma can work its way through the crust is to react with the enclosing rocks. The magma could, by chemical reaction, dissolve or partially dissolve the rocks above the magma. Some melting or partial melting is also possible. For melting to occur, a similar amount of crystallization would have to occur elsewhere in the magma because magmas do not have enough excess heat energy to do much melting. Excess heat energy would have to be in the form of high temperature, but lava at the surface is generally very close to its melting point, indicating that, in general, magmas do not have excessively high temperatures. This process would tend to change the composition of the magma by assimilation of parts of the enclosing rocks. Differentiation would probably also be going on in the magma.

Emplacement of Batholiths

The emplacement of batholiths is a major problem of geology. The great size of a batholith like the Sierra Nevada immediately raises the question: What happened to the large volume of rocks that the batholith displaced? Traditionally this question has been answered by two opposing points of view. One school of thought holds that the batholith made room for itself by a combination of stoping and pushing aside the pre-existing rocks. The many included blocks, presumably of prebatholith rocks, found in some batholiths, and the zone of the deformed earlier rocks that surrounds most batholiths are the main evidence for this viewpoint. The opposing viewpoint is that the batholith was created more or less in place by the metamorphism of the pre-existing rocks that accompanied the mountain-building event during which the batholith was emplaced. This possibility is included in the rock cycle, Fig. 3-2. The evidence for this theory is the metamorphic rocks that surround most batholiths. In some cases these zones of high-grade metamorphic rocks are very wide, and in other cases the heat from the crystallizing magma is enough to account for the smaller halo of metamorphism. However, recent studies show that most batholiths have different isotope compositions than upper crustal rocks, and so could not have formed by the metamorphism of such rocks.

It seems unreasonable to think that a huge batholith the size of the Sierra Nevada was all emplaced at the same time. Recent studies have shown that this batholith is composed of a number of much smaller plutons. The finer-grained margins of these plutons that resulted from more rapid cooling indicate that at no time was the entire batholith molten. Radioactive dates indicate that the emplacement of the batholith took place over a few tens of millions of years. (See Fig. 3-39.)

Because of their large surface area, batholiths have traditionally

Figure 3-36. Salt domes showing some common shapes and the fractures caused in the surrounding rocks by their rise.

Figure 3-37. Granite with many dark inclusions in Yosemite National Park, California. The glove shows the scale. Photo by F. C. Calkins, U.S. Geological Survey.

Pegmatites and Quartz Veins

The water and other volatiles (gases) in magma may also be involved in the formation of pegmatite and ore veins. When a deep intrusive body, such as a batholith, is emplaced, the rim is quickly chilled and solidifies first. This is commonly recorded by the fine-grained margins of an intrusive body. The volatiles are thus trapped inside the intrusive body; and as it solidifies, presumably from the margin inward, the volatiles are confined to a smaller and smaller volume in the center of the body. The residual fluid is rich in the elements that form the last minerals on the Bowen series, especially quartz, feldspar, and mica, as well as less common elements that do not fit into the rock-forming minerals. This water-rich residual fluid eventually escapes, perhaps as a result of the increase in pressure that can result when the volatiles are concentrated in a small volume. The increase in pressure may cause cracks that are filled by this material (Fig. 3-41). Contraction due to cooling may also help to cause the cracks. The fluid has a low viscosity because of the volatiles, and this increases the mobility of the ions. As a result, very

been considered to be thick bodies extending deep into the crust. Some recent seismic and gravity studies suggest that batholiths may be relatively thin features. Figures 3-27 and 3-40 show these two interpretations of batholiths.

All of these data have been incorporated into one recent theory. This theory holds that the batholith is formed by the bubble-like rise of the detached pods of magma generated in the upper mantle or deep crust. These pods of magma form a thin batholith near the surface. At some places the magma reaches the surface and forms extensive volcanic rocks, and some batholiths intrude volcanic rocks that are similar in composition to the main batholith. These magma pods tend to spread out near the surface and so might account for the deformation that surrounds many batholiths. In passing through the crustal rocks, the magma would react with those rocks and could assimilate some of them by reaction and partial melting. In any case, the rocks through which the magma passes would be metamorphosed. Erosion of the thin layer of granitic rocks would expose an extensive metamorphic terrane similar to that found in many mountain ranges. Thus this theory would explain most of the data of batholiths and also explain the origin of extensive areas of metamorphic rocks. Like batholiths, these metamorphic terranes are difficult to explain because the known rise in temperature with depth is not enough to account for the amount of metamorphism.

Many aspects of the batholith problem are related to the structural significance of batholiths and will be discussed further in Chapter 14.

Figure 3-38. Stoping. Forceful injection of the magma breaks fragments from the roof that sink through the magma, reacting with it.

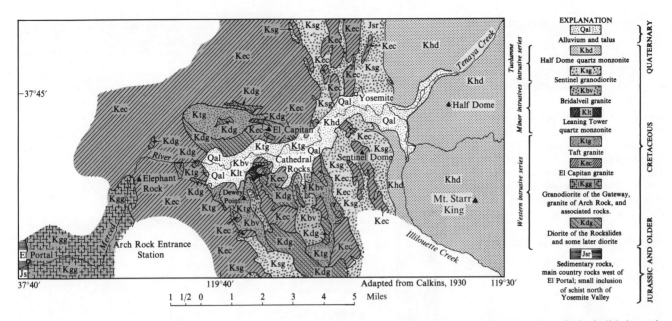

Figure 3-39. Generalized geologic map of Yosemite Valley in the Sierra Nevada, California. The Sierra Nevada batholith is made of many smaller intrusive bodies as shown on this map. Granodiorite and quartz monzonite are types of granitic rocks included under the term granite in this text. From F. C. Calkins and D. L. Peck, California Division of Mines and Geology, *Bulletin* 182, 1962.

large crystals form. The resulting very-coarse-grained rocks are called *pegmatites,* and they may have the same composition as the enclosing magma or may contain unusual minerals. In the latter case the fluids are believed to concentrate the minor elements to form the unusual minerals. Pegmatites commonly occur as dike-like masses in or near intrusive bodies. They are actively sought because they contain large and often beautiful minerals, such as beryl, tourmaline, topaz, fluorite, all of the rock-forming minerals, and many more.

The residual fluids are probably also responsible for the formation of quartz-metal veins, but the process is not yet understood. The last fluid in a batholith would be rich in quartz and in the elements such as copper, lead, zinc, gold, silver, and sulfur, that do not fit into the crystal structures of the rock-forming minerals.

A batholith may contain only a very tiny percentage of these elements, but the total amount of metal may be quite large. Quartz veins containing metals in the form of sulfide minerals are found around some intrusive bodies. (See Fig. 3-41.) These veins do not grade into pegmatites,

so they must form by a somewhat different, although probably related, process.

In still other cases these fluids may cause widespread alteration of the intrusive body and/or the surrounding rocks. Much more study is needed to understand the actions of

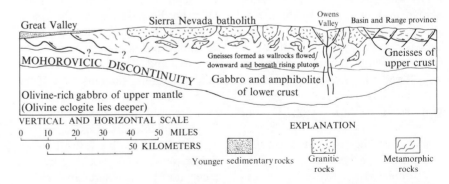

Figure 3-40. Hypothetical cross-section through the Sierra Nevada, based on the idea that the granitic rocks of the batholith moved through the metamorphic rocks to their present positions. This cross-section is based in part on seismic data. Compare with Figure 3-27 that shows a thicker batholith. From Warren Hamilton and W. B. Myers, U.S. Geological Survey, *Professional Paper* 554C, 1967.

Figure 3-41. Cross-section of a crystallizing batholith showing origin of veins. Crystallization has proceeded inward from the fine-grained, quickly chilled margin. The volatiles are concentrated in the still molten center. The increased pressure causes cracks to form that are filled by the metal-bearing quartz-rich volatiles.

the magmatic volatiles that are believed to cause all of these diverse effects.

Identification and Interpretation

Identification of the coarse-grained igneous rocks will present no serious problems if one is able to identify the minerals and estimate their relative amounts. The first clue comes from the color of the rock which, following the charts (Figs. 3-42 and 43), will suggest which minerals should be looked for. In the use of the term *light color,* it should be understood that any shade of red or pink, no matter how dark, must be considered a light color. In the same way, any shade of green should be considered a dark color. Overall color should not be used in place of good mineral determination, because exceptions are common enough to be troublesome (especially on examinations).

In a sense, the fine-grained, or volcanic, rocks are more difficult to identify than are the coarse-grained rocks because less data, in the form

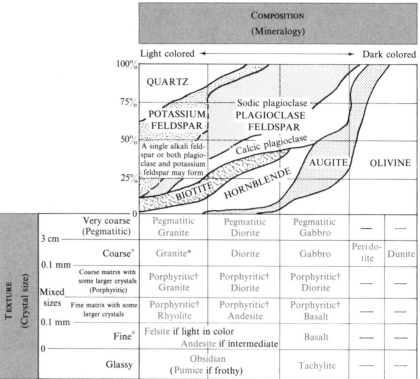

Figure 3-42. Chart showing the classification of the igneous rocks. Composition is indicated vertically, and texture is indicated horizontally. The upper part of the figure shows the range in mineral composition of each rock type.

of identifiable minerals, are available; but because of this, the classification is less rigid. The procedure is similar in that color is used as the first step and, if no porphyritic crystals are present, the only step. If the porphyritic crystals can be identified, they are used in the same way as are the identifiable minerals in coarse-grained rocks; but because all potential minerals may not be present, color must be considered in deciding on a name. Care must be used not to mistake filled gas holes for porphyritic crystals. In many cases, it is necessary to have a chemical analysis in order to name a fine-grained rock

accurately. Also, the volcanic rocks are not simply fine-grained equivalents of the coarse-grained rocks, as the chart might lead one to believe, but contain some small but distinct variations in composition, which chemical analyses will detect.

On the chart shown as Fig. 3-42, granite is used for most of the coarse-grained, quartz-bearing rocks. A number of other names are also in use for these rocks, depending on whether they contain one feldspar, two feldspars, or mixed feldspar (perthite). Diorite generally has more dark minerals than does granite, and the feldspar is plagio-

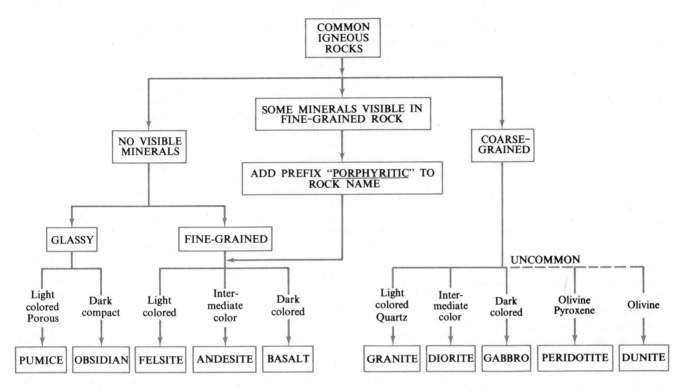

Figure 3-43. Flow diagram showing the identification of igneous rocks.

clase. Quartz diorite is distinguished from granite in that it contains only plagioclase feldspar. The latter distinction, however, is not always easy to make.

Understanding the formation of a rock is much more important than simply naming it. It is necessary to discuss classification so that the meaning of rock names and the re-lationships among the various rock types are understood. Therefore, when studying a rock, try to interpret its origin, following mainly the discussion in the section on texture and mode of origin. Thus the coarse texture of a granite implies that the specimen came from a large, slowly cooled, intrusive body that crystallized deep in the earth and has been exposed by uplift and deep erosion. A pumice sample must have come from a volcano, perhaps from an explosive eruption, and study of other nearby exposures may enable reconstruction of the volcano. When actual outcrops are studied, all of the information discussed here can be used in deciphering the origin and history of the rocks.

Summary

The rocks of the crust are classified by origin into three types: *igneous*, solidified from magma; *metamorphic*, changed from another rock, generally by high temperature and pressure; and *sedimentary*, deposited in layers on the earth's surface by water, wind, and ice.

Magma is a natural, mobile, hot melt composed of rock-forming materials (mainly silicates) and volatiles (mainly steam) that are held in solution by pressure. It may or may not contain suspended solids.

In the sudden eruption of low viscosity magma, the expanding gases form a froth, resulting in the light, glassy rock *pumice*. Coarser bubble holes result in the rock *scoria*.

In eruption of more viscous magmas, the gas pressure shatters the rock into *pyroclastic* fragments: *volcanic ash* if sand-size or smaller, *volcanic blocks* if larger. The *pyroclastic rocks* formed by this process are named according to the size of the fragments. Consolidated ash is called *tuff;* a pyroclastic rock with mainly larger fragments is called *volcanic breccia*.

Volcanic bombs are formed by pyroclastic material that is still molten when thrown out.

Present-day volcanoes occur in areas, such as the rim of the Pacific Ocean, the Mediterranean Sea, and the Mid-Atlantic Ridge, that are geologically active.

The quietest eruptions are very fluid basalt lava flows from dikes or dike swarms that can produce great thicknesses and cover huge areas.

A *shield volcano* is formed by eruption of very fluid basalt lava from a central vent.

Fluid basalt forms *ropy* or *pahoehoe* lava flows.

More viscous basalt lava breaks into blocks, forming *blocky* or *aa* lava flows.

Cinder cones are formed by pyroclastic material thrown out of a vent by the eruption of still more viscous magma.

Composite volcanoes are the very high volcanic mountains made of a combination of pyroclastic material and lava. If the lava is so viscous as to plug the vent, the pressure of the volatiles in the magma may build up until an explosion results. The more viscous the magma, the more numerous or destructive the explosions.

Very explosive eruptions of hot gas and shattered glass fragments are called *nuées ardentes* (*glowing clouds, glowing avalanches*); this fluid pyroclastic material forms the rock *welded tuff* when the hot glass fragments come to rest and weld together. Welded tuffs are relatively abundant and may or may not be associated with dome volcanoes.

Pillow lava forms when lava flowing into water bodies is chilled relatively rapidly.

Eruptions cannot yet be predicted, but magma movements are detected by many shallow, weak earthquakes, by tilt-meter measurements, and other studies.

Volcanic destruction is of two kinds: direct damage by the material erupted; and indirect damage by mudflows of pyroclastic material, and by floods from melted glaciers and breached lakes.

Bowen's reaction series shows the sequence in the crystallization of a basalt magma.

Separation of early-formed crystals may form rocks with a composition different from the original melt; this process is called *differentiation.*

In a *continuous mineral series,* the full range of compositions of that mineral group occurs in sequence from one end member to another.

In a *discontinuous mineral series,* the change is from one mineral group to another in separate, discrete steps.

With falling temperature, Bowen's reaction series has the following sequence of crystallization: a discontinuous series of olivine to pyroxene to amphibole to biotite, concurrently with a continuous series of calcic to sodic plagioclase, followed by the sequence potassium feldspar, muscovite, and, lastly, quartz.

Igneous rocks are classified by composition and texture. The composition classification follows the reaction series. Texture refers to the grain size, and indicates the mode of occurrence and cooling history of the rock.

Intrusive igneous rocks are solidified below the surface of the earth; *extrusive* igneous rocks are solidified on the surface.

Intrusive igneous rocks are *discordant* if they cut across the bedding of the rocks they intrude, and *concordant* if they are more or less parallel to the bedding.

Discordant intrusives are *batholiths,* very large, coarse-grained bodies; *stocks* or *plutons,* smaller, coarse-grained bodies; and *dikes,* tabular, usually parallel-walled, linear, fine-grained bodies.

Concordant intrusives are *sills,* thin, usually fine-grained bodies intruded between sedimentary layers without noticeable deformation; and *laccoliths,* thicker, fine- to coarse-grained bodies which up-arch the sedimentary layers above them.

Extrusive rocks result from lava flows or other volcanic activity. They are fine-grained or, with very rapid cooling, glassy.

The textural terms are:

Glassy — no crystals

Fine-grained — grains can be seen in sunlight with a good hand lens

Coarse-grained — crystals easily seen

Pegmatitic — crystals over one centimenter

Porphyritic — coarse crystals in a fine-grained or glassy matrix, or some very coarse crystals in a coarse-grained rock. Porphyritic texture may record a change in conditions during the cooling history of the rock.

Magma may originate near the bottom of the earth's crust and may rise because the melted rock expands and is less dense than the surrounding rocks.

Salt domes from buried salt beds are also less dense than the surrounding rocks, and may illustrate another case of gravitational rise of lighter material.

Expansion of the volatiles in magma or erosion by expanding gas bubbles may help the magma to rise.

How huge batholiths are emplaced is a major problem. Most large batholiths are composed of a number of smaller plutons. Traditionally batholiths have been thought to extend deep into the crust and to have been emplaced by *stoping* and pushing aside the pre-existing rocks.

Stoping is a process by which blocks of pre-existing rocks above the magma fall through the melted magma, allowing the magma to rise.

Another theory is that batholiths may be relatively thin features. They may be emplaced by the bubble-like rise of detached pods of magma, which metamorphose the rocks through which they pass.

Quartz veins and very coarse-grained *pegmatite* masses may result from the residual fluid in magma, which is composed of trapped volatiles, elements to form the last minerals of Bowen's series, and less common elements which do not fit into the common rock-forming minerals and therefore may form unusual minerals.

In general, coarse-grained igneous rocks are identified by identifying minerals and estimating their relative amounts. The type of feldspar is important.

Fine-grained igneous rocks are identified primarily by color or chemical analysis.

Study of the texture and composition of the rock enables interpretation of its formation.

Questions

3-1. Define igneous rock.

3-2. Describe the formation of pumice.

3-3. Describe the types of volcanoes. Include how they are formed and which is steepest and why.

3-4. What hazards do volcanoes pose?

3-5. Describe the crystallization of a basaltic melt.

3-6. Describe the crystallization of plagioclase in a magma.

3-7. What happens when a crystal of calcic plagioclase is melted?

3-8. Is the classification of most igneous rocks genetic? Discuss.

3-9. How can a sill be distinguished from a flow?

3-10. What cooling history does a porphyritic texture imply?

3-11. What is a batholith; what is the usual composition and what problems do batholiths present?

3-12. Name a concordant igneous body.

3-13. In what type of occurrence (sill, flow, batholith, etc.) would you expect to find glassy, fine-grained, porphyritic, and coarse-grained igneous rocks?

3-14. Describe the formation of ore veins.

3-15. Igneous rocks are classified on the bases of _____ and _____.

3-16. What minerals are most likely to be visible in a felsite porphyry?

3-17. How does pegmatite form?

3-18. What are the main volcanic gases?

Supplementary Reading

General

Ernst, W. G., *Earth Materials*. Englewood Cliffs, N.J.: Prentice-Hall, Inc., 1969, 150 pp. (paperback).

Pough, F. H., *A Field Guide to Rocks and Minerals*. Boston: Houghton Mifflin Co., 1955, 349 pp.

Igneous Principles

Mason, Brian, *Principles of Geochemistry* (3rd ed.), Chapter 5, "Magmatism and igneous rocks," pp. 96-148. New York: John Wiley & Sons, Inc., 1966.

Tuttle, O. F., "The Origin of Granite," *Scientific American* (April, 1955), Vol. 192, No. 4, pp. 77-82. Reprint 819, W. H. Freeman and Co., San Francisco.

Volcanoes

Aspinall, W. P., H. Sigurdsson, and J. B. Shepherd, "Eruption of Soufrière Volcano on St. Vincent Island, 1971–1972," *Science* (July 13, 1973) Vol. 181, No. 4095, pp. 117-124.

Bullard, F. M., *Volcanoes: In History, in Theory, in Eruption*. Austin, Texas: University of Texas Press, 1962, 441 pp.

Crandell, D. R., and D. R. Mullineaux, *Volcanic Hazards at Mount Rainier, Washington*, U.S. Geological Survey Bulletin 1238, 1967, 26 pp.

Eaton, J. P., and K. J. Murata, "How Volcanoes Grow," *Science* (October 7, 1960), Vol. 132, No. 3432, pp. 925-938.

Macdonald, G. A., "Volcanology," *Science* (March 10, 1961), Vol. 133, No. 3454, pp. 673-679.

Macdonald, G. A., *Volcanoes*, Englewood Cliffs, N.J.: Prentice-Hall, Inc., 1972, 510 pp.

Williams, R. S. Jr., and J. G. Moore, "Iceland Chills a Lava Flow," *Geotimes* (August, 1973), Vol. 18, No. 8, pp. 14-17. Describes the use of sea water in an attempt to save a fishing port from the hazards of a lava flow.

Weathering and Sedimentary Rocks

4

The study of sedimentary rocks began very early. Da Vinci around 1500 and Steno in 1668 noted that sedimentary rocks had originally been deposited in horizontal layers, that the fossils that they contain were once living animals, and that seas had once covered much of what is now land. These were major observations, and had they been generally recognized, geology would have developed much earlier than it did. During the eighteenth century a number of attempts were made to classify and organize the sedimentary layers. Neptunism, discussed in Chapter 3, was one of the poorer of these attempts. Study of the sedimentary layers and their contained fossils ultimately led, in the early nineteenth century, to a geologic time scale and to deciphering the geologic history of the earth. These topics will be covered in Chapter 17.

The study of sedimentary rocks themselves, their composition and formation, came much later. Study of sedimentary rocks may have lagged behind study of igneous rocks because they seemed to present fewer problems than controversial rocks such as basalt. Study of how sedimentary rocks form probably started in earnest when uniformitarianism was generally accepted. Uniformitarianism, as was described in the last chapter, in this case means that to understand ancient rocks, we must study how similar rocks are forming today. Uniformitarianism is central to all geology, and so it is a theme to which we will return in later chapters. It began with James Hutton, late in the eighteenth century, but it did not become widely accepted until it was popularized by Charles Lyell (1797–1875). Lyell wrote a textbook that was first published in 1830 and went on through many editions. He extended and illustrated uniformitarianism and was responsible for its acceptance.

Even though the scene was set for study of the sedimentary rocks with the acceptance of uniformitarianism and the development of microscope techniques, study of sedimentary rocks developed slowly. The microscope was used first on igneous and metamorphic rocks as described in Chapter 3. Study of these rocks gave insight into internal processes of the earth, and so was an important research area. The practical geologists of the nineteenth century were engaged in mining, and so they, too, were mainly interested in igneous and metamorphic rocks. A few geologists did do important work on sedimentary rocks, but it took a new economic incentive to give the study of sedimentary rocks its rightful place in geology. Petroleum is found in sedimentary rocks, and its importance in the twentieth century encouraged detailed study of all aspects of sedimentary rocks.

Sedimentary rocks are formed at or near the surface of the earth. They are composed of rock fragments, weathering products, organic material, or precipitates. Most are deposited in beds or layers. They comprise a very small volume of the

The north wall of the Grand Canyon. This U. S. Geological Survey field party was photographed about 1920 by L. F. Noble.

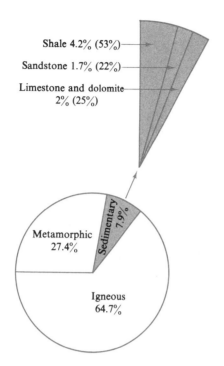

Shale 4.2% (53%)
Sandstone 1.7% (22%)
Limestone and dolomite 2% (25%)

Metamorphic 27.4%

Sedimentary 7.9%

Igneous 64.7%

Figure 4-1. Relative abundance of sedimentary rocks in the earth's crust. For example, shale forms 4.2 per cent of all rocks, and 53 per cent of the sedimentary rocks. Data from Ronov and Yaroshevsky, 1969.

earth, only about 8 per cent of the crust. (See Fig. 4-1.) In spite of their small volume, however, they cover about 75 per cent of the surface. They are formed in a number of different ways, but because the raw materials come from weathering, we will begin there.

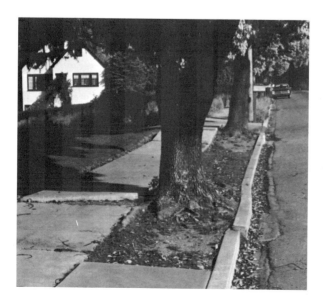

Figure 4-2. Sidewalk and curb have been moved by tree roots.

Weathering

Mechanical weathering breaks up a pre-existing rock into smaller fragments, and *chemical weathering,* acting on these small fragments, rearranges the elements into new minerals.

Mechanical weathering is done mainly by the forces produced by the expansion of water due to freezing, and to a lesser extent by such things as forces of growing roots (see Fig. 4-2), worms and other burrowing animals, lightning, and work of man. The most important of these methods is the result of the 9 per cent expansion that water undergoes when it freezes. Tremendous forces are produced in this way, and their efficacy is shown by the thick layers of broken rock that form in areas that go through a daily freeze-thaw cycle, many times a year. (See Fig. 4-3.)

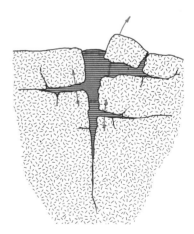

Figure 4-3. Frost wedging. Water in crack freezes from top down because ice is less dense than water (floats) and because the water in a crack is cooled most at the surface. This restricts the water in the crack. Because ice occupies more volume than water, further freezing creates forces due to this expansion. These forces cause further cracking of the rock.

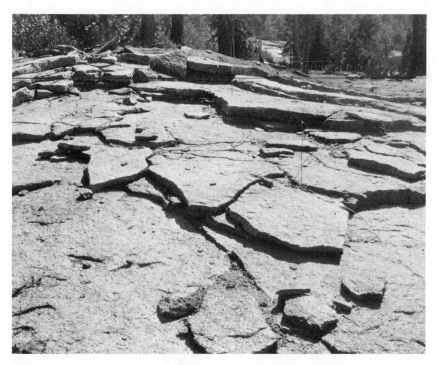

Figure 4-4. Cracks or joints parallel to the surface are believed to be caused by expansion. The expansion occurs as erosion removes the overlying rocks. The granitic rocks in this view crystallized at least a mile beneath the surface. Jointing of this type on a larger scale is conspicuous in many areas of granitic rocks. Photo by G. K. Gilbert, U.S. Geological Survey.

and the shape of the fragments in such a rock can reveal some of the history of the rock. Thus if the fragments are sharp and angular, they probably were buried quickly; but if the fragments are rounded, they were abraded during long transportation. (See Fig. 4-6.)

Water is important in most types of weathering. The role of water is largely determined by its structure. The water molecule is composed of an oxygen atom and two smaller hydrogen atoms that are attached to the oxygen about 105 degrees apart, as shown in Fig. 2-6. The hydrogen has a slight residual positive charge, and the oxygen a similar weak negative charge. Because of these charges the molecules tend to form in groups of two to eight. More groups are formed near the freezing point than in liquid water, and because the groups take up more space, water expands on freezing. These residual charges also make water an excellent

Another type of mechanical weathering is caused by the expansion of rock when erosion removes the weight of the overlying rocks. The outermost layer of the rock may expand enough so that it breaks from the main mass, forming cracks or joints parallel to the surface. (See Fig. 4-4.) This may also occur in boulders released by frost action. In this case layers like onion skin may form. (See Fig. 4-5.) However, most weathering of boulders in this way is caused by chemical weathering, as noted below.

The products of mechanical weathering include everything from the huge boulders found beneath cliffs to particles the size of silt. The erosion and transportation of these fragments will be discussed in Chapters 7 through 11. These fragments can form rocks such as sandstones,

Figure 4-5. Expansion of surface layers can produce spherical masses with onion skin layers. Such weathering probably begins on joint cracks; and although splitting-off of the layers is mechanical weathering, the expansion of the layers is probably caused largely by chemical weathering. This photo of basalt in Puerto Rico by C. A. Kaye, U.S. Geological Survey.

Figure 4-6. Sand grains showing different degrees of roundness. The rounded sand grains were probably shaped by abrasion during longer transportation than the angular grains. If the fragments are of different materials, the softer will be rounded first.

solvent. Removal of material in solution by ground water is called *leaching.*

The rate of chemical weathering depends in general on the temperature, the surface area, and the amount of water available. Except in cold or very dry climates, it generally keeps up with mechanical weathering, and the two can be separated only in concept. Mechanical weathering provides the large surface area necessary for chemical activity to take place at the surface of the earth. (See Fig. 4-7.)

In following chemical weathering, we will see what happens to the elements in the main rock-forming minerals. The main reactions involved are *oxidation, hydration,* and *carbonation.* Oxidation is reaction with oxygen in the air to form an oxide. Hydration is reaction with water, and carbonation is reaction with carbon dioxide in the air to form a carbonate. The carbonation reaction begins with the uniting of carbon dioxide (CO_2) and water (H_2O) to form carbonic acid (H_2CO_3), or

$$CO_2 + H_2O \rightarrow H_2CO_3$$

This acid plays an important role in many weathering reactions. The air in soil, because of biologic processes, contains up to ten per cent carbon dioxide, much more than the small amount in the atmosphere. In all of these reactions, substances are added

so that the total volume is increased. The details of how these reactions take place are not well understood and are an important area of current research in soil science, a relatively new branch of science. The reactions take place slowly and are probably influenced by the composition of the water films that surround the particles and by the organic materials present (particularly the acids). The importance of organic materials in weathering and soil formation is covered in the next section.

Air pollutants such as carbon dioxide and sulfates, released by burning fossil fuels and industrial processes, cause accelerated weathering of building stone and other materials in cities. The ultimate fate of many pollutants is not known, but it appears that some are absorbed and decomposed by soil.

The rain water involved in most weathering is not pure water. In addition to carbon dioxide, it contains dissolved salts. Atmospheric water

vapor, in general, will not condense to form clouds or rain without a nucleus around which to condense. Dust particles provide some of these nuclei, but most are tiny crystals of salts from the ocean. Spray from ocean waves evaporates, and the tiny crystals of salts formed in this way are carried away by the wind. The salts formed when sea water is evaporated are described in a later section. Near the ocean, rain water has the same chemical composition as sea water, but is very much diluted. Farther inland, the composition is variable because the winds transport the various salts from the ocean differently. The chemical composition of rain water does have an effect on the clays produced by weathering, but little more can be said on the complex subject of clay mineralogy here. However, the salts in rain water accumulate in lakes that have no outlet to the sea, such as Great Salt Lake.

Chemical weathering is most intense on grain surfaces and because of the increase in volume during chemical weathering, this trait can cause disaggregation of a rock. This process may account for fragmentation of rocks in areas where mechanical weathering by frost action does not occur. Laboratory experiments show that expansion and contraction from daily heating and cooling are

6 square feet
864 sq. inches
(6 × 12 × 12)   1728 sq. in.   3456 sq. in.   6912 sq. in.

Figure 4-7. Surface area versus particle size. Shows how mechanical weathering increases the surface area and thus helps to increase chemical weathering. A cubic foot is shown in each case.

not enough to cause disintegration, even on a desert.

Climate affects the topography produced by weathering. Dry areas where mechanical weathering predominates generally have bold, angular cliffs (see Fig. 4-8), and humid areas have rounded topography. Limestone, which is made of the soluble mineral calcite, is readily dissolved by carbonic acid. Thus it is rapidly weathered and tends to form valleys in humid regions, but it is a resistant rock and forms bold outcrops in dry areas. In warm, moist climates, weathering may extend to depths of over 400 feet. Maximum weathering probably takes place just below the ground surface—the depth at which fence posts tend to break.

Chemical weathering can be de-fined also as changes taking place near the surface that tend to restore the minerals to equilibrium with their surroundings. The primary minerals form, in most cases, at elevated temperature and pressure and are in equilibrium under those conditions. When these minerals are exposed at the surface, they are in conditions far from those under which they formed; therefore, they react to form new minerals which will be in equilibrium. One can then predict that the highest-temperature minerals will probably be affected first in chemical weathering. Thus it turns out that Bowen's reaction series (Fig. 3-23), which shows the sequence of mineral formation with falling temperature, also shows the relative stability to weathering. Olivine is the first to be affected and quartz the least affected, although biotite is easily weathered. Organic soil acids can in some cases upset these relationships.

The weathering of potassium feldspar will illustrate chemical weathering. An overall reaction can be written as

2 potassium feldspar
$$2KAlSi_3O_8$$
+ carbonic acid + water
$$H_2CO_3 \qquad H_2O$$
→ clay + potassium carbonate
$$Al_2Si_2O_5(OH)_4 \qquad K_2CO_3$$
+ 4 silica
$$4\ SiO_2$$

The feldspar is broken down by a process in which hydrogen replaces potassium. The hydrogen can come from water, from carbonic acid, or from the hydrogen ions that surround

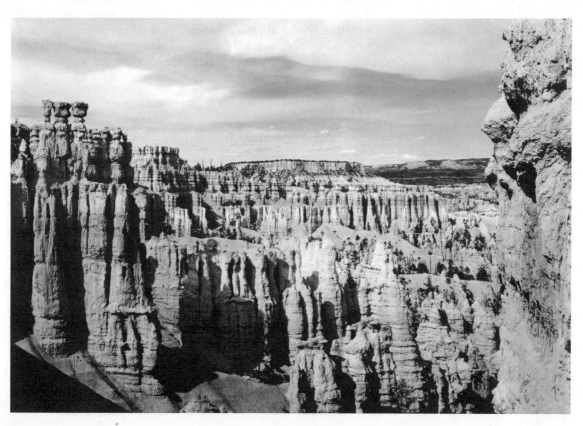

Figure 4-8. Angular topography in an arid area. Weathering has accentuated minor differences among the sedimentary layers. Bryce Canyon National Park, Utah. Photo from Union Pacific Railroad.

plant roots. The first two sources are inorganic, and the last is an example of biological weathering. The potassium released in the latter way is taken up by the plant. The clay mineral shown in the illustrative reaction is not the only clay mineral that can form. Potassium and magnesium clays are also common and form in some weathering environments. Weathering reactions of other feldspars are similar, differing only in that sodium and calcium are involved in place of potassium.

The weathering of the ferromagnesian minerals is also similar, and magnesium-bearing clays may form in this way. The ferromagnesian minerals are more easily weathered because the iron is easily oxidized, breaking down the mineral.

The products of chemical weathering of the common minerals can be summarized as follows:

Quartz — little chemical weathering, sand grains produced.

Feldspars—

Potassium feldspar ($KAlSi_3O_8$). K by carbonation produces K_2CO_3.
Al,Si,O by hydration produce clay ($Al_2Si_2O_5(OH)_4$).
SiO_2 is released as soluble silica and colloidal silica.

Plagioclase feldspars—produce the same end products except that $CaCO_3$ and Na_2CO_3 are produced instead of K_2CO_3.

Muscovite — produces the same products as potassium feldspar.

Ferromagnesian minerals — depending on composition, produce the same products plus iron oxides — hematite Fe_2O_3 (red iron oxide) or limonite $FeO(OH)$ (yellow iron oxide). Magnesium forms soluble mag-

Table 4-1. Weathering products and their disposition.

Weathering Product	Disposition
Clay	Forms shale.
Silica	Forms chert and similar rocks.
Potassium carbonate (K_2CO_3)	Some is transported to ocean, some is used by plant life, and some is adsorbed on or taken into certain clays.
Calcium carbonate ($CaCO_3$)	Forms limestone.
Sodium carbonate (Na_2CO_3)	Dissolves in ocean.
Iron oxides	Form sedimentary deposits of iron oxide.
Magnesium	Some replaces calcium in limestone to form dolostone, and some is transported to ocean.

nesium carbonate, and some is used in producing clay minerals.

The products of weathering and their ultimate disposition are summarized in Table 4-1. (See also Fig. 4-9.)

From this brief outline of chemical weathering it is possible to make some interpretation of sedimentary rocks. As might be expected, most sandstones are composed largely of quartz because it is the mineral most resistant to weathering. Some sandstones contain much feldspar, and such rocks formed under conditions that did not permit chemical weath-

ering to complete its job. This situation could have developed several ways. Accumulation might have been too rapid to permit complete weathering, or the climate may have been too dry or too cold. Certainly there are other possibilities, but this serves as an example of how sedimentary rocks can be interpreted.

Soils

Soils are of great economic importance. They can be said to be the bridge between terrestrial life and

Figure 4-9. Simplified chart showing origins of sedimentary rocks.

A horizon

B horizon

C horizon

Soil

Bedrock

A.

B.

Figure 4-10. Soil profiles. A. Residual soil with well-developed horizons or layers. B. Transported soil without layering.

the inanimate world. Soil, as the term is used here, is the material that supports plant life, and so supports all terrestrial life. Only lichen and primitive plants, such as silver sword in Hawaii, can get nutrition directly from rock without soil. The development of soil is a complex interplay of weathering and biologic processes.

The importance of soil to life is shown by the composition of some foods. Steak contains about 12.5 per cent phosphorus, iron, and calcium by weight. Thus each pound of steak contains about two ounces of rock material obtained from soil by plants. A much larger amount of rock had to be weathered to release these two ounces. Plants use much weathered rock material. For example, an acre of alfalfa yields about four tons of alfalfa per year. To produce this alfalfa, over two tons of rock must be weathered to obtain the phosphorus, calcium, magnesium, potassium, and other elements in the alfalfa.[1] This corresponds to about 0.007 inches of rock weathered. The average rate of erosion is estimated to be about one inch per 9,000 years, so this is about 5.5 times the annual erosion rate. This "uses" soil at a very high rate, so fertilizers must be added.

Thus soils can be used faster than they form, a fact not always taken into account in agricultural economics.

There are many different types of soils. At first it was thought that the parent rock determined the soil type. We now know that climate is more important than parent rock; but the situation is complex, and other factors, especially the type of vegetation, are involved. Soils can be described as mature and immature. On hillslopes, erosion may prevent soils from reaching maturity and the transported soil may accumulate in valleys. (See Fig. 4-10B.) Many soils form on surface materials that have been transported. It apparently takes at least a few hundred years for a soil to develop. For this reason soil conservation is very important. The range in rate of soil formation is great. In 45 years, 14 inches of soil formed on pumice from the eruption of the volcano Krakatoa. At the other extreme, in many places no soil has formed on glacially polished surfaces in about 10,000 years. (See Fig. 10-9.)

In most soils three layers, or *horizons,* can be recognized. They are the topsoil, the subsoil, and the

partly decomposed bedrock. In addition, in some areas, especially forests, a layer of organic material may form at the surface. (See Figs. 4-10 and 4-11.) The material that gives the upper parts of many soils their dark color is *humus.* Humus is formed by the action of bacteria and molds on the plant material in the soil. Chemical weathering releases some of the materials used by plants, and humus also provides food for plants.

Soils form as a result of the weathering processes just discussed and the effects of the organic material present, mainly humus. One of the most important processes in the formation of soils is the leaching of material from the top soil and the deposition of this material in the subsoil. Water reacts with humus to form acid that is very effective both

Figure 4-11. Soil profile revealed in a pit near Ladson, South Carolina. Photo by H. E. Malde, U.S. Geological Survey.

[1]W. D. Keller, "Geochemical weathering of rocks: source of raw materials for good living," *Journal of Geological Education,* Vol. 14, No. 1, February, 1966, pp. 17-22.

in leaching and in causing chemical weathering. Earlier, the similar role of carbonic acid, formed by carbon dioxide and water, was discussed. Carbon dioxide is also formed in soil from the decay of organic material.

We can now consider a few of the more important types of soils. In the United States the soil of the humid east is different from that of the drier west. The dividing line is approximately at the 25-inch annual rainfall line and runs almost north-south near 97° longitude. (See Fig. 4-12.) The characteristics of the soils about to be discussed are summarized in Table 4-2.

The soil that predominates in eastern United States is called *pedalfer* (*pedon*, Greek for soil, and Al, Fe,

the symbols for aluminum and iron). Humus is well developed in the temperate, moist climate. The abundant rain water becomes strongly acid by reaction with the humus and leaches the topsoil. Aluminum and iron leached from the topsoil are deposited in the subsoil, giving it a brown (limonite) color. Soluble materials such as calcite or dolomite are leached from this type of soil and are generally removed by the ground water. These soils are quite acid, and clays are well developed. Ground water carries the clay to the subsoil so that topsoil may be somewhat sandy, due to resistant quartz fragments. Pedalfers generally are best developed under forests. Conifer forests produce a thick layer of litter on

Figure 4-12. Map of major soil types in the United States. The dividing line is approximately at the 25-inch annual rainfall line except in the northern Rocky Mountains where low temperature affects the soil type.

the forest floor and so develop more humus. Therefore, the soil is very acid and its development is accentuated; the resulting soil is light gray

Table 4-2. Generalized summary of soil types. See text for details.

CLIMATE	TEMPERATE HUMID (>25" RAINFALL)	TEMPERATE DRY (< 25" RAINFALL)	TROPICAL SAVANNA (HEAVY SEASONAL RAINFALL)		EXTREME ARCTIC, DESERT (HOT OR COLD)
VEGETATION	Forest	Grass-brush	Grass and tree		Almost none, so no humus develops
TYPICAL AREA	Eastern U.S.	Western U.S.	—		—
SOIL TYPE	Pedalfer	Pedocal	Laterite		
TOPSOIL	Sandy; light colored; acid.	Commonly enriched in calcite; whitish color.	Enriched in iron (and aluminum); brick red color.	Zones not developed	No real soil forms because there is no organic material. Chemical weathering is very slow.
SUBSOIL	Enriched in aluminum, iron, clay; brown color (limonite).	Enriched in calcite; whitish color.	All other elements removed by leaching.	Zones not developed	
REMARKS	Extreme development in conifer forests because abundant humus makes ground water very acid. Produces a light gray soil because of removal of iron.	*Caliche* is the name applied to the calcite-rich soils.	Apparently bacteria destroy the humus so no acid is available to remove iron. Tropical rain forests develop soils similar to pedalfers.		

in color. It is in the pedalfers that calcite, generally called agricultural lime or limestone, is spread on the soil to combat its acidity.

In the drier west, less humus develops so the limited amounts of ground water are much less active chemically. Chemical weathering is also much slower in this climate so there is less clay than in pedalfers. Under these conditions leaching is not complete, and the soluble materials, especially calcite, are deposited in the subsoil because the ground water evaporates before it can remove them. Such soils are called *pedocals* (*pedon* for soil, and *cal* for calcite). The calcite forms as whitish material in the soil, called *caliche* (Spanish for lime). Caliche may accumulate in the topsoil as well as in the subsoil. This can occur when the ground water moves upward through capillary openings because of the evaporation at the surface. Pedocals tend to develop under grass- or brush-covered areas.

Soil formation in tropical climates is not fully understood, in part because few areas have been well studied. The soil in rain forests is somewhat similar to the pedalfers. The soil of the grass- and tree-covered savannas, however, is much different. Apparently because of the seasonal heavy rainfall, leaching is the dominant process. Everything, including silica and clays, is leached from the soil and carried away by ground water except iron and aluminum, which form hydrated oxides. Apparently, at the high temperatures, bacteria destroy the humus so that the ground water is not acid and cannot remove iron and aluminum. The soils produced in this way are called *laterites*. (See Fig. 4-13.) Laterite is from the Latin word for brick, and its reddish-brown color is similar to that of a brick. Laterites do not form distinct layers as other soils do. As might be expected, they are poor soils on which to grow crops, as shown by the many unsuccessful attempts to develop new farmlands in emerging tropical countries. Most laterites are iron-rich; however, in some areas they are mainly aluminum. Such a mixture of hydrated aluminum oxides is called *bauxite* and is a valuable ore. Bauxite probably forms only from the weathering of aluminum-rich rocks that have little iron, such as some feldspar-rich igneous rocks. At some places laterites are mined for iron ore and, at a few places, for their nickel or manganese; here the parent rock probably was rich in these elements.

In the extreme climates, such as arctic and desert, very little chemical weathering occurs and there is almost no organic material; therefore, no real soil can develop.

Trace Elements in the Soil and Health

Some elements present in the environment in only trace amounts have a large influence on health. A few obvious examples of good and bad effects of trace elements are widely known. The relationship between dental caries and fluorine in the water supply, and selenium poisoning of cattle by "loco weed" (*Astragalus racemosus*) are known to almost everyone.

Although some trace elements, such as iodine, apparently move through the atmosphere, most come from the soil and enter the life cycle either through water supply or plants. Although climate is more important in soil formation than the type of bedrock, the bedrock does influence the trace elements in the soil. Because most of these elements are metals, mining areas are commonly enriched in one or more of them. Many other areas are also either enriched or deficient in some or many of the trace elements. The amount of chemical weathering seems important, and both arid and tropical areas are generally deficient. Not all trace elements are beneficial, as will be

Figure 4-13. Simplified summary of weathering showing how laterites form.

shown. Insecticides, sprays, and fertilizers also add materials to plants and soils.

Deciding which elements should be considered trace elements is not a simple decision. Most living tissue is composed of hydrogen, carbon, nitrogen, oxygen, sodium, magnesium, phosphorus, sulfur, chlorine, potassium, calcium, and in those animals with hemoglobin, iron. In general the trace elements regulate enzyme actions. The elements that do this are calcium, magnesium, iron, manganese, cobalt, copper, zinc, and molybdenum. Zinc seems to be necessary for growth and to reach maturity. Obviously there is overlap between the two lists. In addition, there is another group of elements called age elements because they accumulate during life. This group includes beryllium, aluminum, silicon, titanium, vanadium, chromium, nickel, arsenic, selenium, tin, lead, silver, cadmium, barium, gold, and mercury. The role of this last list is not clearly understood, and future studies may show that some of them belong in the other lists. Excess or deficiency of some of these can cause great changes or even death. Some can cause tumors and so may cause some cancers. The role of metals in both the cause and treatment of cancer is an area of current research. Some of the evidence showing the effect of trace elements on health is comparison of maps showing trace elements in soils with maps showing either mortality or morbidity of disease. Such evidence is subject to various interpretations and so is somewhat controversial. In the following paragraphs, a number of examples of the effects of trace elements will be described.

Fluorine aids in reducing dental caries, and many municipalities add it to drinking water for this reason. Geographic areas where teeth are good also differ from areas where teeth are poor in having higher calcium, molybdenum, aluminum, titanium, and calcium-magnesium ratio, and lower copper, manganese, barium, and strontium. The evidence that fluorine reduces caries is excellent, but in spite of this, many people object to adding it to their water. Low-fluorine areas have higher rates of both osteoporosis and aortic calcification, so the benefits of fluorine may go far beyond sound teeth. In addition, other studies suggest that vanadium, boron, strontium, barium, lithium, and titanium may prevent caries, and lead, copper, zinc, and chromium may increase them.

In Georgia, highland areas with soils containing many trace elements have lower cardiovascular death rates for middle-aged white males than coastal-plain areas where soils are deficient in trace elements. (See Fig. 4-14.) However, the amounts of trace elements in plants from these two areas are not too different, suggesting that if trace elements are the cause, they may enter the life cycle through the water supply or other

Figure 4-14. Relationship between cardiovascular disease and geologic provinces in Georgia. The areas with both high and low death rates for cardiovascular disease among white males of ages 35–74 are shown. From U. S. Geological Survey, *Professional Paper 574 C,* 1970.

means.

Goiter is well known to be caused by a deficiency in iodine. We avoid the enlarged neck of goiter by adding iodine to salt in places where iodine is low. Iodine is in sea water, so those living near the sea or eating much sea food generally get enough. Mountain areas and recently glaciated areas tend to be low in iodine. The distribution of iodine is erratic, so many exceptions can be found.

Cancer may also be related to trace elements. Low-iodine areas have a suspicious relationship to breast cancer. Cancer is less common near the Gulf of Mexico than in northern states, but this might be related more to climate than to trace elements. Diabetes is less common in the tropics, and insulin doses are smaller there; again this may be climatic.

Cattle fed forage low in selenium have a very high incidence of muscular dystrophy, which can be cured by selenium injection. The soils of part of the northern great plains are rich in selenium; some plants, such as loco weed, concentrate the selenium, and cattle eating it are poisoned. Apparently the selenium enrichment of this plant is seasonal or otherwise variable, because ranchers report conflicting observations. Thus too little and too much selenium are both detrimental to animal life.

In a Florida study, salt-sick disease in cattle and anemia in children were related to deficiencies in iron and copper in the soil. Addition of these two elements to diet promptly caused recovery. In addition, cobalt aided in copper retention, and molybdenum accentuated copper deficiency and also interfered with phosphorus metabolism.

Anthrax, another cattle disease, is controlled by soil type. Apparently the spores live in organic-rich calcium soils, and the disease reaches cattle after such infected soils dry out.

Some other studies suggest that soils in western United States contain toxic amounts of vanadium, molybdenum, and selenium. Some of the peat used in gardens contains high amounts of uranium, and some natural fertilizers add cadmium and zinc to soils. Mercury is high in many soils. All of these points have a bearing on claims of the value of foods raised on organic versus chemical fertilizer, and apparently every case must be considered individually. Thus some food fads may be based unknowingly on trace elements. Medicinal plants in many cases also depend on trace elements for their value.

Lithium is sometimes used in the treatment of certain types of mental illness. Cities with trace amounts of lithium in their water supply have fewer admissions to mental hospitals than those that do not, according to one preliminary study.

Lithification

The products of weathering generally are eroded, transported, and deposited before they are transformed into sedimentary rocks. These processes are described in Chapters 7 through 11.

The transformation of a sediment into a rock is called *lithification*. Several processes are involved. *Cementation* is the deposition by ground water of soluble material between the grains and, as might be expected, is most effective in coarse-grained, well-sorted, permeable rocks. The main cementing agents are

Calcite—recognized by acid test. (Do not mistake the effervescence of cement for the overall effervescence of limestone.)

Silica—generally produces the toughest rocks.

Iron oxides—color the rock red or yellow.

Other cements, such as dolomite, are possible. In certain poorly sorted rocks, clay, which generally colors the rock gray or green-gray, may be thought of as forming the cement.

Compaction is an effective lithifier of fine-grained rocks such as shale and siltstone. Compaction generally comes about from the weight of overburden during burial, and the reduction in volume due to compaction and squeezing out of water may amount to over 50 per cent in some shales.

Recrystallization probably is important in producing chert and in transforming calcite-bearing mud into limestone. Recrystallization produces an interlocking texture.

Classification and Formation

As with the igneous rocks, it is more important to interpret the formation of these rocks than merely to name them. Knowing the origin of the materials that compose a sedimentary rock and understanding the origin of its sedimentary features will permit such interpretation.

According to the way they were formed, sedimentary rocks can be classified as marine, lacustrine (lake deposited), glacial, eolian (wind deposited), fluvial (river deposited), etc. They can be classified also by type as limestone, chert, quartzose sandstone, etc., or by their mode of origin as clastic, chemical precipitate, or organic. In practice, these are blended to produce a practical classification as follows.

Clastic Rocks

These are composed of rock fragments or mineral grains broken from any type of pre-existing rock. (See Fig. 4-15.) They are subdivided according to fragment size. Commonly, sizes are mixed, requiring intermediate names such as sandy siltstone. They are recognized by their clastic texture. The fragments originate from mechanical weathering. The agents of erosion, transportation, and deposition that bring these fragments together and so form the clastic sedimentary rocks are described in Chapters 7 through 11.

Composition is not used in the general classification of clastic sedimentary rocks because the composition may be affected by several factors. The examples of this are almost endless, but a few simple ones will illustrate. A quartz sandstone may be formed if the source area is mainly quartz (perhaps an older quartz sandstone), or a quartz sandstone may be the result of prolonged weathering, leaving only quartz fragments. Large amounts of

Table 4-3. Clastic sedimentary rocks.

SIZE	SEDIMENT	ROCK
Over 2 mm	Gravel	*Conglomerate* — generally has a sandy matrix. May be subdivided into round-stone and sharpstone conglomerate (*breccia* is a synonym) depending on the fragment shape. (See Fig. 4-16.)
$^1/_{16}$–2 mm	Sand	*Sandstone* — recognized by gritty feel. (See Figs. 4-17 and 4-18.) Generally designated as coarse, medium, or fine if well sorted. They have also been subdivided on the basis of composition. The more important types are *Quartzose sandstone* (mainly quartz) *Arkosic sandstone (arkose)* — over 20 per cent feldspar. *Graywacke* — poorly sorted, with clay or chloritic matrix.
$^1/_{256}$–$^1/_{16}$ mm	Silt	*Siltstone (mudstone)*—may be necessary to rub on teeth to detect grittiness, thus distinguishing it from shale.
Less than $^1/_{256}$ mm	Clay	*Shale* — distinguished from siltstone by its lack of grittiness and its *fissility* (ability to split very easily on bedding planes). Rocks composed of clay that lack fissility are called *claystone.*

feldspar in a sandstone may imply rapid deposition and burial before chemical weathering could decompose the feldspar, or it might imply a cold climate in which chemical weathering is very slow. In many studies composition is used to subdivide sandstones as indicated in Table 4-3.

The actual formation of shale is somewhat more complex than indicated in Table 4-3. Much of the clay that gives a shale its fissility apparently develops or is mechanically reoriented after deposition. This is suggested by the lack of geologically young shale. Young fine-grained clastic rocks are mainly mudstones.

Non-clastic Rocks

These are formed by chemical precipitation, by biologic precipitation, and by accumulation of organic material. These processes extract specific materials from their surroundings, generally sea water, and

Clastic texture

Interlocking texture. Developed in chemical precipitates such as some limestones and in many igneous and metamorphic rocks.

Figure 4-15. Recognition of textures is very important in the identification and interpretation of rocks.

A.   B.

Figure 4-16. Conglomerate. A. Rounded pebbles in a sandstone matrix. B. Sharpstone conglomerate (breccia). The angular pebbles in this rock were probably not transported far compared with the pebbles in A. Photos from Ward's Natural Science Establishment, Inc., Rochester, N.Y.

Figure 4-17. Sandstone. The bedding is too thick to show in this specimen. Photo from Ward's Natural Science Establishment, Inc., Rochester, N.Y.

precipitate these substances, forming rocks. The rocks are classified mainly by composition. As with the clastic rocks, these rocks are commonly mixed, both among themselves and with the clastic rocks.

Limestone is composed of calcite. It is recognized by effervescence with dilute hydrochloric acid. It is generally of biologic origin and may contain fossils. (See Fig. 4-19A.) A rock composed mainly of fossils or fossil fragments is called *coquina.* (See Fig. 4-19B.)

Limestone can form in many ways as shown in Table 4-4. Most limestone probably originates from organisms that remove calcium carbonate from sea water. The remains of these animals may accumulate to form the limestone directly, or they may be broken and redeposited. Calcite can also be precipitated chemically. The amount of carbon dioxide in sea water controls the amount of calcite that can remain in solution. If the amount of carbon dioxide is reduced by warming the water, as would occur in shallow tropical water, calcite may be precipitated. Marine plants also remove carbon dioxide from sea water and aid in calcite precipitation. Bacteria can also aid chemical precipitation of calcite by making the water more alkaline. Decaying organic material can increase the amount of carbon dioxide and so retard precipitation of calcite. Limestone is presently being formed by chemical precipitation on the shallow Bahama Banks where the factors discussed are favorable.

Chalk is soft, white limestone formed by the accumulation of the shells of microscopic animals. The shells are composed of calcium carbonate, and chalk is recognized by effervescence with acid.

Dolostone (dolomite) is composed of dolomite. It is recognized by effervescence after scratching (to produce powder) with dilute hydrochloric acid; it will also react (with-

Figure 4-18. Sandstone outcrop. Photo from New Mexico State Tourist Bureau.

out scratching) with concentrated or with warm dilute hydrochloric acid. Dolostone is generally formed by replacement of calcite, presumably soon after burial. The reduction in volume in this replacement may produce irregular voids and generally obliterates fossils.

Chert is microscopically fine-grained silica (SiO_2). It is equivalent to chalcedony (see Mineral Table, Appendix A). It is unfortunate that many names have been applied to the fine-grained varieties of silica. These names were based on minor differences such as color and luster that cannot be substantiated by modern methods, and there is little agreement among mineralogists on their naming. Both chert and chalcedony may contain opal and fine-grained quartz. Dozens of names have been applied to chalcedonic silica; of these, a few may be useful, e.g., agate for banded types, and flint for dark gray or black chert.

Chert originates in several ways. Some may precipitate directly from sea water in areas where volcanism releases abundant silica. Most comes from the accumulation of silica shells of organisms. These organisms remove dissolved silica from sea water to form their shells or skeletons. These silica remains come from diatoms, radiolaria, and sponge spicules, and are composed of opal. Opal is easily recrystallized to form chert. Thus much chert is recrystallized, making the origin difficult to discern. In recrystallization the silica may replace other materials, and fossils replaced by silica are common. If this occurs in limestone, beautifully preserved fossils with delicate features intact can be recovered by dissolving

A.   B.

Figure 4-19. Limestone. A. Fossiliferous limestone. A fine-grained white limestone with abundant fossil fragments. B. Coquina, a variety of limestone composed almost entirely of fossil fragments. Photos from Ward's Natural Science Establishment, Inc., Rochester, N.Y.

the limestone with acid. Chert may form either beds or nodules.

Gypsum forms from the evaporation of sea water. Rocks formed in this way are termed *evaporites;* they are discussed below. The identification of gypsum is described in Appendix A.

Rock salt is composed of halite. It is recognized by its cubic cleavage and

its taste. (See Appendix A.) It is deposited when restricted parts of the sea are evaporated.

Diatomite is a soft, white rock composed of the remains of microscopic plants. Because the remains are composed of silica, it is distinguished from chalk by lack of effervescence.

Coal is formed by the accumulation of plant material.

Table 4-4. Origins of the more important non-clastic sedimentary rocks.

| | Biologic | Chemical | | Remarks |
		Evaporite	Other	
Limestone	×	× Rare	×	
Dolostone		× Rare		Forms mainly by replacement
Chert	×		× Precipitation from volcanic silica	Most biologic chert recrystallizes from opal
Gypsum		×		
Rock salt		×		
Chalk	×			
Diatomite	×			
Coal	×			

A.

B.

Figure 4-20. A. Mud cracks, which form as a result of drying mud flats, can be used to indicate top and bottom. B. Mud cracks in Death Valley, California. Photo by J. R. Stacy, U.S. Geological Survey.

Evaporite deposits are formed by the evaporation of sea water. Gypsum and rock salt are the main rocks formed in this way. When sea water is evaporated at surface temperatures, such as in a restricted basin, the first mineral precipitated is calcite. Dolomite is the next mineral precipitated, but only very small amounts of limestone and dolomite can be formed in this way. Evaporation of a half-mile column of sea water would only produce an inch or two of limestone and dolomite. After about two-thirds of the water is evaporated, gypsum is precipitated; and when nine-tenths of the water is removed, halite forms. During the last stages of precipitation, potassium and magnesium salts form. Thick

beds of rock salt imply evaporation of large amounts of sea water. This is discussed in Chapter 6.

The rocks described above are the most important, but by no means the only, sedimentary rocks. Any mixture of types is possible and, because any weathering product may form a sedimentary rock, endless variety is possible. Some less abundant types include economically important deposits, described in Chapter 6, such as iron oxides, phosphorous rocks, bauxite (aluminum ore), and potash.

Shale, limestone, and sandstone make up over 99 per cent of all sedimentary rocks. Because feldspars are the most abundant minerals, their main weathering product, shale, is the most abundant sedimentary rock.

Limestone is next in abundance, and sandstone is last. (See Fig. 4-1.) The abundance of shale is somewhat less than is predicted from the abundance of clay-forming silicate minerals, suggesting that some clay is deposited in the deep sea basins. Clay particles are very small and sink slowly; they can be carried thousands of miles by gentle currents.

Features of Sedimentary Rocks

The most noticeable feature of an outcrop of sedimentary rocks is the bedding, which records the layers in the order of deposition with the oldest at the bottom. In some instances the beds are too thick to show in a small outcrop, and, for the same reason, many hand specimens do not show obvious bedding.

Study of the type of fossils and of sedimentary features can tell much about the environment of deposition and, in some cases, the post-depositional history. Mud cracks on bedding planes record periodic drying and suggest shallow water and, perhaps, seasonal drying (Fig. 4-20). Ripple-marks also suggest shallow water with some current action, but are also known to form in deep water. Detailed study of ripple-marks can show type and direction of current. (See Figs. 4-21 and 4-22.) Current action can also cause localized scouring or cutting to produce cut-and-fill features. (See Fig. 4-23.) Another type of current feature is cross-bedding. (See Figs. 4-24 and 4-25.) These features can also be used to recognize beds that have been overturned by folding or faulting. (See Fig. 4-26.) Currents can also align mineral grains and fossils

Table 4-5. Composition of sea water.

Element	Per Cent	Origin	Rock Formed
Chlorine	0.019	volcanoes	rock salt (halite)
Sodium	0.011	chemical weathering	
Magnesium	0.001	chemical weathering	dolomite
Sulfur	0.0009	volcanoes	gypsum
Calcium	0.0004	chemical weathering	limestone (calcite)
Potassium	0.0004	chemical weathering	potassium salts

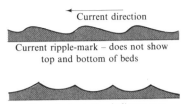

Current direction

Current ripple-mark – does not show
top and bottom of beds

Wave ripple-mark – indicates top
and bottom of beds. Formed by
back-and-forth slosh

Figure 4-21. Ripple-marks form where currents can act on sediments. Ripple-marks form at right angles to the current direction, and asymmetrical ripple-marks indicate which way the current flowed. To see why only wave ripple-marks can be used to determine top and bottom of beds, turn the page upside-down.

and so impart linear internal structures to rocks that can be used to determine the current direction.

If a mixture of sediments is suddenly deposited into a sedimentary basin, then the large fragments sink faster than the small ones. This does

Figure 4-22. Oscillation ripple-marks, Erie County, Ohio. Note the presence of ripple-marks in successive layers. Photo by D. F. Demarest, U.S. Geological Survey.

not violate Galileo's famous experiment, because here we are concerned with falling bodies in a viscous medium in which velocity of descent is controlled mainly by the size. The type of bedding produced by this kind of sedimentation is called *graded bedding* (Fig. 4-27); we can demonstrate this experimentally by putting sediments of various sizes in a jar of water, shaking, and then allowing the contents to settle. Graded beds can form in many ways, such as the sudden influx of sediment by a storm or seasonal flow of rivers. Their formation at the base of the continental slope is described in the next section.

The sorting of a sediment determines other of its textural features. The *porosity* of a sediment is a measure of its empty or void space, and its *permeability* is a measure of the interconnections of the pore spaces. These properties become quite important when one is concerned with recovering oil or ground water from an underground reservoir. The amount of pore space in sedimentary

A. B. C.

D.

Figure 4-23. Scouring produced by localized currents. A. Initial deposition. B. Channel cut by current. C. Deposition is resumed, filling the channel. Such a feature records an interruption in deposition and can be used to distinguish top and bottom. D. Channel filled with conglomerate. Note that the conglomerate is more resistant to weathering than the sandstone. Channel is about ten feet deep. Daggett County, Utah. Photo by W. R. Hansen, U.S. Geological Survey.

rocks is quite varied but surprisingly large. If we consider packing of uniform spheres (and most sand grains are rounded), we find that the closest possible packing contains 26 per cent void space, and the loosest, 48 per cent. (See Fig. 4-28.) The permeability will vary depending on the amount of pore space, the sorting, the cementation, and the size of the particles. A shale may have over 30 per cent porosity, but the surface tension of water will prevent flowage through the tiny openings.

Slump features as shown in Fig. 4-29 form if sediments are deposited on a slope or if they are tilted while still soft. Another type of deformation of soft rock is the vertical movement of sediment shown in Fig. 4-30.

Deposition

Most sedimentary rocks are deposited in oceans, and these marine rocks, easily dated by their contained fossils, record much of the earth's history. In the present section a few of the more important marine environments will be considered.

Figure 4-24. Development of cross-beds. A. Initial deposition. B. Channel cut by current. C. Channel filled by deposition from one side. D. Normal deposition after channel is filled. Cross-beds can also be used to distinguish top and bottom.

One of the interesting observations of marine sedimentary rocks is that at many places individual sedimentary units, such as sandstones, can be traced for hundreds of miles. At other places the beds are irregular and have limited horizontal extent. To see how this comes about we will consider how marine sedimentary rocks are formed. For simplicity, we will limit the discussion to clastic rocks, sandstones and shales. As will be seen, the origin of the fragments that compose these rocks is weathering and erosion. These fragments are delivered to the ocean by rivers.

Because most marine sediments consist of material delivered by rivers, most sedimentary rocks form near the continents. Sampling and drilling in the oceans show this to be true. If we wish to interpret the thick accumulations of old sedimentary rocks exposed in today's mountain ranges, we should study the continental shelves and the continental slopes. The structure of these areas is described in Chapter 15. (See Fig. 4-31.)

The continental shelf is part of the continent, and its extent is determined by seismic studies of the crustal structure. However, the 100-fathom, or 600-foot, depth line is generally close to the edge of the continental shelf. The continental shelf is almost flat; it has an average slope of about 10 feet per mile. The break in slope between the shelf and the continental slope generally begins at a depth of about 400 to 500 feet, and the range is from near sea level to 1,500 feet. The average depth of the shelves is near 200 feet. The continental slope extends on the average to a depth of two miles or more, with a slope of two to six degrees. This is a very steep slope compared to the mountain ranges of the earth.

The continental shelves are cov-

A.   B.

Figure 4-25. Cross-bedding. A. In river-deposited coarse sandstone. Photo by H. E. Malde, U.S. Geological Survey. B. In wind-deposited sandstone in Zion National Park, Utah. Photo by H. E. Gregory, U.S. Geological Survey.

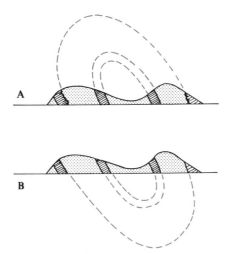

Figure 4-26. Alternate interpretations of the structure of outcrops of folded sedimentary rocks. The ripple-marks in A. show that it is the correct interpretation. In the absence of ripple-marks, or other sedimentary features that indicate top and bottom, it would not be possible to choose between the two interpretations.

Figure 4-27. Graded bedding. Produced by the more rapid settling of coarse material than fine material.

Close packing   Open packing

Figure 4-28. The packing controls the porosity, as illustrated here in the idealized case of spheres.

Figure 4-29. Soft sediments can slide down even very gentle slopes, developing slump features.

ered with clastic sediments for the most part. One would expect that, in general, the coarsest material should be found near shore or near river mouths and the finer silt and clay should be farther from the source. (See Fig. 4-32.) This is probably true in a very general way, but the many departures suggest that the situation is not simple. (See Fig. 4-33.) One very obvious reason is that during the recent glacial time the oceans were lower because much water was held in the glacial ice. Apparently, too, tidal and other currents are fairly effective in moving and sorting clastic material on the relatively shallow shelves, as will be shown in Chapter 11. This type of reworking of the sediments may pos-

sibly account for the uniform beds extending over large areas that are found on the present continents. In any event, pebbles and conglomerate are not uncommon far out on the shelf.

The present coastal plains on the Atlantic and Gulf coasts are underlain by what are apparently continental shelf deposits. Uplift has exposed the landward parts of these sedimentary accumulations, and seismic studies and deep drilling indicate that they extend onto and thicken on the present continental shelf. On the Gulf Coast these deposits are several miles thick, and their features and fossils indicate shallow-water deposition. The latter suggests slow subsidence.

Much less is known about the structure and formation of the continental slopes. They will be discussed more fully in Chapter 15. Limited sampling shows the surface, at least, to be covered mainly with fine sediments, presumably swept off the shelves. A more important process in the building of the slopes may be submarine landslides or mudslides. Following earthquakes, submarine telephone cables may be broken. The cables do not all break at the time of the earthquake but at later times, farther down the slope. In some cases many cables have been broken and the time of breaking is accurately known. This, plus the burial of some of the cables reported by the repair ships, strongly suggests that submarine landslides move down the continental slopes.

Another feature of continental shelves and slopes is the submarine canyons. These submarine canyons have many of the features of river canyons, and some are larger than such river canyons as the Grand Can-

A.

B.

Figure 4-30. A. Photo of deformed bedding. B. Diagram showing how deformation of soft rock forms the structure shown in photo.

yon. The origin of such features has been the subject of lively speculation, and not everyone is yet satisfied with the explanations offered. At first they were thought to have formed by ordinary river erosion at a time when sea level was lower than it is at present.

Many of the submarine canyons have heads near the mouths of large rivers, making this a reasonable hypothesis. However, many of them extend to depths of over two miles, and sea level changes of this magnitude in the recent past seem impossible. The

submarine landslides mentioned above offer the most likely explanation. Such landslides consist of sediment-laden water. Because they are heavier than the surrounding water, these landslides or mudflows move along the bottom and are capable of eroding the soft bottom material. These muddy waters are called *turbidity currents*. They can move down very gentle slopes and can transport fairly large pebbles. Landsliding of unconsolidated material on the continental slopes can account for the canyons there; and turbidity currents initiated by stream discharge, storms, or other currents can account for the submarine canyons on the continental shelves. As well as explaining the canyons, these processes can also account for the building of the continental slopes, for seismic studies show the slopes to be composed of sedimentary rocks.

Turbidity currents and submarine

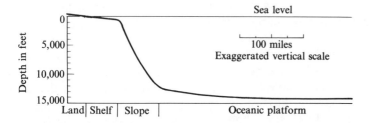

Figure 4-31. Continental margin showing typical continental shelf, continental slope, and oceanic platform. The vertical scale is exaggerated.

Figure 4-32. Idealized diagram showing coarse material being deposited near the shore and finer material further out.

Figure 4-33. Bottom sediments on part of the continental shelf in northeastern United States. The many departures from the idealized case shown in Figure 4-32 are probably caused by bottom currents and by deposition when sea level was lowered during glacial times. Adapted from F. P. Shepard and G. V. Cohee, *Bulletin,* Geological Society of America, Vol. 47, 1936.

landslides can also account for the smooth ocean bottoms at the foot of the continental slope at many places. (See Fig. 4-34.) These smooth surfaces are apparently similar to the fans that develop where rivers enter broad valleys. These submarine fans contain pebbles and shallow-water shells at some places. These fans do not occur everywhere at the foot of the continental slope.

To further evaluate the possible existence and importance of turbidity currents in the geologic past, we will look at their depositional features. Turbidity currents, because of their density and speed, can transport large fragments. These larger frag-

ments and the main mass of fine material are deposited fairly rapidly. The larger, heavier fragments settle first, forming graded beds. The eroding ability of turbidity currents suggests that they may erode or otherwise deform the top of the underlying bed. This is shown in Fig. 4-35.

In the deep ocean, far from land, the sediments consist mainly of wind-carried dust, largely of volcanic origin, and the fallen shells of tiny animals that live near the surface. Sediments of this type that are composed of more than 30 per cent biologic material are called *oozes.* (See Fig. 4-36.) The rate of accumulation

of oozes is very slow compared with other sedimentary rocks. It is estimated that it requires between 1,000 and 10,000 years to deposit one millimeter. The volcanic dust settles so slowly that most of it is oxidized while sinking to the bottom, and so it is termed red clay. Fig. 4-37 shows some features of deep ocean sediments.

Interpreting Sedimentary Rocks—Some Simple Examples

Interpreting geologic history from the study of sedimentary rocks is an important activity carried on by many geologists. These studies are carried on at many levels, both in the field and in the laboratory. The few examples given here may help you to interpret the rocks in your area. The examples here are simple, but a little imagination will reveal the many other possibilities and the many complexities that are possible. The following examples will be limited to the clastic sedimentary rocks, especially sandstones.

In studying a hand specimen of sandstone, the composition of the grains may reveal the rock type that was eroded to form the sandstone. The amount of weathering influences the mineralogy of the grains so that the grains of rock that could reveal the source are, in general, only present if very little weathering has occurred. Such a lack of weathering could be caused by a dry or cold climate, rapid or short transportation, or rapid burial, perhaps caused by rapid deposition. A quartz sandstone could reveal the opposite conditions. Study of the size, shape and

Figure 4-34. Turbidity current in Lake Ontario at mouth of Niagara River. The turbidity current is caused by the sediment-bearing river water. Note that part of the river water is carried by long-shore currents.

sorting of the grains may help in deciding which conditions were most important in the formation of the sandstone. Large grains suggest nearness to the source or short distance of transportation. Angular fragments, unless very resistant to abrasion, have not traveled very far. The sorting can reveal, among other things, the mode of transportation. Wind-transported sediments are very well sorted, and glacially deposited rocks have an extreme size range. The modes of transportation are described in Chapters 7 through 11.

All of these features can be seen with a hand lens, but the complete study of a hand specimen in the laboratory requires the use of the microscope and other instruments.

Much more can be learned at the outcrop of the rock layer. Here the bedding can be seen. The thickness and uniformity of the beds can reveal much about their origin. The irregular bedding of river deposits contrasts with the regularity of many shallow-water marine sedimentary rocks. The many features of sedimentary rocks, such as ripple-marks,

cross-beds and graded bedding, can also be studied at the outcrop. These features and their interpretation were described earlier. Fossils can also be collected at the outcrop. The environment of deposition can, in many cases, be determined from study of fossils, because the types of animals that live near shore are different from those in deep water; they also differ between warm and cold water. Perhaps the most important use made of fossils is to determine the age of the sedimentary layers which is the topic of Chapter 17. The de-

A.

B.

Figure 4-35. Depositional features of turbidity currents. A. Photo. The bottom of a bed is shown. B. Diagram. The lobe-like casts point in the opposite direction to current movement. The feature that crosses the specimen is the cast of a groove, probably caused by a boulder carried by the turbidity current. The specimen shown is about two feet in diameter.

formation, if any, of the beds can also be seen in many outcrops.

The next level of study of a sedimentary layer is to trace its areal extent by mapping its outcrops. By combining the interpretations made at each outcrop it is possible to reconstruct the basin in which the beds were deposited. An example of such an interpretation is shown in Fig. 4-38. In this example, the source area for the Morrison Formation is revealed from the current directions shown by the cross-bedding, the increase in thickness near the source, and the coarseness of the grains near the source. Note also that the Morrison Formation is exposed in only a small part of the basin and everywhere else it has either been removed by erosion or is covered by younger rocks. Most geologic interpretations must be made on the basis of such fragmentary evidence.

Figure 4-36. Radiolarian ooze from a depth of 17066 feet, about 600 miles south of Hawaii. Photo from D. R. Horn, Lamont-Doherty Geological Observatory.

Figure 4-37. Ocean bottom photographs. A. Hydrographer Canyon off the east coast of the United States. Soft coral and other bottom organisms at a depth of 2,400 feet. B. Hydrographer Canyon showing brittle stars and other organisms. C. Deep sea bottom in Romanche Trench. Photos courtesy of R. M. Pratt.

Summary

Sedimentary rocks are composed of rock fragments, weathering products, organic material, or precipitates; and most are deposited in *beds* or layers. They compose only about 8 per cent of the earth's crust, although they cover about 75 per cent of the surface.

Mechanical weathering breaks up a rock into smaller fragments; *chemical weathering* rearranges the elements into new minerals.

Mechanical weathering is caused mostly by the forces produced by the expansion of freezing water, and to a lesser extent by growing roots, burrowing animals, lightning and man.

Leaching is the removal of material in solution by ground water.

Chemical weathering is controlled by temperature, surface area, and amount of water available. Mechanical weathering provides the large surface area needed by chemical weathering.

Chemical weathering increases volume by adding substances. *Oxidation* is reaction with oxygen in air to form an oxide; *hydration* is reaction with water; and *carbonation* is reaction with carbon dioxide in air to form a carbonate.

Figure 4-38. Reconstruction of an ancient sedimentary basin from interpretation of the features in scattered outcrops. A. Map of outcrops of the Morrison Formation. B. Current direction inferred from cross-bedding in the Salt Wash Member of the Morrison Formation. The arrows show the down-slope direction of cross-beds. Length of the arrows is proportional to the consistency. C. Thickness and sediment type in the Salt Wash Member. Note that the coarser sediments are closer to the source area inferred from both thickness and cross-bedding. From L. C. Craig and others, U.S. Geological Survey, *Bulletin* 1009E, 1955.

Carbonation involves uniting carbon dioxide with water to form carbonic acid.

Dry climates tend to have bold cliffs because of the predominance of mechanical weathering. Humid areas have rounded topography. Maximum weathering probably takes place just below ground surface.

Bowen's reaction series, in general, shows relative stability to chemical weathering; olivine is first to be affected and quartz least affected, although biotite is easily weathered.

Quartz—very little weathering, sand grains produced.

Feldspars and muscovite — potassium, calcium, and sodium by carbonation form carbonates. Some potassium carbonate goes to ocean, some used by plants, some to clays. Calcium carbonate forms limestone. Sodium carbonate dissolves in ocean. Aluminum, silicon, and oxygen produce clay which forms shale. Silicon and oxygen produce silica which forms chert and similar rocks.

Ferromagnesian minerals produce same products plus iron into *hematite* (red iron oxide) or *limonite* (yellow iron oxide), and magnesium into some clay minerals or magnesium carbonate which replaces calcium in limestone to form dolostone.

Most sandstones are composed largely of resistant quartz fragments; sandstones with much feldspar formed under conditions which did not permit complete chemical weathering.

Soil is the material that supports plant life. Climate is more important than parent rock type in determining soil type.

Most soils have three layers or *horizons*: topsoil, subsoil, and partly decomposed bedrock. *Humus,* dark material in the upper part of soil, is formed by action of bacteria and molds on plant material. Soils form by weathering processes, leaching and

carbonation, and effects of an acid formed by water and humus.

Soils in humid areas are called *pedalfers* (*pedon*, Greek for soil, and Al, Fe for aluminum and iron). Pedalfers are quite acid because of abundant rainfall and humus. Aluminum and iron are leached from topsoil and deposited in subsoil.

Soils in drier areas (less than approximately 25 inches of precipitation annually) are *pedocals* (*pedon* for soil and *cal* for calcite). Leaching is not complete, and calcite is deposited rather than removed by solution in ground water. Such calcite deposits are called *caliche.*

Laterites are formed in tropical savannas; everything is leached away except iron and aluminum which form reddish brown hydrated oxide deposits. *Bauxite,* aluminum ore, is such a deposit, weathered from an aluminum-rich parent rock.

Trace elements in soils, which enter the life cycle through water supplies or plants, have a large influence on health.

Lithification is the transformation of sediment into a rock. *Cementation* is deposition of soluble material, such as calcite, silica, and iron oxides, between the grains by ground water. *Compaction* lithifies fine sediments by weight of overburden during burial. *Recrystallization* produces the interlocking texture of chert and limestone.

Clastic rocks are composed of fragments or minerals broken from any type of pre-existing rock; clastic rocks are generally subdivided by fragment size, not by composition except for some sandstone types.

Conglomerate — fragments over 2 mm; may be subdivided into *roundstone* or *sharpstone conglomerates* (*breccia*) if fragments rounded or angular respectively.

Sandstone — fragments between $1/16$ and 2 mm. Some compositional types

are *quartzose sandstone* (mainly quartz), *arkosic sandstone* or *arkose* (over 20 per cent feldspar), and *graywacke* (poorly sorted, clay or chloritic matrix).

Siltstone (*mudstone*) — fragments $1/256$ to $1/16$ mm, gritty.

Shale—fragments less than $1/256$ mm (clay), not gritty, has *fissility* (ability to split on bedding planes). Non-fissile clay rocks are *claystone.*

Non-clastic rocks are formed by chemical or biologic precipitation, and by accumulation of organic material.

Limestone is composed of calcite, in some cases from chemical precipitation but usually of biologic origin. If composed mainly of fossils it is called *coquina.*

Chalk is soft, white limestone formed by accumulation of the calcium carbonate shells of microscopic animals.

Dolostone (*dolomite*) is formed by replacement of calcite with dolomite.

Chert is fine-grained silica; some precipitated directly from sea water, most from accumulation of silica shells of organisms.

An *evaporite* is a deposit formed by evaporation of sea water. The precipitation sequence is calcite, dolomite, gypsum, rock salt, and lastly potassium and magnesium salts. *Gypsum* and *rock salt* (*halite*) are the main rocks formed this way.

Diatomite is soft, white rock composed of silica remains of microscopic plants.

Coal is formed by accumulation of plant material.

Shale, limestone, and sandstone make up over 99 per cent of all sedimentary rocks. Shale is most abundant because of the abundance of feldspars.

Bedding of sedimentary rocks records the layers in order of deposition with the oldest at the bottom.

Mud cracks record periodic drying

and can show top and bottom of deformed beds.

Ripple-marks can show type and direction of currents, and wave ripplemarks can show top and bottom.

Cut-and-fill features are caused by localized scouring, and can show top and bottom.

Cross-bedding, another current feature, also shows top and bottom.

Graded bedding is formed by larger fragments settling faster than smaller fragments; bottom is indicated by the larger fragments.

Porosity is a measure of the amount of void space in a sediment; *permeability* is a measure of the interconnections of the void spaces.

Continental shelves, the almost flat portions of the continent below sea level, are generally covered with clastic sediments.

Continental slopes, the steep slopes between the continental shelves and the deep ocean platforms, are mainly covered with fine sediments.

Slopes and shelves have some large *submarine canyons.* These may be caused by *turbidity currents,* underwater landslides consisting of heavy, sediment-laden water which can erode the soft bottom material.

In the deep ocean, the sediments are composed of slowly settled, wind-carried dust, largely of volcanic origin, and fallen shells of tiny animals. Sediments that have more than 30 per cent biologic material are called *oozes*.

Sedimentary rocks are important in the interpretation of geologic history through study of grain size and shape, depositional features, size and shape of depositional area, fossil study including dating, and deformation of the area.

Questions

4-1. Describe two methods of mechanical weathering.

4-2. Describe two methods of chemical weathering.

4-3. List the common minerals in the order of their stability to chemical weathering.

4-4. In what climate is chemical weathering most effective and why?

4-5. In what climate is mechanical weathering most effective and why?

4-6. Why is quartz a common mineral in sandstones?

4-7. What effects would you expect the increased carbon dioxide in industrial areas to have on stone buildings?

4-8. Why is mature soil zoned? Describe the zones.

4-9. Discuss the importance of climate and rock type on the soil produced.

4-10. List the trace elements that are harmful and those that are helpful to man.

4-11. What is the origin of chert?

4-12. What are the common cements in sandstones?

4-13. Clastic sedimentary rocks are classified on the basis of _____.

4-14. What rocks would form from the weathering of a granite in a cold climate? In a warm, moist climate?

4-15. How can top and bottom of the beds be determined in a group of vertical sedimentary beds?

4-16. How might sediments deposited in shallow water differ from those deposited in deep water?

4-17. What features could be used to distinguish turbidity-current-deposited sediments?

4-18. What geologic history might be inferred from each of the following rocks?
a. A feldspar-rich sandstone.
b. A well-rounded conglomerate.
c. A sandstone composed almost entirely of quartz.
d. A limestone with abundant fossils.

4-19. What kinds of fossils would you expect to find in rocks deposited in the following environments?
a. Tropical shallow water.
b. Deep water.
c. Temperate near shore.
d. Marsh.

Supplementary Reading

General

Ernst, W. G., *Earth Materials*. Englewood Cliffs, N.J.: Prentice-Hall, Inc., 1969, 150 pp. (paperback).

Pough, F. H., *A Field Guide to Rocks and Minerals*. Boston: Houghton Mifflin Co., 1955, 349 pp.

Weathering

Boyer, R. E., "Field Guide to Rock Weathering," *ESCP Pamphlet Series, PS-1*. Boston: Houghton Mifflin Co., 1971, 37 pp.

Foth, Henry, and H. S. Jacobs, "Field Guide to Soils," *ESCP Pamphlet Series, PS-2*. Boston: Houghton Mifflin Co., 1971, 38 pp.

Keller, W. D., "Geochemical Weathering of Rocks: Source of Raw Materials for Good Living," *Journal of Geological Education* (February, 1966), Vol. 14, No. 1, pp. 17-22.

Keller, W. D., *Principles of Chemical Weathering*. Columbia, Mo.: Lucas Brothers, 1957 (revision), 111 pp.

Kellogg, C. E., "Soil," *Scientific American* (July, 1950), Vol. 183, No. 1, pp. 30-39. Reprint 821, W. H. Freeman and Co., San Francisco.

Sedimentary Rocks

Freeman, Tom, "Field Guide to Layered Rocks," *ESCP Pamphlet Series, PS-3*. Boston: Houghton Mifflin Co., 1971, 43 pp.

Heezen, B. C., "The Origin of Submarine Canyons," *Scientific American* (August, 1956), Vol. 195, No. 2, pp. 36-41. Reprint 807, W. H. Freeman and Co., San Francisco.

Hsü, K. J., "When the Mediterranean Dried Up," *Scientific American* (December, 1972), Vol. 227, No. 6, pp. 26-36. An excellent article on the interpretation of climate and environment from sedimentary rocks and other geologic features.

Kuenen, P. H., "Sand," *Scientific American* (April, 1960), Vol. 202, No. 4, pp. 94-106. Reprint 803, W. H. Freeman and Co., San Francisco.

Laporte, L. F., *Ancient Environments*. Englewood Cliffs, N.J.: Prentice-Hall, Inc., 1968, 116 pp. (paperback).

Pettijohn, F. J., and P. E. Potter, *Atlas and Glossary of Primary Sedimentary Structures*. New York: Springer-Verlag, 1964, 370 pp.

Potter, P. E., and F. J. Pettijohn, *Paleocurrents and Basin Analysis*. Berlin: Springer-Verlag, published in United States by Academic Press, Inc., New York, 1963, 296 pp.

Shrock, R. R., *Sequence in Layered Rocks*. New York: McGraw-Hill Book Co., 1948, 507 pp.

Zenger, D. H., "Dolomitization and Uniformitarianism," *Journal of Geological Education* (May, 1972), Vol. 20, No. 3, pp. 107-124.

Trace Elements and Health

Cannon, H. L., and D. F. Davidson, "Relation of Geology and Trace Elements to Nutrition," Special Paper No. 90, Geological Society of America, 1967, 68 pp.

Gillie, R. B., "Endemic Goiter," *Scientific American* (June, 1971), Vol. 224, No. 6, pp. 92-101.

Scherp, H. W., "Dental Caries: Prospects for Prevention," *Science* (September 24, 1971), Vol. 173, No. 4003, pp. 1199-1205.

Shacklette, H. T., H. I. Sauer, and A. T. Miesch, *Geochemical Environments and Cardiovascular Mortality Rates in Georgia*. U.S. Geological Survey Professional Paper 574C, Washington, D.C.: U.S. Government Printing Office, 1970, 39 pp.

Van Ness, G. B., "Ecology of Anthrax," *Science* (June 25, 1971), Vol. 172, No. 3990, pp. 1303-1307.

Warren, H. V., "Medical Geology and Geography," *Science* (April 23, 1965), Vol. 148, No. 3669, pp. 534-536.

Metamorphic Rocks

5

Metamorphic rocks are rocks of any type or composition that have been changed in either mineral composition or texture.

The study of metamorphic rocks developed later than the study of igneous rocks, although the same methods of study are used for both rock types. Metamorphic rocks are the more difficult of the two to study because of their complex structures and their variations in composition and mineralogy. Werner had included the metamorphic rocks with the granites in his oldest group of Neptunist precipitates. Hutton was the first to recognize clearly that rocks could undergo changes, although his concept of metamorphism was crude in the light of later knowledge. By the middle of the nineteenth century, French geologists had established the two main types of metamorphism, contact metamorphism and regional metamorphism. These terms are described below.

Even before Werner, the idea developed that metamorphic rocks are the oldest rocks. Their complex structures and their locations in the cores of many mountain ranges and under the oldest sedimentary rocks suggest their antiquity. We now know that although the oldest rocks are metamorphic rocks, not all metamorphic rocks are old. The discovery of deformed fossils in metamorphic rocks showed this as early as 1814 when De Charpentier found fossils in mica schist in the Alps. This discovery helped to convince some Neptunists that Werner's age assignment was wrong. In 1881 Reusch found fossils in metamorphic rocks in Norway, and this discovery accelerated the study and understanding of metamorphic rocks. At the present time, the mapping of metamorphic areas is aided by the use of radioactive dating. Mapping complex structures without some means of recognizing the ages of the rocks is a very difficult undertaking.

Around the turn of the century, studies of the intensity or grade of metamorphism were made from field mapping and microscopic study. Studies were made by Barrow in Scotland in 1893, Becke and Grubenmann in the Alps in 1904, Goldschmidt at Oslo in 1911, and Escola in Finland in 1915. These men established the principles involved in the mineralogical changes that occur in metamorphism. Refinements of this work, especially using the results of more recent laboratory studies, is an on-going type of research.

Agents of Metamorphism

Metamorphic rocks are rocks that have been changed while in the solid state either in texture or in mineral composition by any of the following — heat, pressure, directed pressure (stress), shear, or chemically active solutions. Because any rock of any

Complexly folded quartz vein in mica schist. Potomac River gorge near Washington, D. C. Photo by J. C. Reed, Jr., U. S. Geological Survey.

type or composition can be subjected to any or all of the above agents, the variety of metamorphic rocks is endless. Note that the changes that take place in the lithification of a sedimentary rock are, according to the definition, metamorphic. Such changes, however, are not considered metamorphic, but are arbitrarily excluded.

The changes that occur during metamorphism are the result of an attempt to reestablish equilibrium with the new conditions to which the rock is now subjected. Again note the similarity to weathering. Weathering occurs when rocks that formed deep in the earth are subjected to surface conditions. However, when surface-formed rocks, such as shales, are metamorphosed, the changes that occur proceed in the opposite direction and high-temperature minerals are formed.

Many factors promote or retard metamorphic reactions. Large surface area promotes chemical reaction so fine-grained rocks react faster than coarse-grained rocks. Glass is less stable than crystalline material and so also reacts faster.

Heat is an important agent in causing metamorphism. It is well known that as one goes deeper into the earth in mines or drill holes, the temperature increases. Thus, deep burial alone can cause the temperature to rise, but most metamorphic rocks probably form at places where the temperature rise with depth is greater than normal.

Pressure also is the result of burial. The pressure is due to the weight of the overlying rocks. At the increased temperature and pressure deep in the earth, rocks lose much of their strength and so behave plastically. This causes the pressure due to the

weight of overlying rocks to act uniformly in all directions, just as the pressure due to the weight of the water in a swimming pool acts in all directions.

Directed pressure (stress) is also commonly present during metamorphism. The effect of this is to fold the more or less plastic rocks. Stress may deform rocks, and flattened or stretched pebbles or fossils record some of this deformation.

Shear results when the rocks are broken and moved by directed pressure, as in faults. In metamorphic rocks the movement is commonly distributed on many closely spaced shear surfaces. (See Fig. 5-1.)

Chemically active solutions are very important in metamorphic reactions. Even if temperature and pressure are high enough to cause a metamorphic reaction to occur, some water is generally necessary to allow the reaction to take place. Without water to act as a catalyst, the reaction may actually be too slow to occur. Water is generally present in most rocks, and only a thin film on the grain boundaries is necessary. The water may contain other materials that cause or promote the reactions. It can also bring in or remove elements from the rock. Small amounts of water may also greatly reduce the strength of rocks undergoing metamorphism.

The hot waters that cause low-grade metamorphism have apparently been found in some areas in the quest for geothermal power. Hot springs are relatively common and are spectacular at places such as Yellowstone National Park. Recent study of the geothermal area near the Salton Sea in southern California shows that the hot waters there have caused metamorphism in the sur-

rounding sediments. Geothermal waters are described in the next chapter.

Types of Metamorphism

Metamorphic rocks can be subdivided into three genetic groups, although there is gradation among the groups.

1. *Thermal* or *contact metamorphic rocks.* These rocks generally are found at the margins of intrusive igneous bodies such as batholiths. The rocks are termed *hornfels.*
2. *Regional metamorphic rocks* are so-called because they generally occupy large areas. These rocks form deep in the crust, and their presence at the surface reveals much uplift and erosion. The rocks generally have recrystallized under stress so that the new minerals grow in preferred orientation. These rocks are said to be *foliated,* and the most common are *gneiss* (pronounced "nice") and *schist.*
3. *Dynamic metamorphic rocks* result from breaking and grinding without much recrystallization.

Formation and Identification of Metamorphic Rocks

For classification and identification, metamorphic rocks can be subdivided into two textural groups:

1. Foliated—having a directional or layered aspect.
2. Non-foliated — homogeneous or massive rocks.

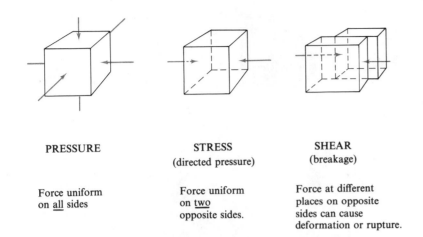

PRESSURE

Force uniform on <u>all</u> sides

STRESS
(directed pressure)

Force uniform on <u>two</u> opposite sides.

SHEAR
(breakage)

Force at different places on opposite sides can cause deformation or rupture.

Figure 5-1. Pressure, stress, and shear.

Figure 5-2. Hornfels, showing bedding and graded bedding. This polished specimen is just over two inches in diameter. Photo by F. C. Armstrong, U.S. Geological Survey.

Non-foliated Metamorphic Rocks

Because the non-foliated rocks are the simpler, we will consider them first. There are two types of non-foliated metamorphic rocks. The first type consists of thermal or contact metamorphic rocks called *hornfels* (singular). They generally occur in narrow belts around intrusive bodies and may originate from any type of parent rock. Sometimes these are called *baked rocks;* the name is appropriate, because their formation is quite similar to the baking of clay pottery in a kiln. They are generally fine-grained, tough rocks that are difficult to identify without microscopic study unless the field relations are clear. They range from being completely recrystallized with none of their original features preserved to being slightly modified rocks with most of the original features preserved. (See Fig. 5-2.) In some of these rocks new minerals grow. At the early stages of growth the irregular growths produce what are called spotted rocks. Calcite-bearing contact rocks are generally quite spectacular, containing large crystals of

garnet and other minerals.

The second type of non-foliated metamorphic rock develops if the newly formed metamorphic minerals are equidimensional and so do not grow in any preferred orientation. The best examples of this are the rocks that result when monomineralic rocks such as limestone, quartz sandstone, and dunite are metamorphosed. In the case of limestone, which is composed of calcite, no new

mineral can form; thus, the calcite crystals, which are small in limestone, grow bigger, develop an interlocking texture, and become *marble*. (See Fig. 5-3.) Marble can be distinguished from limestone only by its larger crystals and lack of fossils; being composed of calcite, both effervesce in acid. Marble may have any color; pure white calcite rocks are more apt to be marble than limestone. Impure limestones, as noted above, develop new minerals,

Figure 5-3. Marble showing coarse interlocking texture. Photo from Ward's Natural Science Establishment, Inc., Rochester, N.Y.

particularly garnet. Dolomite marble also occurs.

Because quartz sandstone similarly can form no new minerals, the quartz crystals enlarge and intergrow to form *quartzite*. Quartzite is distinguished from chalcedony, chert, and opal by its peculiar sugary luster. The term quartzite is best reserved for the metamorphic rock, although quartz sandstone is called quartzite by some and the metamorphic rock is called metaquartzite.

Both marble and quartzite may form under stress, and so the crystals may be aligned. However, because the crystals are equidimensional, this preferred orientation can be detected only by microscopic examination. Thus, although we call these rocks non-foliated, some may have a hidden foliation.

So far we have been concerned with *progressive* metamorphism, which is metamorphism caused by increased temperature, pressure, and other agents of metamorphism. The next example is one of *retrogressive* metamorphism, which is metamorphism at a temperature lower than that of the rock's original formation. Dunite is composed mainly of olivine, which is the highest-temperature mineral in Bowen's series. When it is subjected to the same metamorphic conditions that produce the rocks discussed earlier, it too is metamorphosed — but at a temperature lower than that of its formation. The olivine of dunite is changed to the mineral *serpentine,* and the metamorphic rock produced is called *serpentinite.* Its identification is discussed in Appendix A. In other instances, instead of serpentine, *talc* is the metamorphic mineral that forms. If non-foliated, the talc rock is called *soapstone;* if foliated, it is called *talc schist.* Identification of

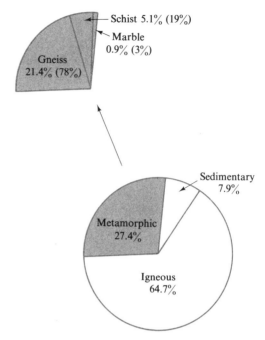

Figure 5-4. Relative abundance of metamorphic rocks in the earth's crust. For example, gneiss forms 21.4 per cent of all rocks and 78 per cent of the metamorphic rocks. Data from Ronov and Yaroshevshy, 1969.

talc is discussed in Appendix A. Slight differences in the composition of the parent rock or of the water solutions that cause this type of metamorphism probably determine which mineral is formed.

Foliated Metamorphic Rocks

Foliated rocks are the more common of the metamorphic rocks (see Fig. 5-4) and are the product of regional metamorphism. Regional metamorphism generally takes place in areas undergoing deformation, and these very active areas are generally exposed in the cores of mountain ranges. Regional metamorphic rocks are produced by the same type of forces that make mountain structures. The forces that fold and fault the shallow rocks exposed on the flanks of mountain ranges can provide the stress fields under which the deep-seated metamorphic rocks, now exposed in the uplifted cores of mountain ranges, were recrystallized.

One example of the rocks pro-

Original flat-lying sedimentary rocks.

Mica, shown diagrammatically by short dashes, grows perpendicular to the forces that may fold the rocks during metamorphism. Note that the foliation may form at an angle to the bedding.

Figure 5-5. Development of foliation.

duced by increasing metamorphic grade will show most of the foliated metamorphic rock types. These same rock types will form when many other parent rocks are metamorphosed; the differences in original chemical composition will change the relative amounts of the metamorphic minerals only somewhat.

Shale is composed mainly of clay. During lower-grade metamorphism, the clay is transformed to mica, and at higher grades of metamorphism, to feldspar. Which micas and which feldspars form depends on the bulk chemical composition. Compare this with chemical weathering and note that weathering is, in a sense, retrogressive metamorphism. Because mica is a flat, platy mineral, it tends to grow with its leaves perpendicular to the maximum stress, forming a preferred orientation. (See Figs. 5-5, 5-6, and 5-7.) Microscopic examination of some foliated rocks shows that the bedding planes are slightly displaced in the plane of the foliation. This suggests that foliation, at least at some places, forms in shear directions. The shear direction is in general at about 45 degrees to the maximum stress in uniform materials. Thus foliation may form perpendicular to the direction of maximum stress, or at about 45 degrees to that direction.

The sequence in the metamorphism of a shale is given in Table 5-1. The rock names are applied for the overall texture of the rock and not strictly for the mineral transformations which occur over a range in temperature and pressure. (See Figs. 5-8, 5-9, and 5-10.) Most schists contain feldspar, but the name gneiss is reserved for rocks with much more feldspar than mica. Other rocks that have the same general bulk chemical composition, thus producing the

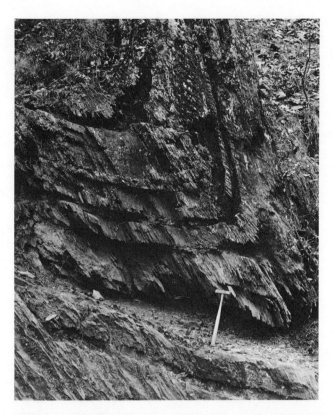

Figure 5-6. Foliation and bedding in folded rocks near Walland, Tennessee. The rock type is slate. Compare with Figure 5-5. Photo by A. Keith, U.S. Geological Survey.

Figure 5-7. Typical exposure of slate on Prudence Island, Rhode Island. Bedding is nearly vertical and is shown by color differences. Foliation is perpendicular to the bedding. Photo by J. B. Woodworth, U.S. Geological Survey.

Figure 5-8. Outcrop of foliated rock. The rock is an impure marble, and the foliation is parallel to the bedding. The light-colored pods above the knife are quartzite, and the pods formed when a sandstone bed broke brittlely when the rocks were folded. Photo by J. A. Donaldson, Geological Survey of Canada, Ottawa.

same metamorphic rocks, are certain pyroclastic and volcanic sedimentary rocks, many sandstones, arkose, granite, and rhyolite. The coarse-grained rocks in this list are rarely the parents of slate or fine-grained schist but, in general, remain more or less unaffected until medium-grade metamorphism is reached. Occurrence of slate in areas where the non-shale rocks are not metamorphosed suggests that slate may be

formed by a mechanical reorientation of clay particles during folding, together with limited recrystallization.

Only one other group of regional metamorphic rocks must be considered among the most common metamorphic rocks. These rocks are formed by the metamorphism of basalt, certain pyroclastic and volcanic sedimentary rocks, gabbro, graywacke, and some calcite-bearing or dolomitic sedimentary rocks. The

sequence is given in Table 5-2.

Dynamic metamorphism. This third type of metamorphism results when rocks are broken, sheared, and ground near the surface where the temperature and pressure are too low to cause any significant recrystallization. Thus, these rocks are commonly associated with fault zones, but there are all gradations between these rocks and ordinary schist, especially lower-grade schist formed from coarse-grained rocks. The fine-grained, ground rock in a fault zone is called *fault gouge* or *fault breccia* if it has little cohesion, and similar cohesive rocks are generally called *mylonite.* As the amount of recrystallization of such rocks increases, they become gradational with schist and gneiss. The origin of these rocks is clear when they are encountered in the field, but they are very difficult to identify in hand specimen.

An unusual type of dynamic metamorphism can occur at impact craters caused by meteorites, as decribed in Chapter 18.

The origin, texture, and mineralogy of the common metamorphic rocks are summarized in Figs. 5-11, 5-12, and 5-13.

Interpretation

Study of metamorphic rocks, coupled with laboratory experiments, yields much information about conditions and processes deep in the crust. That is, if we know the conditions of formation of a mineral or, better, an assemblage of minerals, when we find rocks composed of such minerals, we know the conditions under which the rocks formed. The problem is complicated by the diversity

Figure 5-9. Garnet-mica schist. The large garnet crystals grew during metamorphism. Foliation does not show clearly in the orientation pictured. Photo from Smithsonian Institution.

of natural rocks in that they contain many more elements, which are the variables in experiments, than can be easily studied in experiments. It is also not possible to reproduce the effects of geologic time in the laboratory, nor is it generally possible to detect the effects of solutions or volatiles that may have catalyzed the reactions in natural rocks. Thus, although we are very rapidly gaining an understanding of metamorphic processes, many problems remain.

The purpose of this chapter is to give the student some ability in identifying metamorphic rocks. He will not, except in a very general way, be able to interpret the detailed geologic history of these rocks without knowledge of chemistry, mineralogy, and field and microscope study. To interpret metamorphic rocks fully, one must know whether they contain equilibrium mineral assemblages or whether conditions changed before the metamorphic reactions were complete; whether material was added or subtracted during metamorphism; and whether the rocks were subjected to one or more periods of metamorphism. One can think of many more possibilities that must be considered: for example, if the reactions take place during deep

Table 5-1. Metamorphism of shale.

SEDIMENTARY ROCK	LOW-GRADE METAMORPHISM	MEDIUM-GRADE METAMORPHISM	HIGH-GRADE METAMORPHISM
Shale ⟶	Slate ⟶	Schist ⟶	Gneiss
Clay	Clay begins to be transformed into mica. Mica crystals are too small to see, but impart a foliation to the rock. May also form by mechanical rearrangement of clay alone.	Mica grains are larger so that the rock has a conspicuous foliation.	Mica has transformed largely to feldspar, giving the rock a banded or layered aspect.

burial, will the rocks revert to something like their original state during slow uplift? No doubt some such effect may operate in some instances, but most of the chemical reactions involved go very slowly in the reverse direction. If this were not so,

nothing but weathering products would be visible at the surface.

Metamorphic Grade

The concept of metamorphic grade was introduced above in discussing

Table 5-2. Metamorphism of basalt.

PARENT ROCK	LOW-GRADE METAMORPHISM	MEDIUM- AND HIGH-GRADE METAMORPHISM
Basalt ⟶	Greenschist ⟶	**Amphibolite** (Amphibole schist)
	Fine-grained, foliated Chlorite (green "mica") Green amphibole Quartz	Coarse-grained, foliated Plagioclase Dark amphibole

A.

B.

Figure 5-10. Gneiss. A. The foliation is shown by discontinuous layers of dark and light colored minerals. B. The layers in this gneiss are more irregular. Photos from Ward's Natural Science Establishment, Inc., Rochester, N.Y.

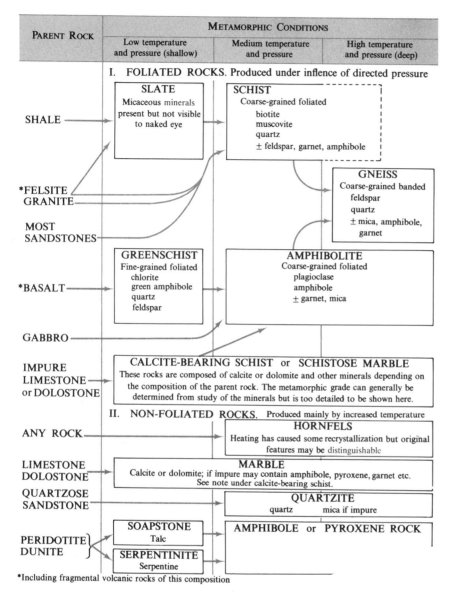

Parent Rock	Metamorphic Conditions		
	Low temperature and pressure (shallow)	Medium temperature and pressure	High temperature and pressure (deep)

I. FOLIATED ROCKS. Produced under inflence of directed pressure

SLATE
Micaceous minerals present but not visible to naked eye

SCHIST
Coarse-grained foliated
biotite
muscovite
quartz
± feldspar, garnet, amphibole

GNEISS
Coarse-grained banded
feldspar
quartz
± mica, amphibole, garnet

GREENSCHIST
Fine-grained foliated
chlorite
green amphibole
quartz
feldspar

AMPHIBOLITE
Coarse-grained foliated
plagioclase
amphibole
± garnet, mica

SHALE

*FELSITE
GRANITE

MOST
SANDSTONES

*BASALT

GABBRO

IMPURE
LIMESTONE
or DOLOSTONE

CALCITE-BEARING SCHIST or SCHISTOSE MARBLE
These rocks are composed of calcite or dolomite and other minerals depending on the composition of the parent rock. The metamorphic grade can generally be determined from study of the minerals but is too detailed to be shown here.

II. NON-FOLIATED ROCKS. Produced mainly by increased temperature

ANY ROCK

HORNFELS
Heating has caused some recrystallization but original features may be distinguishable

LIMESTONE
DOLOSTONE

MARBLE
Calcite or dolomite; if impure may contain amphibole, pyroxene, garnet etc.
See note under calcite-bearing schist.

QUARTZOSE
SANDSTONE

QUARTZITE
quartz   mica if impure

PERIDOTITE
DUNITE

SOAPSTONE
Talc

SERPENTINITE
Serpentine

AMPHIBOLE or PYROXENE ROCK

*Including fragmental volcanic rocks of this composition

Figure 5-11. A generalized chart showing the origin of common metamorphic rocks.

the formation of metamorphic rocks. Metamorphic grade is an interpretation based on field and laboratory evidence and should be further examined. The first attempts to map and study grade of metamorphic rocks consisted of mapping mineral zones. This was done by restricting the study to a single parent rock type, such as shale, and recording the first occurrence of certain minerals as the metamorphic grade increased. (See Fig. 5-14.) This simple method has been improved by considering assemblages of several minerals and by comparing such equilibrium assemblages in rocks of many compositions. Many problems are encountered in recognizing equilibrium assemblages, and much field

and laboratory work has been required to find widespread equilibrium assemblages that can be used in this way.

Problems

The origin of metamorphic rocks presents many geologic problems. The regional metamorphic rocks are the most difficult to explain. These rocks formed under high temperature and pressure. The depths at which these temperatures and pressures can occur, assuming reasonable temperature gradients, are in the range of 10 to 20 miles. (See Fig. 5-15.) The amount of erosion in areas where high-grade regional metamorphic rocks are exposed is probably less than this, indicating that the temperature gradient must have been greater. A higher temperature gradient is not unreasonable because such rocks form in geologically active areas. The origin of such high-temperature gradients is a problem. A possible solution to this problem is the passage of magma through the rocks, as suggested in Chapter 3 in connection with the origin of batholiths. When viewed on a large scale, as in Fig. 5-14, the narrow transition zone between high-grade metamorphic rocks (sillimanite-bearing rocks) and low-grade rocks (chlorite-bearing rocks) is also difficult to explain. That is, if the heat comes from deep burial, the heat conduction of rocks is such that the surrounding rocks should have been much hotter.

Another problem is the origin of granite. As shown in Fig. 5-15, at the highest grades of metamorphism some melting may occur, and so

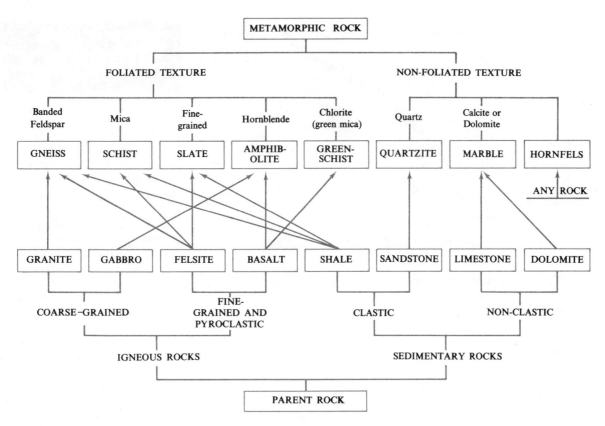

Figure 5-12. Origin of common metamorphic rocks. Not all possible parent rocks are shown in this simplified diagram.

metamorphic processes may merge with igneous activity. Many of the small lenses of granitic rocks found in high-grade gneiss probably originate from such partial melting. (See Fig. 5-16.) The origin of larger bodies of granite, such as batholiths, in high-grade metamorphic terranes is a problem. Some show chilled, fine-grained margins and other features that are clearly igneous. Others have completely gradational contacts and could be metamorphic rocks or could have intruded the gneiss just after the gneiss formed and was still hot. Some of the latter granites are probably the result of metamorphism, but, as pointed out in Chapter 3, the isotope composition of most batholiths indicates an origin in the lower crust or upper mantle.

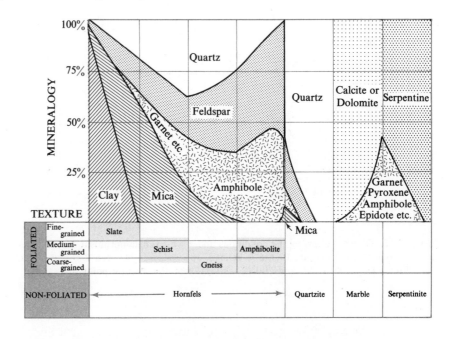

Figure 5-13. Texture and mineralogy of common metamorphic rocks.

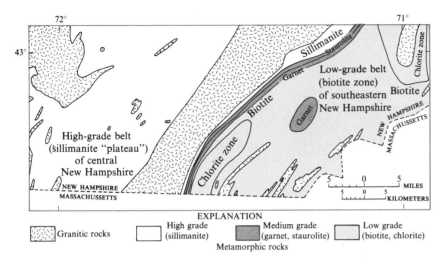

Figure 5-14. Metamorphic grade as interpreted from metamorphic minerals in shaly rocks in southeastern New Hampshire. From Warren Hamilton and W. B. Myers, U.S. Geological Survey, *Professional Paper 554C*, 1967.

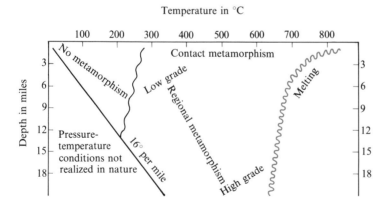

Figure 5-15. Diagram showing the conditions under which metamorphism occurs. Pressure is proportional to depth. Biotite is believed to crystallize between 200 and 300°C and sillimanite between 600 and 700°C, or very near the melting point of granite.

Summary

Metamorphic rocks are rocks which have been changed, either in texture or mineral composition by heat and pressure, usually as a result of burial, directed pressure (stress), shear, or chemically active solutions, especially water.

There are three genetic groups with gradation between them:

1. *Thermal* or *contact metamorphic rocks*, called *hornfels*, generally found at margins of intrusive igneous bodies.
2. *Regional metamorphic rocks* occupy large areas, and reveal much uplift and erosion after their formation. They are usually *foliated* (the new minerals have preferred orientation because of recrystallization under stress). The most common are *gneiss* and *schist*.
3. *Dynamic metamorphic rocks* result from breaking and grinding without much recrystallization.

Textural groups are (1) *foliated* — directional or layered aspect and (2) *non-foliated* — homogeneous or massive.

There are two types of *non-foliated* rocks:

1. Thermal or contact metamorphic rocks, called *hornfels*, fine-grained, tough *baked rocks* in narrow belts around intrusive bodies.
2. The development of equidimensional new minerals, usually from monomineralic parent rocks. Limestone into

marble; quartz sandstone into *quartzite*; dunite into *serpentinite* if serpentine is the new mineral, *soapstone* if talc is the new mineral (foliated talc rock is *talc schist*).

Progressive metamorphism is caused by increased pressure and temperature, etc., as in cases of marble and quartzite. *Retrogressive* metamorphism is at lower temperature, etc., than parent rock's formation; the olivine in dunite is the highest temperature mineral in Bowen's reaction series, and it is metamorphosed at a temperature lower than that of its formation.

Foliated metamorphic rocks are caused by the stresses that fold mountains in regions undergoing deformation. Rocks are named largely for texture and not strictly for minerals. Examples are given in Tables 5-1 and 5-2.

Dynamic metamorphism occurs in fault zones or, rarely, at meteorite impact craters. Non-cohesive ground rock in fault zone is *fault gouge* or *fault breccia*. Similar cohesive rocks are *mylonite*. These grade into schist and gneiss.

Interpretation of metamorphic rocks gives valuable insight into earth processes and history, but is complex and difficult. High temperatures required by regional metamorphic rocks pose problems. Metamorphic processes may merge with igneous as in the origin of some granite bodies.

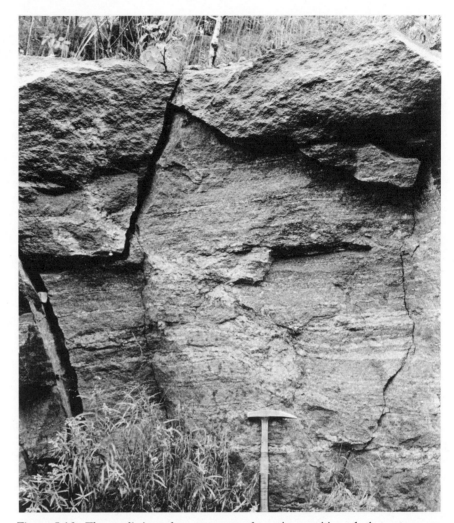

Figure 5-16. The small, irregular occurrence of massive granitic rock that cuts across the foliation may have formed by partial melting. The granitic rock is localized in a minor fault zone where the foliation is disturbed. Dannemora Mountain, New York. Photo by A. W. Postel, U.S. Geological Survey.

Questions

5-1. Define metamorphic rock.

5-2. What are the types of metamorphism, and which agents are most important in each?

5-3. What is foliation, and how does it form?

5-4. What is contact metamorphism, and what types of rocks are formed?

5-5. How does a shale differ from a slate?

5-6. What is the low-grade metamorphic equivalent of a basalt?

5-7. Serpentinite forms from metamorphism of _____.

5-8. Soapstone forms from metamorphism of _____.

5-9. How is the metamorphic grade established?

5-10. How might the field relationships differ between an igneous and a metamorphic granite?

Questions
Covering All the Rocks and Minerals in Chapters 2, 3, 4, and 5

1. Rockhounds collect many different rocks and minerals that they cut and polish. What physical properties must such rocks and minerals have? List the various minerals and rocks that have these qualities.

2. Name and interpret as completely as possible the significance of the following rocks:

   a. A coarse-grained rock with interlocking texture composed mainly of quartz and feldspar with some dark minerals.

   b. A fine-grained, dark-gray rock with a few well-formed plagioclase and olivine crystals.

   c. A foliated rock with much muscovite.

   d. A fine-grained, white rock composed of calcite with some clam shells.

   e. A clastic rock composed of well-rounded quartz grains about 1 mm in diameter.

   f. A clastic rock composed of quartz and feldspar with some hornblende and mica. The grains are mainly between 1 and 2 mm in diameter and are angular.

   [*Note:* After reading Chapters 7 through 11 you may want to review your interpretations of the last two rocks.]

3. Before starting to identify rock specimens, it may help to list all of the rocks mentioned in the text in a form like the one suggested below. This will show the importance of texture in recognition and interpretation of rocks.

Check which type:

	IGNEOUS	SEDIMENTARY	METAMORPHIC
COARSE-GRAINED Clastic texture			
Interlocking texture Massive			
Foliated			
FINE-GRAINED			
Glass			
Frothy			

Supplementary Reading

General

Ernst, W. G., *Earth Materials*. Englewood Cliffs, N. J.: Prentice-Hall, Inc., 1969, 150 pp. (paperback).

Pough, F. H., *A Field Guide to Rocks and Minerals*. Boston: Houghton Mifflin Co., 1955. 349 pp.

Metamorphic Rocks

Fyfe, W. S., F. J. Turner, and John Verhoogen, *Metamorphic Reactions and Metamorphic Facies,* Chapter 1, "Concept of metamorphic facies," pp. 3-20; Chapter 7, "Mineral assemblages of individual metamorphic facies," pp. 199-239. Geol. Soc. of America, 1958, Memoir 73.

Higgins, M. W., "Cataclastic rocks," U.S. Geological Survey, Professional Paper 687, 1971, 97 pp.

Mason, Brian, *Principles of Geochemistry,* (3rd ed.), Chapter 10, "Metamorphism and metamorphic rocks." New York: John Wiley & Sons, Inc., 1966, pp. 248-282.

Mineral Resources

6

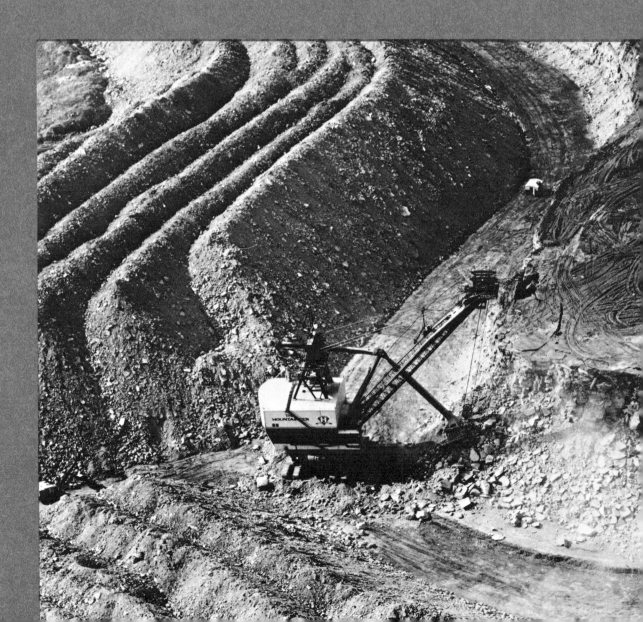

Many of the materials that form the crust of the earth are useful to man. The term *mineral resources* is used to mean natural resources that come from the earth. Note that the term "mineral" used in this sense does not follow the definition of mineral used earlier because many of the materials discussed in this chapter, strictly speaking, are not minerals. The most important mineral resources are metals and energy-producing materials, and this chapter will be limited mainly to these. Many other mineral products are mined, however, such as sand, gravel, building stone, clay, diatomite, and limestone.

Scientific understanding of how economic mineral deposits form developed very slowly. In 1556, Agricola's book *De Re Metallica* was published, and, as noted earlier, it was an important, useful book that helped the miners of that time. Agricola recognized that the veins that contain the ore minerals were deposited by solutions in channels that formed later than the surrounding rock. In these observations he was far ahead of his time; however, no further progress was made for almost 200 years. In that interval a number of fanciful ideas developed, such as veins are the twigs of a golden tree whose roots are at the center of the earth, and veins develop as a result of celestial influences, especially the sun's rays.

By the middle of the eighteenth century, more scientific study was progressing, particularly in Germany at the Mining Academy at Freiberg where Werner was active. In the nineteenth century, two important theories of vein formation were put forward. The first is that veins are deposited by hot water solutions, and the second is that solutions moving through rocks leach the metals that are then deposited in veins. Apparently both processes cause ore deposits. The late nineteenth century was a time of great interest in mining, and many geologists contributed to our understanding of mineral deposits. In the early twentieth century, much of this work was used by Waldemar Lindgren to formulate a classification of mineral deposits. The classification used in this chapter is adapted from Lindgren.

This chapter is concerned with the origin of mineral resources. Few new principles are involved, so it will, in a way, serve as a review of minerals and rocks. Before discussing formation of mineral deposits, we will consider some of their unique features and their importance.

Distribution and Importance

Mineral resources are not spread evenly around the continents, but are scattered. (See Fig. 6-1.) This creates the have and the have-not nations. A mine or other mineral deposit is a rare feature, geologically speaking.

The importance of mineral deposits to man cannot be overemphasized. The wealth of nations is dependent on their natural resources.

Strip mining coal in Ohio. Photo from U. S. Bureau of Mines and Marion Power Shovel Co.

Although not always emphasized, mineral deposits have played a very important role in history. In ancient times, wars were fought and civilizations prospered or failed because of gold and silver deposits. In recent times, metals and energy-producing materials, such as iron, coal, and petroleum, have determined the course of history.

Exhaustibility

Every mineral deposit has finite dimensions. When the ore is mined out, the deposit has been used. Thus all mineral deposits are exhaustible, and there is no way to replace them. This is not news to anyone who has traveled through the west and seen the many ghost towns. The importance of the exhaustibility of mineral deposits is seen when rates of mining are considered. In the last fifty years more metal has been mined than in all preceding time. Thus, we are using our natural resources at an unprecedented rate. Unless new sources are found, we face the prospect of nations decaying like the ghost towns of the west. This drastic possibility will be averted by increased exploration (prospecting), using new, sensitive chemical, physical, and biologic methods; by finding substitutes, such as plastics; and by developing new refining methods that are effective with very low-grade ores. In any case, it seems quite evident that we will have to pay more for our metals in the future.

An ore is anything that can be mined at a profit. Therefore, what is ore in the United States may not be ore in the Canadian Arctic or Central Australia because of the cost of mining and the distance to market.

It also follows that what is worthless today may be ore tomorrow because of changing economic conditions or technological changes.

Water, both surface and underground, and soil are the most important mineral resources. Life would not be possible without both. These two basic resources are different from all other mineral resources in that they are replenishable, although the time required to replenish them may be very long by human standards.

Resources and Man

The mineral resources of the earth are finite, and once mined, no more new metal can be had. This is one aspect of the spaceship earth analogy. Some metals, such as aluminum, will probably never run out, but some others, especially silver and mercury, and perhaps lead, zinc, gold and tin may very well run out.

Because the distribution of metal mines is random, no industrial country is self sufficient in all mineral resources. Obviously, the industrial countries are using mineral resources in much greater quantities than the undeveloped countries. The undeveloped or non-industrial countries are striving to become industrial countries and raise their standards of living. Thus the rate of use of minerals is bound to increase, and the increase will be rapid because most of the earth's population lives in the undeveloped countries.

Some examples of metal usage will show the magnitude of the problem. The per capita use of iron in the United States in 1967 was about one

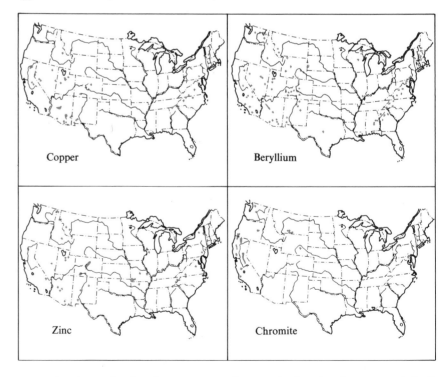

Figure 6-1. Maps showing the occurrence of economic deposits of some metals in the United States. Note the similarities between gold and silver, and among silver, lead,

ton, and for the total world population the figure was .17 tons. The United States' usage was about six times that of the world average, or, said another way, iron production would have to be increased six times if the whole earth is to have the same standard of living that the United States had in 1967. It is estimated that the population of the earth will double by the year 2000. To keep the same world per capita use of iron, production will have to be doubled; or if everyone is to have the United States' 1967 standard of living, production must be increased 12 times. The world production of iron in 1967 was 550 million tons. Of course, not only iron but all other mineral resources will be needed in similar ratios to achieve these goals. See Table 6-1. These projections are probably too high for a number of

reasons, such as export of manufactured goods by the United States. On the other hand, per capita consumption in the United States and elsewhere has been rapidly accelerating, suggesting that the projections may be too low. It seems clear that the potential demand for metals in the year 2000 will be at least several times that of today. The question is, can this potential demand be met?

The differences in mineral usage reveal a number of other differences among countries. The United States is a highly industrialized country. With only 6 per cent of the world's population and 6 per cent of the earth's land surface, in 1966 it used 16 per cent of the total earth's production of coal, 25 per cent of the petroleum, 26 per cent of the steel, 35 per cent of the copper, and 53 per cent of the aluminum. It must

be remembered, too, that these amounts add to the already large quantities of these metals that are now in use in buildings, machinery, and the like. The amounts of mineral resources used are increasing in all countries; but although the amounts used per capita are increasing in the industrial countries, they are decreasing in many undeveloped countries. The rapid growth of population in the undeveloped countries is the cause. The industrial countries are increasing their standards of living, while in the undeveloped countries the people are becoming poorer, at least in mineral resources. See Fig. 6-2.

The United States, like most industrial countries, is not self sufficient in many mineral resources. It must import large quantities. A question of great importance is, what is a fair price to pay an undeveloped country for its unreplenishable mineral resources? All people look forward to the day when they will have a standard of living like ours. This will be difficult to achieve if their resources are gone. On the other hand, they cannot develop or even use their resources without capital, and their only ready way to get capital is to sell their resources. It takes many millions of dollars to find a resource, and many more to develop it, and, of course, much technical know-how. The historical background to the present world situation shows how undeveloped countries have been exploited in the past. The colonial system developed with the industrial revolution. The mother country was generally an industrial country, and the colonies supplied the mother country with raw materials. The political breakdown of this system began with the American

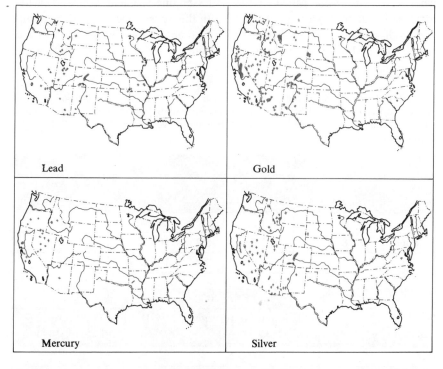

Lead   Gold

Mercury   Silver

and zinc. Most of these deposits are in the geologically active areas that will be discussed in Chapter 14. Data from Mineral Resource Maps of the U.S. Geological Survey.

Revolution and was pretty well completed after World War II with the formation of many new independent countries in Africa and Asia. Most of the colonialism of today is economic colonialism, and revolves around the question of the price paid for mineral resources. Prices may

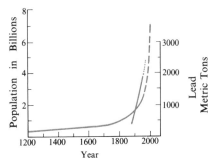

Figure 6-2. Projected increase in the use of copper, lead, and iron, compared to projected increase in population. Data from C. F. Park, Jr., *Affluence in Jeopardy*, 1968.

have to increase, and this may change the standards of living of the whole world. The constant upward revision of the royalties paid for oil in the Near East is an example of the kinds of change that seem likely to occur. Anyone who does not believe the importance of mineral resources in world politics needs only to study the history of Europe for the last 200 years or so. The territorial settlements after each war involved mainly the important mineral resource areas, such as the iron of Alsace-Lorraine, the coal of the Saar, and the lead-zinc of Poland.

Waste Disposal — "Garbage is a resource for which we have not yet found a use." The amount of material discarded is alarming. In the United States it is about 5.5 pounds per person per day — almost twice what it was in 1920. Much of the increase is due to modern packaging and to disposable items. Most cities either bury or burn their garbage —

processes that have been in use since before recorded history. Some cities have tried to make fertilizer out of garbage but have been unable to sell it. Several cities are compacting trash and either shipping it away or using it locally for land fill. Experiments are underway using trash as fuel to run electric generators and converting garbage into a nutritious protein, or to petroleum.

One way to eliminate part of the disposal problem is to recycle as much trash as possible. Bottles and aluminum cans are currently being recycled by consumers. Commercially reclaimed metals include about 40 per cent of our lead and 25 per cent of our copper. About three million tons of iron are discarded each year in the United States. Recycling trash can greatly reduce the need to find and mine metal ores, and so lessen one problem discussed earlier. In this sense, one can look at dumps and automobile wrecking yards as national resources although they still

Table 6-1. Current and predicted use of some metals. The amounts needed to bring world usage up to United States standards are enormous for the present population of the world, and staggering if population doubles by the year 2000. From C. F. Park, Jr., *Affluence in Jeopardy*.

	YEAR 1967			
METAL	UNITED STATES PER CAPITA	WORLD PER CAPITA	WORLD PRODUCTION	BRING WORLD TO U.S. STANDARD
Iron	1 ton	.17 ton	550×10^6 tons	$\times 6$
Copper	18 lbs.	3.2 lbs.	5.25×10^6 tons	$\times 5.5$
Lead	12 lbs.	1.5 lbs.	2.44×10^6 tons	$\times 8$

	YEAR 2000	
METAL	DOUBLE POPULATION AS TODAY	DOUBLE POPULATION WORLD TO U.S. STANDARD
Iron	$\times 2$	$\times 12$
Copper	$\times 2$	$\times 11$
Lead	$\times 2$	$\times 16$

Table 6-2. Concentration factors for some common metals.

METAL	PER CENT IN EARTH'S CRUST	PER CENT IN ORE	CONCENTRATION FACTOR
MERCURY	0.000008	0.2	25,000
LEAD	0.00125	5.0	4,000
NICKEL	0.0075	1.0	133
COPPER	0.0055	0.5	90
IRON	5.63	30.0	6
ALUMINUM	8.23	38.0	5

may be odorous eyesores.

It is estimated that the United States each day produces enough waste to cover 400 acres to a depth of ten feet. The cost to government for collection and disposal is $3000 million per year, a sum only exceeded by funds for schools and roads. We may yet be ". . . standing knee-deep in garbage firing rockets at the moon . . ." (*Newsweek,* June 29, 1970, p. 77).

Formation of Mineral Deposits

Mineral deposits occur at places where unusual conditions caused the concentration of some elements far in excess of their normal abundances. Table 6-2 shows the average amounts of some elements in the earth's crust. Copper, for instance, has an average abundance of 0.0055 per cent, but copper ore in the United States must average 0.5 per cent copper. Thus copper must be concentrated by at least 90 times its average amount to form copper ore. Many mines produce more than one metal, thus reducing the percentage of any one element that must be present to make the deposit economic. In any case, to form a metal mine, an element must be concentrated at least several to a few thousand times its average abundance. We will now see how this can

be done. (See Tables 6-3 and 6-4.)

Many mines are associated with igneous rocks, and we will consider them first. Because we do not know how igneous rocks form, we have little knowledge of where the metals come from, except that they are part of the magma. The different magma types, such as granite and gabbro, each have typical metals that are associated with them. The distribution of ore metals is probably inherited from the original distribution of metals that developed when the earth formed.

Disseminated Magmatic Deposits are the simplest of the magmatic deposits. The valuable mineral is disseminated or scattered throughout the igneous body. In the diamond deposits of South Africa, for example, the diamonds are disseminated in an unusual rock, somewhat similar to peridotite. (See Fig. 6-3.) In this example, the valuable mineral is unique to this rock type and, be-

Table 6-3. Genetic classification of mineral deposits.

TYPE	EXAMPLE
Magmatic	
Disseminated	Diamonds — South Africa (Fig. 6-3)
Crystal settling	Chromite — Stillwater, Montana (Fig. 6-4)
Late magmatic	Magnetite — Adirondack Mountains, New York (Fig. 6-5)
Pegmatite	Beryl and lithium — Black Hills, South Dakota (Fig. 6-6)
Hydrothermal	Copper — Butte, Montana (Fig. 6-7)
Contact-metamorphic	Lead and silver — Leadville, Colorado (Fig. 6-9)
Sedimentary	
Evaporite	Potassium — Carlsbad, New Mexico
Placer	Gold — Sierra Nevada foothills, California (Fig. 6-11)
Weathering	Bauxite — Arkansas (Fig. 6-12)

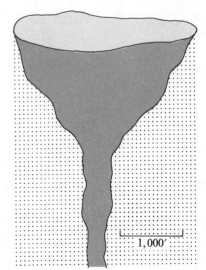

Figure 6-3. South African diamond pipe. Diamonds are scattered through more or less vertical, cylindrical bodies of an unusual igneous rock somewhat similar to peridotite.

cause of its great value, can be mined. In the other types of deposits, the ore mineral or minerals are concentrated rather than disseminated within the body.

Crystal Settling was already described in the discussion of differentiation of magma. (See Figs. 3-25 and 6-4.) Heavy, early-crystallized minerals sink to the bottom of the

igneous body. Many chromite and/or magnetite deposits form in this way, generally in gabbro bodies. (Magnetite crystallizes early under some conditions and late under

Table 6-4. Geologic occurrence of some metals.

METAL	MAIN ORE MINERALS	PRINCIPAL TYPE OCCURRENCE	EXAMPLE
Aluminum	Bauxite $Al_2O_3 \cdot nH_2O$	Weathering	Arkansas
Antimony	Stibnite Sb_2S_3	Hydrothermal	Yellow Pine, Idaho
Beryllium	Beryl $Be_3Al_2Si_6O_{18}$	Pegmatite	Black Hills, S.D.
Cadmium	By-product of zinc ores	Hydrothermal (in sphalerite)	Missouri (Tri-State Region)
Chromium	Chromite $Fe(Cr,Fe)_2O_4$	Magmatic	Stillwater, Montana
Cobalt	Cobaltite $CoAsS$ Smaltite $CoAs_3$	Hydrothermal	Zaire
Copper	Bornite Cu_5FeS_4 Enargite Cu_3AsS_4 Chalcocite Cu_2S	Hydrothermal	Bingham, Utah Butte, Montana
Gold	Tellurides $AuTe_2$ Native gold Au	Hydrothermal Placer	Homestake, S.D. Motherlode, Calif.
Iron	Hematite Fe_2O_3 Magnetite Fe_3O_4	Sedimentary and Weathering	Lake Superior Region
Lead	Galena PbS	Hydrothermal	Missouri (Tri-State Region) Coeur d'Alene, Id.
Manganese	Oxides MnO_2	Weathering	Chamberlain, S.D.
Mercury	Cinnabar HgS	Hydrothermal	New Almaden, Calif.
Molybdenum	Molybdenite MoS_2	Hydrothermal	Climax, Colo.
Nickel	Sulfides NiS	Magmatic	Sudbury, Ontario
Platinum	Native platinum Pt	Magmatic Placer	Ural Mountains, Russia
Silver	Sulfides Ag_2S By-product of zinc-lead mine	Hydrothermal	Comstock, Nev. Coeur d'Alene, Id.
Tin	Cassiterite SnO_2	Hydrothermal Placer	Malaya
Titanium	Ilmenite $FeTiO_3$ Rutile TiO_2	Magmatic	Adirondack Mtns., New York
Tungsten	Scheelite $CaWO_4$ Wolframite $(Fe,Mn)WO_4$	Contact-metamorphic	Small deposits in Nev. and Calif.
Uranium	Pitchblende U_3O_8 Uraninite UO_2 Carnotite $K_2(UO_2)_2(VO_4)_2 \cdot 3H_2O$	Sedimentary (Carnotite)	Colorado Plateau
Vanadium	Carnotite $K_2(UO_2)_2(VO_4)_2 \cdot 3H_2O$ Roscoelite $K(V,Al)_2(OH)_2AlSi_3O_{10}$	Sedimentary (Carnotite)	Colorado Plateau
Zinc	Sphalerite ZnS	Hydrothermal	Missouri (Tri-State Region)

others.) In some cases, the segregation of the early-formed ore minerals is effected by squeezing the liquid magma away, leaving the early-formed crystals.

Late Magmatic Deposits form when the ore-forming constituents of the magma are among the last to crystallize. The metalliferous material is heavy; it tends to settle to the bottom of the igneous body and to crystallize there and in the interstices between the earlier-formed minerals. (See Fig. 6-5.) At the late stage when these deposits form, as pointed out

earlier, the magma may be water-rich.

Pegmatite bodies also form late in the crystallization of a magma. These extremely coarse-grained rocks may form as dikes or irregular pods. (See Fig. 6-6.) Just how the large crystals form is not known, but the water- and volatile-rich solutions must permit the growth of large crystals. Not all pegmatites are igneous, for many are found in metamorphic terranes far from any igneous body. They are sought for their large crystals of mica, which are used in industry, and

for feldspar, quartz crystals, and gems. They are also the main source of columbium, tantalum, the rare earths, beryl (the main beryllium mineral), and lithium. Many other minerals occur in pegmatites, including some high-temperature sulfides.

Hydrothermal, or hot water, deposits are the most common type of metal deposit. Their possible origin from the late magmatic fluids was discussed earlier. They grade from high-temperature to low-temperature deposits. Typical hydrothermal deposits contain sulfide minerals, such as pyrite, chalcopyrite, and galena, in quartz veins. (See Figs. 6-7 and 6-8.) Generally they are associated with batholiths, and the veins are both in the batholith and the surrounding rocks. However, many hydrothermal deposits have no nearby outcropping igneous rocks and the origin of the depositing solutions is not known. In such cases the ores may have been deposited by circulating ground water, as mentioned later in connection with geothermal power. In these instances, the metals might come from deep magmas or might be leached from other rocks by the circulating ground water. Examples of hydrothermal deposits include copper at Butte, silver at the Comstock Lode, lead-zinc at Missouri, and gold at the Mother Lode in California.

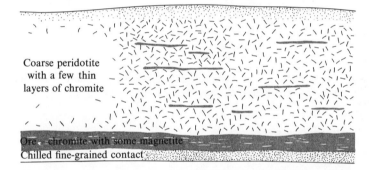

Figure 6-4. Chromite layers form by settling of early-formed chromite crystals. The parent igneous rock is peridotite, dunite, or gabbro. Commonly the igneous body has a more irregular shape than shown.

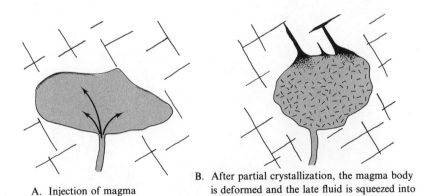

A. Injection of magma

B. After partial crystallization, the magma body is deformed and the late fluid is squeezed into newly-formed cracks.

Figure 6-5. An example of segregation of late magmatic fluid. In this case, the late fluid is separated from the rest of the magma; in other cases, the late fluid may sink and fill the spaces between the early crystals near the bottom of the igneous body.

Contact - Metamorphic Deposits form in the zone of metamorphism that surrounds an intrusive body. The magma and its fluids react with the wall rock and the ground water and replace the wall rock with ore minerals. These deposits are commonly found in limestone near intrusive bodies because the limestone is especially reactive with the fluids. (See Fig. 6-9.) Many deposits formed in this way contain high-temperature sulfide minerals similar to those in high-temperature hydro-thermal deposits. Some iron deposits, a few fairly large, formed in this way: Iron Springs, Utah, is an example.

Sedimentary processes can also concentrate valuable elements.

Evaporites, or chemical precipitates, such as salt beds, especially those containing potassium salts, are sought actively. Such beds come from the evaporation of sea water or of a salt lake. Because sea water contains only 3.5 per cent dissolved material, mostly common salt, several thousand feet of sea water must evaporate to produce 100 feet of salt. Many salt deposits, however, are much thicker than this. The evaporation is thought to occur in shallow, slowly sinking basins, partially dammed by sand bars, and fresh brine from the ocean flowing over the sand bars, perhaps during storms. Recently it has been suggested that salts could accumulate in deep basins with high evaporation and stagnant bottoms. The chemistry of these brines is complex, but the general sequence of precipitation is calcium and magnesium carbonate, gypsum, salt (halite), and finally potassium and bromine salts. Salt beds are mined in the area of an ancient basin in Michigan, Ohio, and New York. Potassium salts are mined near Carlsbad, New Mexico.

At other times and places in the geologic past, generally in restricted basins, deposits of phosphate and iron-rich rocks have occurred. Rather specialized conditions are probably necessary to form such

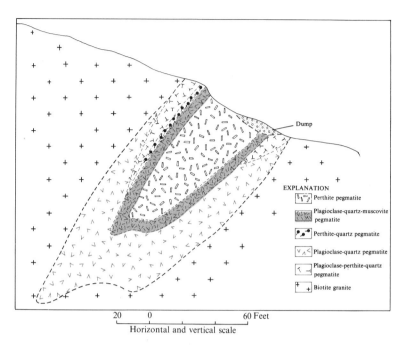

Figure 6-6. Cross-section of the Hyatt pegmatite. Larimer County, Colorado. From W. R. Thurston, U.S. Geological Survey, *Bulletin* 1011, 1955.

Figure 6-7. Underground mining, Kelley Mine, Butte, Montana. Some of the veins in the granitic rock being drilled can be seen. Photo from Anaconda Co.

Figure 6-8. Aerial view of the open pit mine at Chuquicamata, Chile. The veins here are too narrow to mine economically underground. The ore averages 1⅓ per cent copper, and about 142 tons of ore and waste rock must be moved for each ton of copper recovered. This low-grade ore body has been in production over 50 years. Photo from Anaconda Co.

Figure 6-9. Cross-section of part of the Leadville district, Colorado. Most of the ore bodies replace limestone and are believed to have originated from the porphyry. Some veins are present, and the ore bodies along the fault show that it acted as a passage for some of the ore fluids. Replacement ore bodies generally have complex shapes. See Figure 6-10. From F. A. Aicher, U.S. Geological Survey, *Professional Paper* 148, 1927.

rocks. The phosphate deposits of Idaho, Wyoming, and Montana formed in such a restricted sea.

Placer Deposits form by the concentration of heavy mineral fragments by rivers. A heavy, or more properly, a dense mineral fragment is moved less easily than a lighter fragment of the same size. For this reason a river tends to separate, or sort, its transported load. The denser fragments are the first to be deposited and tend to be concentrated on the insides of curves, at places where the gradient is less, and in similar places. These processes will be described in Chapter 8.

Minerals concentrated in this way must be dense and resistant to weathering. Examples include gold, diamond, and platinum. Gold placers were very important in the opening of the west, especially in California in 1849. (See Fig. 6-11.) Placer deposits are "poor man's deposits" because very little equipment is necessary to prospect for them or to exploit them. This easily-obtained wealth started the westward movement. Ocean waves can also produce placers, and beach sands containing valuable heavy minerals are mined at some places.

Weathering Processes also contribute to mineral wealth. Soils are surely our most valuable mineral commodity. In addition, some tropical soils (laterites) are mined for their aluminum (bauxite), iron, or other elements, as was described earlier. (See Fig. 6-12.)

Weathering processes are also active on outcropping mineral deposits. Their effect on many deposits is great and, unless understood, can be very misleading. If an ore or its weathering products are insoluble, the ore will be enriched by weathering unless the weathered material is removed by erosion. This is especially true in mineral deposits in soluble rocks such as limestone. (See Fig. 6-13.) This kind of enrichment has fooled many prospectors and investors who thought that they had a good mine on the basis of near-surface assays, but found the ore too low-grade to be economical when a shaft was sunk. Gold, lead, copper, silver, and mercury enrichments of this type are very common.

At sulfide deposits, the weathering processes can become very complex. The sulfide minerals are oxidized to become soluble sulfates. Pyrite, which is present in almost every case, weathers to sulfuric acid and soluble iron sulfates, all of which aid in the solution of other sulfide minerals. Iron, silver, and copper sulfides are dissolved in this way. The soluble sulfates trickle slowly downward, enriching the lower part of the deposit. The main zone of enrichment occurs at the water table. Here the chemical environment differs, mainly in its lack of oxygen, and precipitation as sulfides occurs. Again, the shrewd prospector should not be fooled by the near-surface shows unless they, by themselves, are rich and large enough to mine.

Weathering of mineral deposits does help the prospector. The oxidation of the ever-present pyrite leaves a surface residue of iron oxides, especially hematite and limonite. These red-brown caps clearly mark the outcrops of sulfide veins, and so they are the prime target of the prospector. (See Fig. 6-14).

Deep Ocean Deposits have recently attracted much attention. Bottom photographs and dredge samples have revealed that parts of the floor of the Pacific Ocean are covered by manganese nodules. (See Fig. 6-15). These interesting rocks have formed by the precipitation of manganese and iron dissolved in sea water onto sand grains and pebbles on the sea bottom. This precipitation is so slow that 1000 to 100,000 years are required to deposit a layer one millimeter thick. The deposition occurs in the form of oxides and hydroxides. These nodules average about 24 per cent manganese and 14 per cent iron, with small amounts of nickel, cobalt, copper, and other elements. Some contain 2 per cent or more copper, nickel, or cobalt. The elements come from chemical weathering of the continents, by leaching of submarine basalt, and from waters associated with volcanic activity. Because of the large areas involved, these nodules constitute a very valuable resource, and techniques for their mining are being developed. However, the nodules also contain large amounts of silica (SiO_2), and current metallurgical techniques do not permit the separation of manganese from silica profitably.

Energy Resources

The main mineral energy sources are coal and petroleum. Doubtless in the future atomic and nuclear power will overshadow these mineral fuels. Geothermal power is currently being produced and also will be more important in the future.

Coal and petroleum are sometimes called fossil fuels because they

Figure 6-10. A model of a mine made by drafting successive cross-sections on transparent material. The complex shapes of ore bodies can be studied by this technique. Photo from Anaconda Co.

originate from once-living material. Coal comes from plant material. To form coal, plant material must accumulate; but under most conditions, even with abundant growth of vegetation as in a tropical rain forest, plant material does not accumulate. Instead, it decays and is attacked by bacteria as soon as it falls. Most coal beds seem to have developed in swamps, where, as soon as a fallen tree becomes waterlogged, it sinks to the bottom and ceases, or nearly ceases, to rot, especially if the water is stagnant as it is in many swamps. The plant material at the bottom of such a swamp becomes *peat.* The peat ultimately becomes coal, largely due to compaction from the weight of overlying sediments. The first stage in this process produces *lignite,* and further pressure produces *bituminous coal* and finally, in rare cases, *anthracite.* (See Fig. 6-16.) In this process the original peat may be compressed to less than one twenty-fifth of its original thickness. To produce a reasonably pure coal, little other sediment can be accumulating in the swamp at the same time.

Petroleum also comes from organic material, but the process is not so clear as it is with coal. Oil and gas come from marine sediments. Dark organic-rich shale is the probable source. Such sediments are accumulating today in the Black Sea where the bottom muds contain up to 35 per cent organic material compared with about 2.5 per cent in ordinary marine sediments. In normal marine sediments the organic material is oxidized, but in stagnant bottom water it can accumulate. The situation is very similar to the formation of coal. As with coal, apparently, deep burial, some heat, and much time are necessary to produce gas and oil. Once formed, the petroleum must migrate to a permeable rock from which it can be extracted. Oil and gas reservoirs and traps will be described in Chapter 12 in con-

Figure 6-11. Map showing the location of placer gold deposits in the Sierra foothills in California. The ancient rivers along which these placers accumulated were in different locations than the present rivers. The older of the rivers are shown by heavy lines and the younger by dotted lines. These rivers flowed into a sea indicated by the stippled area. From P. C. Bateman and Clyde Wahrhaftig, California Division of Mines and Geology, *Bulletin* 190, 1966.

nection with ground water. Not all oil forms in marine rocks. The oil shales of the Rocky Mountain area, which contain a large percentage of our petroleum reserves, formed in fresh-water lakes, probably by the same process. In oil shale, the petroleum material is disseminated through the shale and has not migrated to a reservoir rock from which it can be easily recovered. The petroleum is recovered by heating the rock, which is a more expensive process than production from an oil well. Oil shale of marine origin occurs in the area from Michigan and western Pennsylvania to Oklahoma, as well as in Alaska and elsewhere.

The elements useful for producing atomic power are uranium and thorium. Thorium occurs mainly in hydrothermal veins. Uranium occurs in hydrothermal veins, pegmatites, and disseminated in some sedimentary rocks.

Figure 6-12. Cross-section of the Pruden bauxite mine, Arkansas. The bauxite is the weathering product of the aluminum-rich igneous rock. From Mackenzie Gordon, Jr. and others, U.S. Geological Survey, *Professional Paper* 299, 1958.

Figure 6-13. Residual concentration by solution of the limestone that encloses the deposit. The ore minerals must be insoluble.

Figure 6-15. Manganese nodules on the floor of the Atlantic Ocean. Photo courtesy of R. M. Pratt.

Figure 6-14. Weathering of a sulfide vein deposit. Rain water percolating to the water table leaches metals from the vein and redeposits them near the water table, creating an enriched zone. The reddish iron oxide cap helps to attract the attention of prospectors.

Geothermal power is relatively new in this country but has been used in Italy for some years. Basically, it is simply the use of steam escaping from the earth to generate power. Hot springs and geysers are fairly common, but at only a few places does steam escape from the earth continuously. Yellowstone National Park is a thermal area where super-heated water or steam could be obtained by drilling. Electric power is being generated in this way at The Geysers in north central California. (See Fig. 6-17.) The water escaping from these thermal areas is mostly rain water that has been heated, probably by contact with a cooling magma. Isotope studies show that very little of the water is primary water from a magma. In most areas, these thermal waters contain much dissolved material, a fact that is both an asset and a major problem. After being used to generate power, the waters can be evaporated by solar heat to harvest some of the salts. This is a clear gain. The problems are the disposal of the raw or partially harvested waters, and the corrosion and deposition that these waters cause in pipes. It is interesting to note that at some places these waters contain appreciable amounts of metals such as lead, copper, and iron. Thus these hot waters may closely resemble the hydrothermal solutions that form mineral veins. As we learn more about such hot waters and how to harness them, they may become an important power source. As will be discussed later, we know that the earth's interior is hot, and perhaps some day we can use deep holes to obtain heat from which substantial power can be generated.

Hot brines (133°F), containing large amounts of dissolved metals, have been found in the Red Sea. The sea water that forms these brines is

Peat bog

Peat

Lignite

Bituminous — bright coal

Coal areas of the United States

Bituminous — cannel coal

Anthracite

Bituminous — splint coal

Coke

Figure 6-16. Types of coal. Courtesy U.S. Bureau of Mines.

Figure 6-17. Geothermal steam escaping from wells at The Geysers, California. Photo by Pacific Gas and Electric Co.

believed to have moved large distances through rocks and to have leached the metals during this movement. In the Red Sea and in southern California, just north of the Gulf of California, the hot metal-bearing waters are associated with mid-ocean ridges, a topic discussed in Chapter 15.

Energy and the Future

Modern civilization depends on energy. At first man's only use of energy was in the form of food. Later he used wood fires, and the industrial revolution required the use of coal. The fossil fuels, coal and petroleum, are the main energy sources at present, with nuclear power developing very rapidly.

The energy sources of the earth are:

    Solar radiation
        Food
        Fossil fuels
        Water power
        Direct solar energy

    Tidal energy
    Geothermal energy
    Nuclear energy

The amount of energy that the earth receives from the sun is many times more than the amount used by man. If a practical way to use it could be devised, no other source would be needed. It should be remembered that the earth re-radiates the energy received from the sun. If it did not, the temperature of the earth would increase, and geologic history shows that the average temperature has remained about the same for at least 500 million years. Solar energy is man's most important energy source. It is used by plants, providing the food that supports all of the life on the earth. In the past, plants in the form of wood provided an important source of energy. A very small amount of solar energy is preserved in the form of fossil fuels such as coal and petroleum.

Although the fossil fuels are our present main source of energy, their future is not bright. The earth's coal

supplies are expected to last for two or three centuries, but petroleum will run out in 70 to 80 years. At the present time, about 60 per cent of the world's and 67 per cent of the United States' industrial energy comes from oil and gas. These figures suggest that many changes are due in the relatively near future.

Natural gas is the cleanest of fossil fuels, and it will probably be the first to run out. Coal, which is the most widely used, releases sulfur and mercury. Sulfur is a major problem in some areas where it causes atmospheric pollution. Use of low-sulfur coal or removal of sulfur from the smoke is required in some cities. Low-sulfur coal is also needed in the iron and steel industries. It is estimated that 3000 tons of mercury are also released from burning coal each year. This is about equal to the amount lost in industrial processes, and over ten times the amount released by natural weathering.

Water power is an attractive power source because of the lack of atmospheric pollution. Damming rivers, however, spoils some of their aesthetic value, and the lakes formed by the dams will ultimately fill with silt. Transporting the power from a hydroelectric plant requires transmission lines. In spite of these problems, water power will probably be important in Africa and South America because these continents lack large supplies of coal but have large unused amounts of water power. Water power could supply the earth with energy at the present rate of usage, but because of the silting of the lakes, it could only do this for one or two centuries.

Solar radiation as a source of energy is attractive because no fuel is used. The disadvantages are the large areas required and that only certain areas such as deserts receive enough sunlight to make such systems feasible. These disadvantages are offset somewhat because desert areas are largely unused for any other purpose, although many ecologists would disagree because they feel that deserts, like mountain, forest, and beach areas, should be preserved for recreation, ecologic and aesthetic reasons.

The relatively cloudless deserts of southwestern United States receive about 0.8 kilowatt per square meter for the middle six to eight hours of the day most of the year. A black surface can absorb most of this energy, but a black body is also an excellent radiator. Recently, surface coatings have been developed that allow the sun's radiation to pass through and be absorbed, but these surface films are reasonably opaque to the lower wavelength radiation from the collector. The collector would heat molten salts that would maintain a fairly constant temperature during overnight operation. A large-scale experiment will be necessary to prove the practicality of such a system.

Tidal energy can only supply a small amount of our energy needs, and no really practical method has been devised to utilize it. The advantages are no fuel, no waste, no pollution, and minimum ecologic and scenic damage.

Geothermal energy is a new energy source in the United States, but has been in use since 1904 in Italy.

The number of volcanic areas where geothermal steam is available is small. The life of a steam well is also limited. The best current estimates suggest that about one-fifth of our present energy needs could be met for about fifty years by geothermal power.

Nuclear energy seems most likely to be the energy source of the future. At present, its use is growing. However, two problems must be overcome to insure its use. They are disposal of radioactive wastes, and development of reactors that use the more plentiful ores. Other serious obstacles are public acceptance of nearby nuclear reactors and disposal of unused heat.

Nuclear energy can be released by fusion and fission. Fusion is the process used in the hydrogen bomb, and so far, no method has been found to release this energy slowly. It remains to be seen if this will ever become a practical power source. Fission reactors are in current use. Some of these reactors are burners in the sense that they consume their fuel; others are convertors or breeders. The difference is of supreme importance for the future. Uranium is the main fuel used in reactors. Natural uranium is a rare element and occurs in three isotopes. Uranium 238 is 99.283 per cent, uranium 235 is 0.711 per cent, and uranium 234 is 0.006 per cent of natural uranium. Of these, uranium 234 is so minor it can be neglected. Uranium 235 is the only naturally

fissionable isotope and is used in burner reactors. Almost all of the present reactors are of this type, and they burn uranium 235. There is only enough uranium 235 to last about a century at the present rate of development. It is possible to build breeder reactors that will convert the abundant uranium 238 and thorium 232 into fissionable isotopes, but some uranium 235 is needed for these breeder reactors. Such breeders must be developed soon, or the uranium 235 will be used up and nuclear energy will not be available — and there seems to be no other fuel for the distant future.

Radioactive waste products and similar very dangerous wastes present special problems. Some have been dumped at sea in special concrete and/or steel containers, but their potential danger to the whole ocean makes this practice not acceptable to at least some people. The waste from an atomic reactor is about the same amount as the total fuel. It must be held for about 600 years before it is safe. Salt is a very good material in which to bury atomic wastes. It is stable in earthquakes, has a high melting point, and is a good nuclear shield. For these reasons the Atomic Energy Commission hopes to bury radioactive wastes inside deep salt domes. The ice of Antarctica has also been suggested as a burial site for radioactive waste.

Summary

Mineral resources are natural resources that come from the earth, such as metals, energy-producing materials, sand and gravel, building stone, limestone, etc.

A mineral deposit is a geologically rare feature, and once used it cannot be replaced except for the very long replacement time for ground water and soil. Special conditions concentrate the valuable material.

Ore is anything that can be mined at a profit. Economic conditions or technological changes affect which deposit is considered ore. Mineral resources greatly affect the course of history.

Disseminated magmatic deposits — the valuable mineral is scattered throughout an igneous body.

Crystal settling — heavy, early-crystallized minerals sink to the bottom of magma in a crystallizing igneous body.

Late magmatic deposits — ore-forming minerals are concentrated by being among the last minerals to crystallize in a magma.

Pegmatite deposits—bodies of late-forming, very large crystals. (Some pegmatite bodies are metamorphic.)

Hydrothermal deposits — these deposits may be from late magmatic fluids, and commonly are metal sulfide minerals in quartz veins. The origin of some hydrothermal-depositing solutions may be circulating ground water and leached metals.

Contact-metamorphic deposits—deposits around intrusive bodies in the zone of metamorphism formed when magma reacts with the surrounding wall rocks (in many cases limestone) and ground water.

Evaporites — chemical precipitates formed by the sedimentary processes in the evaporation of sea water.

Placer deposits—a sedimentary process in which heavy mineral fragments are sorted and concentrated by rivers and ocean waves.

Weathering processes provide our most valuable resource, soil. Also, tropical soils *(laterites)* are mined for *bauxite* (aluminum ore), iron, etc. Weathering processes also cause local zones of enrichment on other types of mineral deposits, some of which may be deceiving concerning the deposit's value.

Deep-ocean deposits — manganese and iron oxides and hydroxides are precipitated from sea water into nodules. Not yet economic to separate the manganese from the accompanying silica.

The main *energy resources* are coal and petroleum which originate from once-living material.

Coal forms from accumulation of plant material under conditions which prevent its rot, probably in stagnant swamps. The sequence is *peat*, to *lignite*, to *bituminous coal*, to, in rare cases, *anthracite*.

Petroleum comes from organic-rich marine shales. To be easily recoverable, the petroleum has to migrate to a reservoir rock.

Atomic power is produced from thorium (mainly from hydrothermal veins) and uranium (from hydrothermal veins, pegmatites, and disseminated in some sedimentary rocks).

Geothermal power is the use of steam from hot spring and geyser areas to generate power.

The energy sources of the earth are:

Solar energy — sufficient if practical way to use it directly can be devised.

Food — solar energy through plants.

Fossil fuels — small amount of solar energy is stored in coal and petroleum. Supplies greatly limited.

Water power—eventually limited by silting behind dams.

Solar radiation — advantage is no fuel used. Disadvantage is large areas are required; practicality uncertain.

Tidal energy — no practical method yet.

Geothermal energy — limited number of geothermal steam areas, and steam well's life is limited.

Nuclear energy — likely energy source of future. Breeder fission reactors will need to be developed to use the more abundant uranium 238 before the necessary, less abundant uranium 235 is used up in the current burner reactors. Disposal of radioactive wastes is a very difficult problem.

Questions

6-1. What are the similarities and differences between mineral resources and other types of natural resources?

6-2. What types of mineral deposits are directly associated with igneous rocks? Might some of these be found in sedimentary or metamophic rocks?

6-3. Is the gold pan still a useful tool for the modern prospector?

6-4. What types of mineral deposits are associated directly with sedimentary rocks?

6-5. What parts of the earth are relatively unexplored for mineral resources?

6-6. Name an ore of lead_____.

zinc_____.

mercury_____.

iron_____.

copper_____.

6-7. Discuss whether the earth's mineral resources are sufficient for the next hundred years.

6-8. List the earth's energy sources, and discuss the future of each.

6-9. What are the problems facing us in disposing of our waste material?

Supplementary Reading

Bachmann, H. G., "The Origin of Ores," *Scientific American* (June, 1960), Vol. 202, No. 6, pp. 146-156.

Barnea, Joseph, "Geothermal Power," *Scientific American* (January, 1972), Vol. 226, No. 1, pp. 70-77.

Brobst, D. A., W. P. Pratt, and V. E. McKelvey, *Summary of United States Mineral Resources,* U. S. Geological Survey Circular 682. Washington, D. C.: U. S. Government Printing Office, 1973, 19 pp.

Brown, T. L., *Energy and the Environment.* Columbus, Ohio: Charles E. Merrill Publishing Co., 1971, 141 pp. (paperback).

Cameron, E. N. (editor), *The Mineral Position of the United States 1975-2000.* Madison, Wisconsin: The University of Wisconsin Press, 1973, 100 pp. (paperback).

Committee on Geological Sciences, National Research Council, National Academy of Sciences, *The Earth and Human Affairs.* San Francisco: Canfield Press, 1972, 138 pp. (paperback).

Committee on Resources and Man, National Academy of Science, National Research Council, *Resources and Man.* San Francisco: W. H. Freeman and Co., 1969, 259 pp. (paperback).

Degens, E. T. and D. A. Ross, "The Red Sea Hot Brines," *Scientific American* (April, 1970), Vol. 222, No. 4, pp. 32-42.

De Nevers, Noel, "Tar Sands and Oil Shales," *Scientific American* (February, 1966), Vol. 211, No. 2, pp. 21-29.

Flawn, P. T., *Mineral Resources: Geology, Engineering, Economics, Politics, Law.* Chicago: Rand McNally, 1966, 406 pp.

Hibbard, W. R., Jr., "Mineral Resources: Challenge or Threat?," *Science* (April 12, 1968), Vol. 160, No. 3824, pp. 143-149.

Hubbert, M. K., "The Energy Resources of the Earth," *Scientific American* (September, 1971), Vol. 224, No. 3, pp. 60-70.

Lang, A. H., *Prospecting in Canada,* 4th ed. Canada Dept. Mines and Technology Survey, Geological Survey of Canada, Economic Geology Series No. 7, 1969, 308 pp.

Lovering, T. S., *Minerals in World Affairs.* New York: Prentice-Hall, Inc., 1943, 394 pp.

McKelvey, V. E., "Mineral Resource Estimates and Public Policy," *American Scientist* (January-February, 1972), Vol. 60, No. 1, pp. 32-40.

Park, C. F., Jr., *Affluence in Jeopardy: Minerals and the Political Economy.* San Francisco: Freeman, Cooper and Company, 1968, 368 pp.

Scientific American (September, 1971), Vol. 224, No. 3. Entire issue devoted to energy.

Skinner, B. J., *Earth Resources.* Englewood Cliffs, New Jersery: Prentice-Hall, Inc., 1969, 150 pp. (paperback).

Theobald, P. K., S. P. Schweinfurth, and D. C. Duncan, *Energy Resources of the United States,* U.S. Geological Survey Circular 650. Washington, D. C.: U. S. Government Printing Office, 1972, 27 pp.

U. S. Bureau of Mines, *Minerals Yearbook.* Washington, D. C.: Government Printing Office, published annually.

Wenk, Edward, Jr., "The Physical Resources of the Ocean," *Scientific American* (September, 1969), Vol. 221, No. 3, pp. 167-176.

The Downslope Movement of Surface Material

7

Erosion

The main process going on today on the earth's surface is clearly erosion and is shown by gullies on hillsides, landslides, and similar examples. The agents of erosion, of which rivers are far and away the most important, are constantly wearing down the continents; nevertheless, we still have mountains. The present rate of erosion is such that, theoretically, all topography will be removed in about 12 million years, but we know that erosion and deposition have been going on for a few billion years. Something must interrupt the work of the rivers and uplift the continents. Thus, much of geology is concerned with the struggle between the external forces of erosion and the internal forces which cause uplift.

Some familiar demonstrations of the internal forces which cause uplift are earthquakes, volcanoes, and folded and tilted sedimentary rocks that are now high on mountains, but were originally formed as flat beds at or below sea level. These internal processes will be discussed later. Physical or dynamic geology is the study of this battle between the internal force of uplift and the external force of erosion. We will consider these two forces separately in order to eliminate confusion, but remember that they go hand in hand.

The underlying cause of erosion is downslope movements under the influence of gravity; the agents that cause such movements are water (rivers, etc.), ice (glaciers), and wind. The material that is moved by these agents comes from the weathering processes discussed earlier.

The sun's energy is the ultimate source of the energy used in erosion, as shown in Fig. 7-1. This diagram also shows that erosion is a necessary by-product of the distribution of the sun's energy around the earth. Without this distribution of energy by the oceans and the atmosphere, the equatorial regions would become hotter and the poles colder because most of the sun's energy falls near the equator.

Importance of Downslope Movement

All of the surface material of the earth tends to move downslope under the influence of gravity. The surface material is more or less weathered and is generally termed *overburden* to differentiate it from bedrock. These downslope movements shape the earth's surface and so must be understood to interpret landscapes. The downslope movement generally ends at a river or stream that carries the material away. Thus, rivers and downslope movements work together to shape the landscape. A small example of this, discussed in the following chapter on rivers, shows that although a river downcuts to form a valley, the

Urban landslide in Oakland, California. The slide developed after heavy rains. Only a small part of the slide is shown.

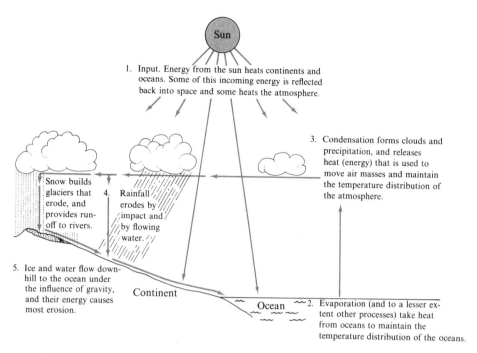

Figure 7-1. The water cycle and the earth's energy balance. Incoming energy from the sun is distributed around the earth by the circulation of the oceans and the atmosphere. The water cycle is part of the energy exchange between the oceans and the atmosphere, and it is the immediate source of the energy used in most erosion.

shape of the valley sides is determined by downslope movements, which widen the narrow cut made by the river. (See Fig. 8-11.) The efficiency of downslope movements is easily shown by analysis of slopes. Such analysis shows that most of the earth's surface slopes less than 5 degrees and that very little of the earth's surface slopes more than 45 degrees. Many forms of downslope movement are slow processes. However, when slopes are disturbed by buildings, roads, or irrigation, more rapid movements may be triggered, causing much damage. (See Fig. 7-2). Thus, knowledge of these processes is important when any type of engineering project is considered on even moderately sloping ground. (See Fig. 7-3.)

Types

Downslope movements may be rapid or very slow; and they may involve only the very topmost surface material, or they may be massive movements that involve the total thickness of overburden or even bedrock. The processes at work on any slope depend on many factors. On steep slopes no overburden generally can develop; therefore, the dominant process on such slopes is *rockfall,* and the slope that develops at the

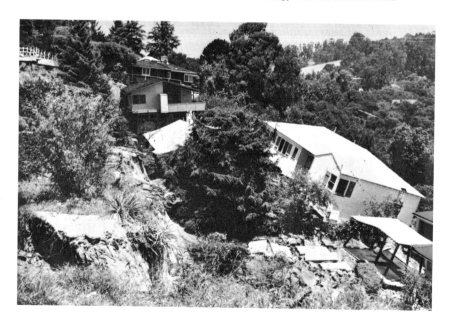

Figure 7-2. Urban landslide in Oakland, California. This slide developed after heavy winter rains. Construction on the top of the hill may have altered the infiltration of rain water. Only a small part of the slide is shown in the photo.

Figure 7-3. Hillside construction can cause earth movements if not properly planned.

foot is called a *talus*. (See Fig. 7-4.) On less steep slopes — how much less depends on the climate which controls the weathering processes — overburden may develop, and this material may move downslope in a number of ways discussed below. Thus weathering, which is controlled by the climate, determines the type of downslope movement, thereby causing the different aspects of the topography in a humid area and a dry area. (See Fig. 7-5.) The shapes of slopes and the theories of how these shapes are developed will be discussed in Chapter 8. The structure and rock types on a hill may also control the erosion processes. Differences in rate or kind of weathering can determine the amount of material available for downslope movement, and so the underlying bedrock structure may be etched in relief.

Erosion by Rain

Falling rain erodes surface material in two ways, by impact and by runoff. The impact of raindrops loosens material and splashes it into the air (see Fig. 7-6); and on hillsides, some of this material falls back at a point lower down the hill. Raindrops hit the ground with a velocity of about 30 feet per second. The resistance of the air controls this velocity. One-tenth of an inch of rain amounts to about 11 tons of water on each acre, and the energy of this water is about 55,000 foot-pounds per second. Ninety per cent of this great amount of energy is dissipated in the impact. Ninety per cent of the impact splashes are to a height of one foot or less, and the lateral splash

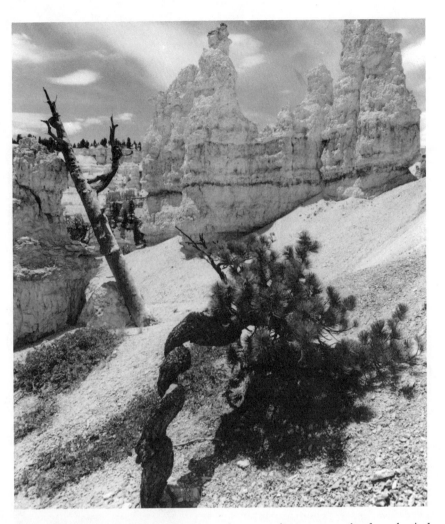

Figure 7-4. A talus develops at the foot of a steep slope as a result of mechanical weathering as shown in this view of Bryce Canyon National Park, Utah. Photo from Union Pacific Railroad.

movement is about four times the height. Because impact erosion is most effective in regions that have little or no grass or forest cover, it is most effective in desert areas (or newly plowed fields) that are subjected to sudden downpours such as thunderstorms. Splash erosion accounts for the otherwise puzzling removal of soil from hilltops where there is little runoff. It can also ruin soil by splashing up the light clay

particles that are then carried away by the runoff, leaving the silt behind.

The rain water that does not infiltrate into the ground runs down the hillside and erodes in the same ways that river water erodes. This water may gather in lower places and cut gullies. The amount of erosion caused in these ways depends on the steepness of the slope and on the type and amount of vegetative cover. The effects of such erosion are best

Figure 7-5. The Appalachian Mountains in North Carolina. Compare the rounded hillsides shown here with the angular topography of arid areas shown in Figure 7-4. Photo by A. Keith, U.S. Geological Survey.

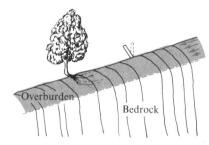

Figure 7-7. The effect of creep on fence posts and trees. This is caused by the slow downslope movement of the overburden.

seen in areas where the vegetation has been removed by fires or man.

Massive Downslope Movements

The massive types of downslope movement that involve appreciable thicknesses of material are creep, earthflow, mudflow, and landslide.

A number of names are applied to these movements as described in the next few pages. They are termed *flows* if they move as viscous fluids, *slides* if they move over the surface, and *falls* if they travel in the air in free fall.

Creep is the slow downslope movement of overburden and, in some cases, bedrock. It is recognized by tilted poles and fence posts if the movement is great enough. Commonly, in such cases, trees will be unable to root themselves, and only shrubs and grass will grow on the slope. In other cases, where the creep is slower, the trunks of trees will be bent as a result of the slow move-

ment. (See Figs. 7-7 and 7-8.)

If the amount of water in the overburden increases, an *earthflow* results. In this case a tongue of overburden breaks away and flows a short distance. An earthflow differs from creep in that a distinct, curved scarp is formed at the breakaway point. (See Fig. 7-9.)

With increasing water, an earthflow may grade into a *mudflow*. The behavior of mudflows is similar to that of fluids. Rain falling on the loose pyroclastic material on the sides of some types of volcanoes produces mudflows, and this is a very

Figure 7-6. Erosion by raindrop. From U.S. Department of Agriculture, 1955 Yearbook, *Water*.

Figure 7-8. Creep in slate at Resurrection Bay, Alaska. Photo by F. H. Moffit, U.S. Geological Survey.

Figure 7-9. Earthflows on a farm in eastern Washington. Note the scarps at the breakaway points. The left flow should probably be classified as a mudflow. Photo from U. S. Department of Agriculture, Soil Conservation Service.

important mode of transportation of volcanic material. Another type of mudflow is common in arid areas. In this case, a heavy thunderstorm produces large amounts of runoff in the drainage area of a stream. The runoff takes the form of rapidly moving sheets of water that pick up much of the loose surface material. Because these sheet floods flow into the main stream, all of this muddy material is concentrated in the main stream course. As a result, a dry stream bed is transformed very rap-

idly into a flood. This flood of muddy material is a mudflow, and it moves very swiftly, in some cases with a steep, wall-like front. The thick mud can carry large boulders, and such a mudflow can cause much damage as it flows out of the mountains. Eventually, loss of water (generally by percolation into the ground) thickens the mudflow to the point that it can no longer flow. (See Fig. 7-10.)

Another type of mudflow, called *solifluction,* occurs especially when frozen ground melts from the top down, as during warm spring days in temperate regions or during the summer in areas of permafrost. (See Fig. 7-11.) In this case the surface mud flows downslope. Solifluction causes many problems in construction, especially in the far north areas of permafrost. Permafrost areas, as the name implies, are areas where the deeply frozen ground does not completely thaw during the summer.

Another type of movement of soil material, called *frost heaving,* is associated with freezing and thawing. In this case the expansion of water as it freezes pushes pebbles upward through the soil or overburden. This phenomenon is well known to farmers in regions such as New England where each spring their fields are covered with pebbles that were not visible in the fall. In the Middle Ages peasants thought that the stones grew in the soil just as crops do. Frost heaving is a complex process, not yet fully understood. The small expansion of water during freezing is not enough to account for the several inches of uplift of pebbles or fence posts common in areas of frost heaving. Most of the soil water is in the pore spaces of the fine material. When this water freezes, it expands and so pushes nearby pebbles up-

Figure 7-10. Mudflow in the San Juan Mountains, Colorado. Its source can be seen on the skyline, and it ends in a lake. The trees growing on the mudflow show its age. Photo by W. Cross, U.S. Geological Survey.

Figure 7-11. Solifluction probably caused the smooth slopes in the middleground in this view of the Sierra Nevada in California. Photo from U.S. Forest Service.

ward. During the process of freezing, soil moisture migrates to these pockets of fine material and forms lenses of ice; it is this additional volume of ice that causes the several inches of uplift. When melting occurs, the surface tension of the water in the fine sediments pulls the fine material together, leaving the larger pebbles at nearly the height to which the expansion pushed them.

In very cold areas, frost heaving causes stone rings or polygons. (See Fig. 7-12.) In such areas, the extreme cold causes contraction cracks to form at the surface. In summer, melt water fills the cracks, and frost heaving proceeds as outlined above. Pebbles rise to the surface, forming the polygons.

Landslides are rapid slides of bedrock and/or overburden. Two types involve bedrock: the *landslide* or *rockslide,* and the *slump.* Rockslides develop when a mass of bedrock breaks loose and slides down the slope (Fig. 7-13.) In many cases the bedrock is broken into many frag-

ments during the fall, and this material behaves as a fluid, spreading out in the valley. It may even flow some

distance uphill on the opposite side of the valley if the valley into which it falls is narrow enough. Such landslides are sometimes called *avalanches,* but this term is best reserved for snow slides. Landslides are generally large and destructive, involving millions of tons of rock. As suggested in the diagrams, landslides are apt to develop if planes of weakness, such as bedding or jointing, are parallel to a slope, especially if the slope has been undercut by a river, a glacier, or men. Thus, landslide danger can be evaluated by geologic study. Slides involving overburden are called *debris slides.*

Slumps tend to develop in cases where a strong, resistant rock overlies weak rocks. Note in Fig. 7-14 the curved plane of slippage, the reverse tilt of the resistant unit that may provide a basin for a pond to

A.

Figure 7-12. Stone polygons formed by frost action. A. Close-up view in Alaska. Photo by H. B. Allen, U.S. Geological Survey.

develop, and the rise of the toe of the slump, which is most pronounced where the underlying rock is very weak and can flow plastically. Unlike rockfalls, slumps develop new cliffs nearly as high as those previous to the slump, setting the stage for a new slump. Thus slumping is a continual process, and generally, in areas of slumping, many previous generations can be seen far in front of the present cliffs.

Triggering

Massive downslope movements may be triggered in many ways. The most common are (1) undercutting of the slope (see Fig. 7-15), (2) overloading of the slope so that it cannot support its new weight and, hence, must flow or slide, (3) vibrations

from earthquakes or explosions that break the bond holding the slope in place (see Figs. 7-13 and 7-17), and (4) additional water. The addition of water is generally seasonal, and this is why many newly made cuts stand until the following spring. The effect of the water is twofold: it adds to the weight of the slope, and it lessens the internal cohesion of the overburden. Although its effect as a lubricant is commonly considered to be its main role, this effect actually is very slight. The main effect is lessening the cohesion of the material by filling the spaces between the grains with water. A simple experiment with sand will illustrate this point. Dry sand can be piled only in cones with slopes slightly over 30 degrees. Damp sand, however, can stand nearly vertically because the small amounts of moisture between the

grains tend to hold the grains together by surface tension. (See Fig. 7-16.) Additional water completely fills the intergrain voids; thus, there is no surface tension to hold the grains together, and a mud forms that flows outward as a fluid.

Slides associated with the 1964 Alaskan earthquake show two other factors. (See Figs. 7-17, 7-18, and 7-19.) Slides may occur under water and create destructive waves, commonly called tidal waves, but better called seismic sea waves because they have nothing to do with tides. (See Chapter 13.) The Alaskan slides also illustrate sliding by a peculiar type of clay that behaves as a fluid when disturbed. This type of clay is termed *quick clay;* it is generally formed from the very fine products of glacial erosion and so is found mainly at northern latitudes. These clays contain much water, commonly 50 per cent or more by weight, and the tiny clay platelets are randomly oriented. Because most quick clays were deposited in the ocean, the interstitial water is saline. The salt ions in this water help to keep the clay particles stuck together, but if fresh ground water moves through the clay, flushing out the salt, the clay becomes unstable. Vibrations may then squeeze the water out, bringing the clay platelets into contact and making the mixture a fluid. (See Fig. 7-20.)

Recognition of any of the types of mass wasting can be of great economic importance when contemplating any construction. The history of many housing developments suggests that even house or lot buyers could benefit by the ability to recognize them. Generally, mass movements are best detected by observation from a distance followed by detailed

B.
Figure 7-12. (cont.) Stone polygons formed by frost action. B. Oblique aerial view near Churchill, Manitoba. Photo from Department of Energy, Mines and Resources, Ottawa, Canadian Government Copyright (A 1426-66).

close study. Generally, a slope in motion will have few trees and may have a bumpy, hummocky appearance. Because some types of glacial deposits also have this appearance, one should know how to recognize glacial topography, which will be discussed in Chapter 10.

Landslides and the Urban Environment

Landslides range from rapid, dramatic events that kill hundreds and cause millions of dollars damage all the way to slow, almost imperceptible creep. Landslides cost hundreds of millions of dollars each year in direct damage, and by delays be-

Figure 7-14. Slump block showing curved plane of slippage and reverse tilt of the slump block.

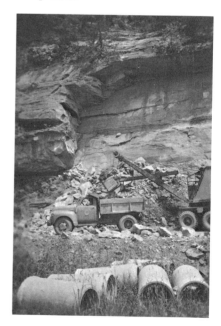

Figure 7-13. Landslide near Hebgen Lake, Montana. A. The slide was triggered by an earthquake and came from the left, leaving a large scar on the mountain side. The debris traveled uphill on the right side of the valley. The slide dammed the Madison River, creating the lake in the foreground. The light-colored area on the slide is a cut made to drain the lake partially. Photo by J. R. Stacy, U.S. Geological Survey. B. This cross-section of the slide was drawn looking in the opposite direction from A. From J. B. Hadley, U.S. Geological Survey, *Professional Paper 435*, 1964.

Figure 7-15. Rockslide caused by undercutting during construction near Quicksand, Kentucky. Photo by J. W. Huddle, U. S. Geological Survey.

Conical pile of dry sand
Angle of repose is about 31°.

Moist sand can form
nearly vertical walls.

Sand grain

Water

Expanded view showing
how surface tension of
the pore water holds
the sand together.

Figure 7-16. The effect of water on the cohesion of sand.

Figure 7-17. Map of Anchorage, Alaska, showing the location of slides caused by the Good Friday Earthquake of 1964. Cross-sections of the L Street and Turnagain Heights slides are shown in Figure 7-18. From Arthur Grantz and others, U.S. Geological Survey, *Circular* 491, 1964.

cause of blocked roads and railroads; but, perhaps more important to the average man, they commonly cause personal financial disaster if a new home is destroyed.

It is difficult to protect oneself from financial loss due to landslides. Landslides are typically excluded from homeowners' insurance, as is earthquake damage. Many policies also exclude damage due to subsidence. These coverages are available, but generally at high cost and should

be investigated by the homeowner.

In many cases the sale of a home is voided if knowledge of hazardous conditions is concealed, but, in general, real estate sales are *caveat emptor*—let the buyer beware. Getting the opinion of a consulting geological engineer is an approach in some areas. The problem with this approach used to be finding a competent person or firm; but licensing laws in several states have made this easier, although many licensed engi-

neers that can legally give such advice know very little about geology. The real problem is that consultants are vulnerable to law suits if their advice is in error; and with the recent rise in malpractice suits in all professional fields, consultants find it almost impossible to get insurance against such suits. Another problem the consultant faces is that he might also be sued by a landowner whose property loses value because of an inaccurate report that it has hazardous geologic conditions. It is difficult to give a geologic opinion on the stability of a single lot because much larger areas have to be studied to determine the geologic conditions. This means that consultants have to charge almost as much to examine a single lot as they do for a whole tract. Drilling and laboratory tests may be needed as well. Thus, the individual cannot do much to protect himself, but must look to government agencies to protect him.

A few cities have laws designed to protect the home buyer. The Los Angeles building codes are commonly cited as examples for the rest of the nation. Los Angeles is endowed with many natural attractions but also with many problems. Legislation dealing with earthquake hazards and air pollution also began in Los Angeles and also has served as a model for the rest of the country. The landslide problem became acute during the very heavy rains of 1952, in part because population pressures and the desire for property with a view led to building homes in the foothills surrounding the city. The January 1952 rains caused about $7,500,000 damage in Los Angeles. The building codes enacted before the following winter rains virtually

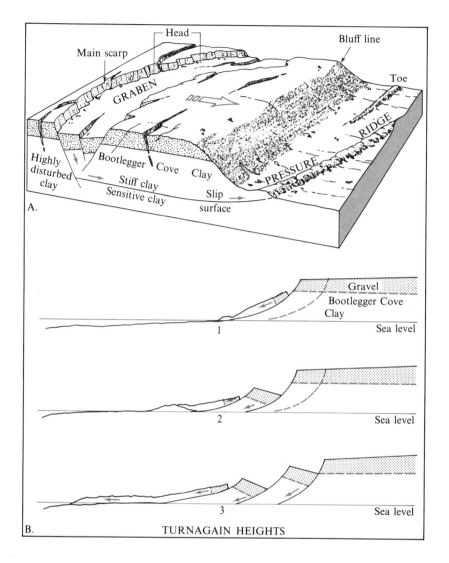

Figure 7-18. Diagrams showing L Street slide (A.) and probable development of Turnagain Heights landslide (B.). See Figure 7-17 for locations. A graben is a down-dropped block. A. From W. R. Hansen, U.S. Geological Survey, *Professional Paper* 542-A, 1965. B. After Arthur Grantz and others, U.S. Geological Survey, *Circular* 491, 1964.

eliminated high, steep cuts and fills, required permits for all hillside grading, and required inspection and geological reports. In addition, drainage has to be conducted to streets and away from cut and fill slopes, and erosion must be controlled by plantings or other devices.

Fills, if not properly compacted, may settle. Fig. 7-21 shows subsidence of an area in San Francisco that was built on filled ground over an old swamp. The city has raised the grade of the streets so that the streets are half way up what was the first floor of the houses.

Landslides can be stabilized in a number of ways. The most important point to make is that all of these methods are done much more economically during the initial construction rather than as a first-aid measure later on. In many cases it is possible, with very minor redesign of any engineering project, to avoid potential landslide areas. One very common misconception is that landslides can be stabilized with planting. Indeed, I have personally met a number of home owners in landslide areas in California who felt that growing either grass or ice plant would stabilize a landslide area. This is a complete misconception, and there are numerous examples of very large trees, even forests, moving as parts of landslide blocks. Many landslides move at depths greater than the normal root depth of most trees.

The main methods of stabilizing landslides are excavation, drainage, construction of retaining walls or similar structures, the use of rock bolts, and the injection of grout into the slide. Highways avoid landslides by bridging over the unstable areas. Most common of all these methods is to drain the landslide. Generally this is done by drilling holes into the slide and providing perforated pipes through which the water can escape. In some cases this is coupled with an impervious layer over part of the landslide to prevent the entrance of surface water into the slide. In a few cases the slides are actually triggered by blasting to stabilize the area. The quick clays (see Fig. 7-20) of Anchorage were subjected to both blasting and vibration in an effort to stabilize them after the 1964 earthquake.

A.

B.

Figure 7-19. Landslides at Anchorage, Alaska, caused by the 1964 earthquake. See Figure 7-17 for locations. A. Looking west at Government Hill School. Note the slumping. B. Aerial view looking south at the hospital on Fourth Avenue. Photos by W. R. Hansen, U.S. Geological Survey, *Professional Paper* 542-A, 1965.

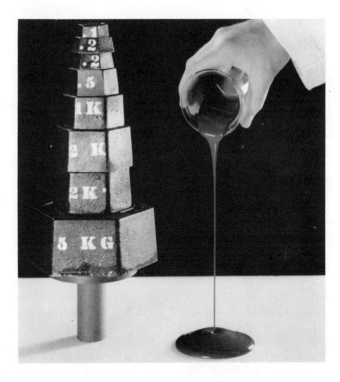

Figure 7-20. A dramatic example of quick clay. On the left an undisturbed sample of quick clay supports 11 kilograms (24 pounds). On the right another sample of the same quick clay pours like a liquid after being stirred. Photo from C. B. Crawford, National Research Council of Canada.

Figure 7-21. Subsidence. The street was regraded and raised after the area subsided so that the garage of this house is unusable.

Erosion is the main process going on on the earth's surface. Rivers are the most important agents. Internal forces uplift the surface so that erosion does not completely level the continents. The underlying cause of erosion is downslope movements under the influence of gravity.

Overburden is the more or less weathered surface material, as opposed to *bedrock.*

On steep slopes no overburden generally develops. *Rockfall* occurs, producing a slope called a *talus* at the bottom.

Weathering, controlled by climate, determines the amount of overburden that develops and the type of downslope movement, thereby causing the different aspects of topography in humid and dry areas. Structure and rock types also control erosion processes.

Falling rain erodes by *impact,* moving material by splashing, and by *runoff* of the rainwater that does not infiltrate the ground.

Types of downslope movement are *flows* if they move as viscous fluids, *slides* if they move over the ground, and *falls* if they travel in the air in free fall.

Creep is the slow downslope movement of overburden and, in some cases, bedrock.

With more water in the overburden, an *earthflow* results, forming a distinct curved scarp at the breakaway point.

With increasing water, a *mudflow* results. Rain on loose pyroclastic material on a volcano, or a heavy thunderstorm in an arid area can produce mudflows.

Solifluction is a type of mudflow that occurs when frozen ground melts from the top down so that the surface mud flows downslope.

Frost heaving results from the expansion of freezing water which pushes pebbles upward through the soil. In cold regions, frost heaving forms stone rings or polygons.

Landslides are rapid slides of bedrock and/or overburden. *Landslides* or *rockslides* involve bedrock; a mass of bedrock breaks loose and slides down the slope. The material is often fragmented and behaves like a fluid in the valley, and the slide is often large and destructive. Although landslides are sometimes called *avalanches,* that term is best reserved for snow slides. Slides involving overburden are called *debris slides.*

Slumps are also rapid slides of bedrock and/or overburden. They tend to develop where a strong, resistant rock overlies weak rocks, and slip down a curved plane, tilting the top unit back, and creating a new cliff that will itself slump.

The most common triggers to massive downslope movements are (1) undercutting of the slope by rivers or man, (2) overloading the slope so that it cannot support its new weight, (3) vibrations from earthquakes or explosions, and (4) additional water which adds to the weight of the slope and lessens the internal cohesion of the material by filling the pore spaces.

Quick clay is a type of clay which behaves as a fluid when disturbed.

Landslides can cause great damage. Landslide potential should be evaluated before engineering projects are done. The main methods of stabilization are draining, excavating, construction of retaining walls or similar structures, rock bolts, injection of grout into slide, or controlled triggering by blasting.

Questions

7-1. Erosion tends to level the land surfaces. Why aren't the continents worn flat?

7-2. Describe the slow means of downslope movement.

7-3. Describe the rapid means of downslope movement.

7-4. Of what importance is raindrop impact in erosion?

7-5. How might rapid mass movements be triggered? What is the role of water in triggering?

7-6. In buying a hillside home, how can you protect yourself against possible landsliding?

Supplementary Reading

Ellison, W. D., "Erosion by Raindrop," *Scientific American* (November, 1948), Vol. 179, No. 5, pp. 40-45. Reprint 817, W. H. Freeman and Co., San Francisco.

Kerr, P. F., "Quick Clay," *Scientific American* (November, 1963), Vol. 209, No. 5, pp. 132-142.

Sharp, R. P., and L. H. Nobles, "Mudflow of 1941 at Wrightwood, Southern California," *Bulletin*, Geological Society of America (May, 1953), Vol. 64, No. 5, pp. 547-560.

Sharpe, C. F. S., *Landslides and Related Phenomena*. New York: Columbia University Press, 1938, 137 pp.

Geologic Work of Rivers and Development of Landscapes

8

Rivers are agents of erosion, transportation, and deposition. That is, they carve their own valleys and carry the eroded material downstream where it is either deposited by the river or delivered into a lake or the ocean. Rivers are the most important agents in transporting the products of erosion: they drain vastly larger areas than do glaciers, and, even in their infrequent times of flowage on deserts, they probably are able to carry more material than the wind can. Thus it is the rivers that receive and transport the products of weathering that are fed into them by runoff from precipitation as well as by downslope movements. Weathering, downslope movement, and rivers work together to shape the landscape. Although the geologic role of rivers has been known for a long time, rivers are complex systems with many variables and much remains to be learned about them. Even more factors are involved in the development of landscape, and this is one of the active fields of research in geology.

Before Hutton's time very little thought was given to erosion and landforms. A few writers did note the erosion and deposition done by rivers, repeating the observation of the Nile made by the ancient Greek Herodotus about 500 B.C. One reason for the lack of interest was that most people believed in a strict interpretation of the Bible. In 1654 Bishop Ussher, from a study of Genesis, said that the earth was created on the night before Sunday, October 23, 4004 B.C., and that the Flood had begun December 7, 2349 B.C. A few thousand years is not enough time for a slow process like erosion to modify the earth's surface to any great extent. As a result, most people thought that the earth was either in the perfect condition in which it had been formed by the

Erosion of plowed field by runoff of rain water. Photo from U. S. Department of Agriculture, Soil Conservation Service.

Creator, or that it had been modified by the Flood or by earthquakes. In the eighteenth century the viewpoint changed somewhat, and near the end of that century Hutton argued that the slow process of erosion was accompanied by deposition, thus perpetuating man's home. Change instead of destruction was Hutton's concept; a benign Deity, instead of a wrathful God, was a current theme. This was the intellectual climate in which modern geology developed.

In the nineteenth century, little more was done in studies of erosion until the last half of the century. The intervening years were active ones for geology, and this was when the geologic time scale was developed, a topic covered in Chapter 17. The scene then shifted to the United States. After the Civil War the West was opening up, and part of the westward interest was geologic exploration. Note that this was the same time that Africa was being explored as described in Chapter 1. On May 24, 1869, at Green River Station, Wyoming, John Wesley Powell, a one-armed Civil War veteran, and nine companions started on an epic boat trip that was to take them more than one thousand miles through the Grand Canyon to Virgin River, Arizona. Powell became a national hero and returned several times to study the Colorado Plateau. From these geological studies, Powell and Grove Karl Gilbert described the erosion and the work of the rivers. The Colorado Plateau, with its flat-lying sedimentary beds, open vistas, and deep canyons, is an excellent place to appreciate the effectiveness of erosion. The work of these men was organized by William Morris Davis in the years between 1880 and 1934 into what he called the cycle of ero-

sion. This concept and the criticism of it by the German father-son team of Albrecht and Walther Penck will be discussed in this chapter.

Flow of Rivers

Rainfall and other forms of precipitation are the source of river water. Only a small fraction of the total precipitation forms the rivers, but the amount of water is great. The fate of most of the rest of the precipitation is described in Chapter 12. The energy of rivers comes from the downward pull of gravity acting on the river water. In physics this is called *potential energy* and is equal to the mass times the acceleration of gravity times the height (mgh). Table 8-1 shows the tremendous amount of energy theoretically available from rivers. Only a small fraction of this energy is actually used to do geologic work. If this were not so, the landscapes of the earth would be far different. To see where the potential energy of rivers goes, and to gain some insight into river processes, the velocity of river flow must be considered.

The average velocity of river flow is less than five miles per hour, and most rivers are in the range from a fraction of a mile per hour to about 20 miles per hour. (Fifteen miles per hour is the fastest ever measured.) These velocities are far less than the 286 miles per hour average velocity (Table 8-1) predicted on the basis of the potential energy of rivers. Apparently the energy of rivers goes largely into heating the water by friction along the bed and banks and by turbulence. It is this friction and turbulence, however, that enable a river to erode and transport, as will

Table 8-1. Calculation of approximate amount of energy available from rivers and the theoretically possible average velocity of rivers.

Potential energy = mass × acceleration of gravity × height
$$P.E. = mgh$$
$$mg = weight$$
$$P.E. = weight × height$$

Average elevation of continents = 2,755 feet (See Chapter 14)
Average runoff of continental U.S. = 1,200,000 million gallons per day (See Chapter 12)

1 gallon of water weighs 8.35 pounds
$$P.E. = 1,200,000,000,000 × 8.35 × 2,755 = 2.76 × 10^{16} \text{ foot-pounds per day}$$

(The total energy used in the U.S. in 1960 was about $3.1 × 10^{19}$ foot-pounds, or about 1,000 times more.) If all of the river's energy were used to heat the water, it would only rise about 3.5°F.

The velocity can also be calculated:
Potential energy = Kinetic energy
$$mgh = ½ mv^2$$
$$v^2 = 2gh$$
$$= 2 × 32 × 2,755$$
$$= 176,000$$
$$v = 418 \text{ feet per second}$$
$$= 286 \text{ mph}$$

be described in the next section. (See Fig. 8-1.)

The flow of a river is determined by many factors. The discharge, or flow, can be defined as the amount of water passing a given cross-section of the river in a time unit (usually given in cubic feet per second, c.f.s.). The discharge, then, is equal to the area of the cross-section times the velocity. The cross-section is determined by the shape of the river's channel, and the velocity is deter-

Smooth channel   Rough channel

Figure 8-1. Idealized diagram showing flow of water. The lines represent paths of water particles. Where the channel is smooth, the water flows in straight paths; but where the channel is rough, the flow is turbulent. Turbulence increases with increasing velocity.

mined by a number of factors, the chief of which are the roughness, shape, and curving of the channel and the slope of the river. The amount of water available for runoff is also a factor, and this is determined by such climatic factors as precipitation, evaporation, vegetation, and permeability of soil and

bedrock. On any given river, all of the factors that control the flow vary from place to place and with season.

The channel is probably the most important feature of a river. However, it is difficult to study because it is generally obscured by the water. In recent years, many soundings have been taken, and much has been learned about channel shape and, as will be seen in the next section, the changes in shape that occur with changes in flow. The shape of a channel greatly influences drag or friction. In general, the smaller the wetted perimeter, the less the drag. This suggests that if this were the only factor, rivers in eroding their channels might tend to form semicircular cross-sections because this shape has the least perimeter for its area. However, many other factors influence channel shape and actual channels are quite irregular. (See Fig. 8-2.) At flood stage, rivers are higher than their channels and spill over into their valleys. This greatly increases

the wetted perimeter, and the resulting drag slows the flow in the flooded valley, reducing the potential damage to the inundated area. The size and amount of the material transported by the river and the erodability of the banks have a great influence on channel shape. (These factors will be described in the next section.) Figure 8-3 shows the distribution of velocity and turbulence in an idealized river channel both on a straight reach and on curves. A river changes in a systematic way from its headwaters to its mouth. Downstream the discharge increases because a larger area is drained; and to accommodate this increased flow, the width and depth both increase, as does the velocity. (See Fig. 8-4.)

The slope of a river also changes systematically from headwaters to mouth as shown in Fig. 8-5. The vertical scale in this figure is greatly exaggerated. A typical river has a slope of several hundred feet per mile near its headwaters and just a few

Figure 8-2. Changes in channel of San Juan River near Bluff, Utah, during flood in 1941. Note the changes in elevation of both the water surface and the channel depth. From L. B. Leopold and Thomas Maddock, Jr., U.S. Geological Survey, *Professional Paper* 252, 1953.

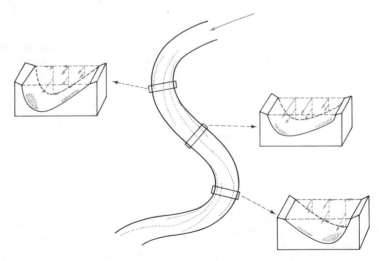

Figure 8-3. Distribution of velocity and turbulence in a river channel. In the cross-sections, the arrows show the direction and magnitude of the velocity. The surface velocity is indicated by two components, and the path of the surface water is shown by the dotted lines in the map view. The areas of most turbulence are indicated by stippling in the cross-sections.

1. Missouri River at **Bismarck**. N. Dak.   4. Missouri River at Kansas City, Mo.   7. Mississippi River at St. Louis, Mo.
2. Missouri River at **Pierre**, S. Dak.   5. Missouri River at Hermann, Mo.   8. Mississippi River at Memphis, Tenn.
3. Missouri River at St. Joseph, Mo.   6. Mississippi River at Alton, Ill.   9. Mississippi River near Vicksburg, Miss.

Figure 8-4. Graphs showing the increase in width, depth, velocity, and discharge downstream in the Missouri River and lower Mississippi River. From L. B. Leopold and Thomas Maddock, Jr., U.S. Geological Survey, *Professional Paper* 252, 1953.

feet per mile near its mouth. The lower 600 miles of the Mississippi River has a slope of less than one half foot per mile. A steep, rapidly flowing river generally has a slope of 30 to 60 feet per mile. A slope of one degree is 92.4 feet per mile. Notice that the profile (Fig. 8-5) is not smooth but has irregularities, especially where tributary streams join. These irregularities are caused by abrupt changes in discharge and load. This suggests that the overall concave upward shape of the long profile is also the result of a balance between discharge, load, and slope, as well as other factors. A brief study

of river processes in the next section will give insight into this balance.

Erosion and Transportation

The main geologic work of rivers is transportation. The load is delivered to the river by downslope movement of material on the valley sides, by erosion by the river itself, and by tributary streams. The load of a river can be considered in three parts: dissolved load, suspended load, and bed load. The total amount of material delivered by the rivers of the world to the oceans is estimated to be about 40,000 million tons per year. Man's activities, especially farming and construction, contribute much of this load.

The dissolved load comes from chemical weathering and from solution by the river itself. Unlike the rest of the stream's load, it has no effect on the flow of the stream. The dissolved load of the world's rivers is estimated at 3,900 million tons per year, just under 10 per cent of the total load. Table 8-2 shows the average dissolved load. About half of the dissolved load is bicarbonate ion (HCO_3^-) that results from the solution of carbon dioxide in water, as described in weathering in Chapter 4.

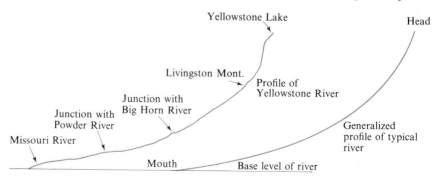

Figure 8-5. Longitudinal river profiles. Vertical scale exaggerated. Yellowstone River profile from U.S. Geological Survey, *Water Supply Paper* 41, 1901.

Table 8-2. Average composition of dissolved material in river waters of the world. From *Data of Geochemistry*, sixth edition, Chapter G, "Chemical Composition of Rivers and Lakes," D. A. Livingstone, U.S. Geological Survey, *Professional Paper* 440-G, 1963.

	PARTS PER MILLION
Bicarbonate (HCO_3)	58.4
Calcium	15.0
Silica (SiO_2)	13.1
Sulfate (SO_4)	11.2
Chlorine	7.8
Sodium	6.3
Magnesium	4.1
Potassium	2.3
Nitrate	1.0
Iron	0.7
	119.9

The suspended load consists of the fine material that, once in the water, settles so slowly that it is carried long distances. As the flow of a river increases, some of the finer bed load becomes suspended load so that the distinction can be made only at a given moment. Suspended material is the cause of muddy river water. Because clay particles settle very slowly, they are carried relatively long distances. Silt settles much more rapidly; it is carried as suspended load only by turbulent water which has enough upward currents to keep the silt from settling. Thus clay is carried farther than silt, resulting in sorting of sediment by rivers. The suspended load is estimated to be almost two-thirds of the total load carried by rivers and amounts to over 25,000 million tons per year for all of the world's rivers.

The bed load is the material that is moved by rolling and sliding along the bottom. It, too, is somewhat sorted by size; also the particles are worn by abrasion, making the pebbles and boulders in a typical stream

bed subspherical. In addition to abrasion and solution, the bed-load pebbles are at times cracked and broken by impacts during movement. Almost all of the movement of most bed loads occurs during the generally very limited times of high flow. Thus the floods that occur seasonally, or even less commonly, mark times that bed loads are moved. In some rivers the ratio of maximum to minimum flow is a few hundred to one. In the summer or fall, it is difficult to believe that the large boulders in the stream bed, particularly a small stream, are part of the bed load and are being moved by the stream. (See Fig. 8-6.) One must remember that the largest boulders may move only during periods of very great flow that may occur a few times each century.

However, if one visits such a stream during the spring when it is filled with melt water, the audible clicks when a large boulder rolls or slides into its neighbor can be heard. The bed load of most streams is about a quarter or less of the total load. The maximum size particle that can be moved by a river is proportional to the square of the velocity. Thus doubling the velocity increases by four times the size of particle that can be moved. In general, doubling the velocity increases the amount of material in transport between four and eight times. The movement of bed load during times of high flow is shown by the changes in size and shape of the channel. (See Fig. 8-2.)

Rivers erode by several processes which were necessarily discussed in

Figure 8-6. A stream at low water level, showing part of the bed load that is moved during times of high discharge. Photo from California Division of Beaches and Parks.

Figure 8-7. Potholes in a river bed cut in granitic rock. The potholes are formed during times of much greater flow. Photo by Mary Hill, courtesy California Division of Mines and Geology.

describing transportation. If parts of the channel are soluble, erosion by solution will occur. Abrasion occurs as a result of the transported material rubbing against the sides and bottom of the channel. A common type of abrasion results in cylindrical holes, called potholes, worn in bedrock, apparently by stones that are spun by turbulent eddy currents. (See Fig. 8-7.) In plunge pools below waterfalls, much of the erosion is caused by the moving water itself. The impact and drag of the water erode and transport material. As noted earlier, the size of particle that can be moved in this way is proportional to the square of the velocity; however, velocities great enough to move sand, and even coarser ma-

A.

B.

C.

D.

Figure 8-8. These two pairs of photographs were taken nearly one hundred years apart and show that little erosion has occurred along the Colorado River in that interval. The early photographs were taken on John Wesley Powell's expeditions. A and B. Sockdologer Rapid, Grand Canyon National Park, Arizona. C and D. Schoolboy Rapids, Split Mountain Canyon, Dinosaur National Park, Utah. The changes in the foreground may have occurred in a single flood. Photo A, 1872, and C, 1871, by J. K. Hillers; B and D, 1968, by H. G. Stephens, all from U.S. Geological Survey.

terial, are necessary to cause erosion of the smooth surface of consolidated clay. The rate of erosion is slow on many rivers and may be noticeable only after major floods. (See Fig. 8-8.)

The rate of erosion of the continents is determined by measuring the amount of material carried by rivers. The average rate of erosion in the United States, determined in this way, is about 2.4 inches per thousand years. Of the major river basins the Columbia River basin is eroding at the rate of 1.5 inches per thousand years, the lowest rate; the Colorado River at 6.5 inches per thousand years; and the Eel River in northern California at between 40

and 80 inches per thousand years, the largest rate. The rate of erosion depends on a complex relationship of many factors, such as rainfall, evaporation, and vegetation, but note that the maximum rate of erosion occurs in areas with about ten inches of rainfall per year. (See Fig. 8-9.)

Erosion tends to deepen, lengthen, and widen a river valley. The ways that a river can deepen its channel were just discussed. At the head of a stream any downcutting lengthens the valley. This process is called headward erosion and is the way that streams eat into the land masses. (See Fig. 8-10.) A stream valley is widened by two methods. The valley side processes, such as creep and

landsliding, widen the valley. (See Fig. 8-11.) Lateral cutting by the stream itself is the other process.

Lateral cutting is most pronounced on the outsides of curves where the valley sides may be undercut. Thus a curving stream widens its valley, and straight stretches of streams longer than ten times the width are not common. Clearly, any irregularity on a stream bottom or bank can deflect the flow and cause a curve to develop. Once formed, a curve tends to migrate, thus widening the valley. (See Fig. 8-12.) Many rivers have distinctive symmetrical curves called *meanders*. (See Fig. 8-13.)

The migration of meanders is shown in Fig. 8-14, which also shows how the meanders may intersect, causing oxbow lakes. (See Fig. 8-15.) Meanders are not formed in the same way as are the irregular curves on a river. The curves that form meanders are not common curves, such as half circles. The meander-forming curve is the one that distributes the river's loss of energy most uniformly. Any other curve would tend to concentrate erosion and deposition at some point on the curve. (The development of river valleys by all of these methods of erosion will be described later.)

The River as a System

Rivers do not occur singly in nature. Each has tributaries and each tributary has smaller tributaries, and so on, down to the smallest rill. All of these parts function together to form a river system, or drainage basin. A change in any part of the system generally affects all of the other parts. One obvious illustration of this balance is that all the tributary streams meet the main stream at the level of

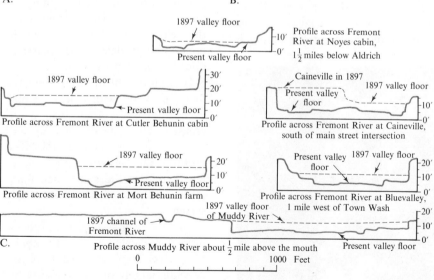

Figure 8-9. Examples of river erosion in Henry Mountains region, Utah, since 1897. Climatic change is the probable reason for this erosion. Parts A and B show Pleasant Creek at Notom. The creek in part B (present day) is 20 feet deep. Part C shows a number of cross-sections of the Fremont River. From C. B. Hunt, U.S. Geological Survey, *Professional Paper* 228, 1953.

Figure 8-10. The two streams in the foreground of this oblique aerial view are lengthening by headward erosion. The bigger one may ultimately capture the stream in the middleground. San Bernardino Mountains, California. Photo by J. R. Balsley, U.S. Geological Survey.

the main stream. This accordance of junctions is too universal to be due to chance. (This accordance is not found in glaciated areas where the stream levels have been changed by glacial erosion.)

The main work of a river is to transport debris, and river systems regulate themselves so that they are able to transport the debris delivered to them. This is generally accomplished by changing the slope, by either erosion or deposition, so that at any given point for the discharge and channel characteristics the velocity is such that the debris is transported. Thus a change in any part of the system will affect the balance of the rest of the system. A stream that has just the right gradient to be able to transport the debris delivered to it by its own erosion, its tributaries, downslope movement of material, and rain wash is called a *graded stream.* This ability is a long-term average so that a given stretch of river at one season may be eroding its bed, and at another time it may be building up its bed. These changes in regimen vary from year to year with such climatic influences as amount and intensity of precipitation.

The long profile of a river (see Fig. 8-5) was mentioned earlier, and the reason for its shape can now be appreciated. The slope near the head must be steep for the river to be able to do its work of eroding and transporting with the small discharge it has available. Farther downstream the velocity and discharge both increase and particle size decreases as does the number of banks (see Fig. 8-16), all of which permit the river to do its work with a lesser slope, even though the total load is increased. Thus the concave upward slope of a river enables the most uniform rate of work per unit of length and at the same time minimizes the work done in the whole system.

In most river systems the profile is more complicated because the ability of a river to erode and transport depends largely on the velocity, the amount of flow, the gradient, and the shape and roughness of the channel. At the point where a tributary joins

Figure 8-11. Widening of a stream valley by a small slide. In this case a forest fire has reduced the vegetation and thus caused the slide. Photo from U.S. Forest Service.

the main stream, the regimen of the stream is abruptly changed by the additional flow and the increased debris from the tributary. The main stream reacts to this change by altering its gradient so that it can handle the new material. These changes are superimposed on the simple profile.

The limit to which the river can erode is, of course, the elevation of its mouth. This limit is called the *base level* of the river. (See Fig. 8-5.) For most big rivers the base level is sea level, but for some that flow into lakes, it is the elevation of the lake. For tributary streams, the base level is the elevation at which they join the main stream.

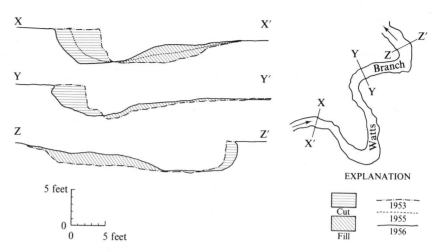

Figure 8-12. Erosion and deposition on Watts Branch near Rockville, Maryland, from 1953 to 1956. From M. G. Wolman and L. B. Leopold, U.S. Geological Survey, *Professional Paper* 282C, 1957.

Deposition

River-deposited sedimentary rocks (fluvial rocks) are relatively rare because rivers deliver most of their loads to lakes or to the ocean. River deposits in the form of deltas develop where rivers enter standing water; even in this case, however, much of the river's load may be reworked by waves and offshore currents, thus becoming marine or lacustrine (lake-deposited) rocks. Deposition occurs because the velocity of river water is slowed so abruptly on entering standing water that the bed load is dropped. The dissolved and suspended loads may be carried farther out. (See Figs. 8-17 and 8-18.) A delta can build outward into the ocean only if the waves and currents are not strong enough to move the deposited material.

Alluvial fans are similar to deltas and are formed where streams flow out of mountains into broad, relatively flat valleys. In this case, the flatness of the valley slows the stream, causing deposition. In addition, especially in arid areas, the mountain stream may seep into the ground in the broad valley. Because the stream constantly shifts its position on the fan, the fan develops a cone shape (Fig. 8-19).

When the flow of a river fills its bed up to the top of the banks, any increase in discharge will cause flooding. The floodwater spills over the river banks, filling part or all of the valley. A river is doing its maximum erosion and transportation at the time of flood, and as the flood-

Figure 8-13. Meanders on the Green River in Utah. Cut-off meanders and old meander scars show the movement of the river on the flood plain. Looking north (upstream) toward the Book Cliffs. Photo from U.S. Geological Survey.

water spreads out, its velocity is quickly reduced, causing deposition near the banks of the river. Such deposits are termed *levees* (or natural levees, to distinguish them from man-made levees built for flood protection). (See Fig. 8-20.) Although the levees help to keep the river within its banks, the area between the levee and the side of the river valley may be lower than the river at flood stage; this area is subject to inundation when the flooding river crests over the levee top. Because the river valleys are prime agricultural areas and because the mechanism of levee building provides a continuing supply of new soil, many people accept periodic flooding and live in the valleys of the large rivers. These people should expect spring flooding, for it is the normal, not the abnormal, behavior of the river. To some extent they can be protected by upstream flood-control dams, which regulate somewhat the flow during flood stage, but this system is expensive and not always effective as discussed later.

Rivers may also build up their beds by deposition. In some cases, a river may deposit material during a long period of time so that its meandering causes deposition covering an entire valley. Such a process produces a smooth surface. Later, the river may, because of some change in its regimen, cut into this surface and so form one or more lower terraces (Fig. 8-21). Terraces can also be formed by lateral cutting of bedrock by the river. One common cause of terrace building is associated with the recent glaciation. In areas where glaciers used to be

Figure 8-14. Development of meanders.

Figure 8-15. Oxbow lakes formed by cut-off meanders of White River, Arkansas. Photo by J. R. Balsley, U.S. Geological Survey.

Figure 8-16. The number of banks decreases downstream, reducing the friction.

Figure 8-17. The Nile delta photographed from Gemini IV. The dark area is cultivated land. Mediterranean Sea to the left and Sinai Peninsula in the middleground. The Nile has a number of distributaries, and this accounts for the large cultivated area. Photo from NASA.

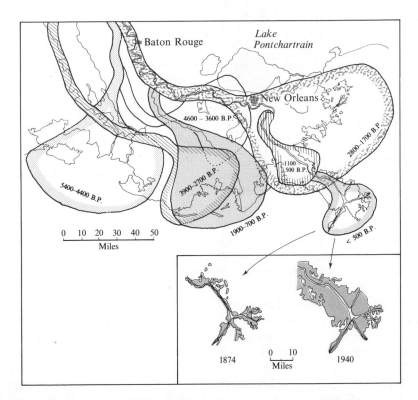

Figure 8-18. Development of the Mississippi River delta. The upper drawing shows the ages of subdeltas in years before present (B.P.). Upper drawing after Kolb and van Lopik, 1966, and lower drawings after Scruton, 1960.

active in the mountains, much debris of glacial erosion was rather suddenly released to the rivers when the glaciers melted. Because the rivers could not transport the material as rapidly as it was delivered, they deposited it in their valleys. As conditions returned to normal, they downcut into these deposits and formed terraces, most of which, in time, will be removed by erosion. As is shown in a later section, the study of terraces can tell much about the history of a river.

Braided streams are an unusual type of stream that forms under some conditions, especially if the load is large and coarse for the slope and discharge, and the banks are easily eroded. In a braided stream the channel divides and rejoins numerous times. (See Fig. 8-22.) Apparently the stream deposits the coarser part of its abundant load in order to attain a steep enough slope to transport the remaining load. This deposition forces the stream to broaden and so erode its banks, adding to the abundant load, perhaps choking the channel. The places where two channels join have increased flow and reduced drag. At such places some of the bed load is

Figure 8-19. Alluvial fans at canyon mouths, Mojave Desert, California. Photo by J. R. Balsley, U.S. Geological Survey.

Figure 8-20. The development of levees. Levees form when silt is deposited beside the river during floods. If, as is generally the case, deposition also occurs on the river bed, the seriousness of later floods increases. The vertical scale is greatly exaggerated.

Figure 8-21. Terraces cut in river-deposited material by the Snake River. Teton Mountains on the skyline. Photo from Union Pacific Railroad.

moved, but this reduces the slope and deposition begins again; however, the load is moved in the process, albeit only a short distance.

River Valleys

River valleys can be described as narrow or broad, although all intermediate steps between the two exist. Narrow valleys are not much wider than the river channel, and broad valleys are much wider than the river.

Narrow valleys, which have V-shaped cross-sections, are generally found where rivers are actively downcutting. On many rivers they occur near the headwaters, but narrow valleys can form anywhere. Many, but far from all, are downcutting, non-graded rivers. Some narrow valleys occur in resistant rocks that slow the lateral cutting of a river; they commonly have rapids and waterfalls where resistant rocks slow the downcutting. Lakes may also form in narrow valleys, but, like falls and rapids, they are only temporary features. (See Figs. 8-23, 8-24, 8-25, 8-26, and 8-27.)

Valley widening begins, in general, after a stream has reached grade and is no longer rapidly downcutting.

This occurs first, on most rivers, near the mouth where the increased flow allows the river to erode more efficiently. Wide flood plains characterize the lower parts of many rivers. (See Fig. 8-13.) In areas of weak rock, widening is also common. Wide valleys are generally fairly flat in cross-section. Meanders are common features of wide valleys, especially in areas with uniform banks composed of fine sediment. Meanders may alternate with areas of irregular curves. Another feature, familiar to fishermen, is the alternation of riffles and pools on most rivers. The riffles form on alternate sides of the river and are composed of the coarser part of the bed load. (See Fig. 8-28.)

River valleys can be used in the interpretation of geologic history. Evidence of uplift can be found in individual river valleys because river valleys widen as erosion progresses. Thus, if a river is cutting a V-shaped, youthful valley within a much wider valley, a change has occurred causing the river to downcut actively. (See Fig. 8-29.) This change could be due to an uplift, a climatic change, or an increase in the flow of the river from a stream capture (discussed below) or other cause. If no further changes occur, the new valley will widen as erosion progresses, removing the evidence. If a new uplift occurs, a new V-shaped valley will be formed. This is one way river terraces are produced (Fig. 8-30). If depression rather than uplift occurs, a river valley will be the site of deposition instead of erosion.

At some places uplift is shown by rivers that have narrow, V-shaped valleys and a meandering map pattern. This happens most commonly when uplift occurs slowly enough that the river that was meandering

Figure 8-22. Braided river. Vertical aerial photograph of North Platte River, Nebraska. Photo from U.S. Geological Survey.

in a broad valley maintains its meandering course, although downcutting. (See Fig. 8-31.)

Environmental Hazards of Running Water

"Floods are as much a part of the phenomena of the landscape as are hills and valleys; they are natural features to be lived with, features which require certain adjustments on our part." *Floods,* by W. G. Hoyt and W. B. Langbein, Princeton University Press, 1955.

Some primitive peoples adapted reasonably well to rivers and accepted seasonal or occasional floods as part of life. As man developed and built bigger and better structures for his comfort, he became less willing to accept the natural behavior of rivers.

A typical river valley consists of a channel, a flat area (called a flood plain) and sloping valley sides. (See Fig. 8-32.) Most of the time the flow is confined to the channel; but at times of high flow, when the channel cannot carry the water, it spills onto the aptly named flood plain. The flood plain is the channel of the river during times of high water. This is the normal, expected behavior of a river. People who live or build structures on flood plains must expect these floods. (See Fig. 8-33.) Flood control is never one hundred per cent effective and can increase the actual damage as we will see later. Floods are not entirely bad events, however, because the silt deposited on the flood plains renews the soil of these generally very productive agricultural areas.

Rivers erode on the outsides of curves and deposit on the insides of curves. (See Fig. 8-14.) This means that the location of a river is always changing. Rivers are commonly used as property boundaries as well as state and international boundaries. Such boundaries are not fixed and lead to many problems. In some cases the boundary is fixed at the position of the river on a certain date. Along the Mississippi, a farmer with land inside a meander loop may wake up some morning and find that his land is separated by the river from the rest of the state and county to which he pays taxes. He may have trouble getting police and fire protection. On the other hand, he now may have an ideal location for moonshining. Recently the United States and Mexico concluded a treaty fixing the international boundary near El Paso. (See Fig. 8-34.)

Storms and mudflows are geologically rapid but normal events. Intense storms that cause floods, landslides, or mudflows can occur almost anywhere. In the United States, such storms are common in the desert areas of the southwest where little note is taken of them because few people are affected. A few cities have expanded into hazardous parts of this region in Arizona and especially in southern California, and the resulting problems will be described later. The hurricanes of the Gulf and southeastern coasts of the United States cause similar problems, although flooding of the coastal areas is the usual result. Some of these storms cause problems as far north as New England. Hurricanes are generally accompanied by heavy rainfall that chokes the rivers, and strong onshore winds drive ocean waters onto the low coastal plain. The low barometric pressure associated with such storms increases the

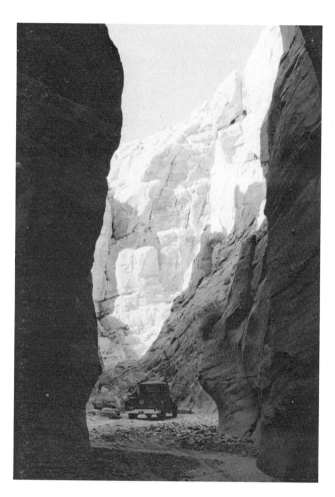

Figure 8-23. An extremely narrow canyon with nearly vertical walls cut in sandstone. Downcutting was apparently faster than downslope movements could shape the valley sides. Photo from California Division of Beaches and Parks.

flooding by allowing much higher than normal tides. Hurricane Camille, which hit the Gulf Coast in August, 1969, had winds up to 200 miles per hour, a low pressure of 905 millibars, and dumped up to 30 inches of rain in a day. The East Pakistan flood of December, 1970, killed between 250,000 and 300,000 people. This flood, erroneously described in the newspapers as a tidal wave, was caused by a hurricane (called a cyclone in that area) that had winds up to about 120 miles per hour and a low pressure of about 950 millibars. The wind and low pressure

caused a storm surge about 18 feet high (normal tides are about six feet), which overran the low, flat area in the Bay of Bengal. In June, 1972, 15 inches of rain fell on Rapid City, South Dakota, causing a flood that killed 237 people and left thousands homeless.

Mudflows caused by relatively rare but intense thunderstorms are not confined to desert areas. They can occur on any steep slope with enough overburden or loose material and not too much underbrush. Forested slopes, especially those with big trees whose shade prevents growth of

underbrush, are susceptible to this type of erosion. The following example is from the Sierra Nevada Mountains near Lake Tahoe on the California-Nevada boundary. The Lake Tahoe area, because of its beauty, its nearness to San Francisco and Sacramento, and its climate that makes it an excellent skiing as well as a summer resort area, is currently undergoing urbanization. In the late afternoon of August 25, 1967, more than an inch of rain fell in the drainage basin of Second Creek. This caused a mudflow of more than 50,000 cubic yards in the 1.5 square-mile basin. The mudflow caused extensive damage to homes, town houses, apartment houses, and roads. A seven-foot-diameter boulder moved in this flow, and at some places, the flow deposited up to 20 feet of debris. Similar events have been observed elsewhere in the Sierra Nevada Mountains, and the results of this type of erosion can be seen in many drainage basins. This particular event was estimated to have reduced the basin by about 0.02 foot and to have a recurrence interval of less than 50 years. This mudflow was a natural event and was not caused by the urbanization, although some conservationists blamed the recent building in the area. This event did affect the building programs in the area because it showed that structures should not be built in the path of potential mudflows and that much larger culverts and other drainage structures had to be installed on structures such as roads that must cross potential mudflow paths. Mudflow paths must be considered when purchasing or developing mountain areas. That no mudflow nor flash flood has occurred in the memory of residents does not mean that one will never occur.

Although not directly related to running water, snow avalanches are another geologic hazard in mountain resort areas. Avalanche chutes are also common in the Lake Tahoe area as well as any mountain area with moderate to heavy winter snows. Avalanche danger is controlled at ski resorts as a result of studies and regulations developed by the Forest Service and the National Ski Patrol System. In lift-served ski areas, the common method of controlling avalanches is to cause them to occur by blasting or other means at times when no one is skiing. Thus many small avalanches may fall harmlessly in a given chute. In uncontrolled areas an avalanche may not fall until a great depth of snow has accumulated. The writer has seen three- to five-foot-diameter trees broken by large avalanches in the Cascade Mountains of Washington. In general, avalanche chutes can be recognized by either a lack of vegetation where the slides occur at grass root level, or by underbrush-covered slopes (slide alder and the like) where sliding occurs after the brush is covered by snow. Slopes that avalanche frequently rarely allow trees to root on them, but it must also be remembered that at many places avalanches occur on forested slopes without disturbing fairly widely spaced trees of any size. Like mudflows and flash floods, avalanches must be considered in buying or developing mountain property. The writer has been offered attractive resort property with all of these hazards clearly shown in the landscape to anyone with an elementary understanding of geologic processes. Unfortunately, land sellers and developers rarely notice or understand these processes.

Figure 8-24. The Yellowstone River in Yellowstone National Park, with falls and rapids. Photo from National Park Service.

Figure 8-25. Block diagram showing how resistant rocks cause falls and rapids.

Figure 8-26. Rogue River, Oregon, has cut a narrow, V-shaped canyon in resistant rocks at this place. Photo from Oregon State Highway Department.

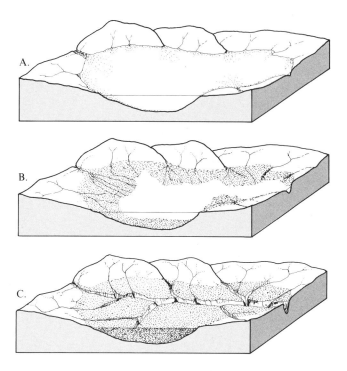

Figure 8-27. Three stages in the life of a lake. A. Initial lake. B. Lake has been partially filled by stream deposition. C. Lake is completely filled, and stream has cut a channel across the fill.

Figure 8-28. A narrow, V-shaped valley in which the river has alternating pools and riffles. Photo from U.S. Forest Service.

Floods—Dammed Rivers and Damned Rivers

Any process that puts more water into a river than the river can carry will cause a flood. Intense precipitation and rapid melting of snow, especially in mountainous headwaters, are common causes. Other natural processes such as dams caused by landslides can cause upstream floods. In this latter case, commonly when the lake behind the slide overtops the slide-dam, the rapidly moving water quickly erodes the slide, emptying the lake rapidly and causing downstream flooding. When the ice breaks up each spring on northern rivers, the floating ice blocks themselves can cause damage to bridges and the like; and ice jams cause floods first upstream and then downstream in the same way as the slide dams just described. Except for landslides, the natural processes described are seasonal and are controlled by climatic processes.

Flood prediction requires much data and cooperation between hydrologists and meteorologists and even at best is not an exact science. The hydrologist must know such characteristics of the drainage basin as how long does it take for a rainfall of a given intensity to cause an increase in discharge of the river. This will occur after a low-intensity rain has saturated the ground, and almost at once in a high-intensity storm, but will also depend on how much moisture was in the soil at the start of the storm. He must also know the characteristics of the river. How much discharge can it carry before it floods? How much area will be flooded as the height of water rises above bank-full? How rapidly will the wave of high water move downstream? The meteorologist must predict the length and intensity of the storm. If melting snow is involved, he must consider the many factors involved in melting snow, such as temperature, humidity, and wind, as well as the amount and condition of the snow. A number of stream gauging stations that report stream height and discharge, and weather stations that report precipitation and other

Figure 8-29. A new valley is being cut into an already existing broad valley in this oblique aerial view of the Snake River on the Oregon-Idaho boundary. Photo from U.S. Geological Survey.

weather elements, are necessary to gather the needed data in a timely fashion.

Data on historical floods have been published by the U. S. Geological Survey in a series of publications called "Hydrologic Investigations Atlases." These atlases show areas flooded and stream gauge data and other data. Flood frequency data show the expected recurrence of floods of different height—that is, the size flood that can be expected every 25 and 50 years, etc. If one of these atlases is available for your area, it should be consulted.

Accurate flood warnings can reduce damage by 5 to 15 per cent and save many lives. The forecasts must be accurate because if too conservative, people will be caught unprepared, and if unnecessary, will be disregarded. Flood preparations are costly and time consuming. Predic-

tions are difficult because a small error in estimating the height of water as flood level or levee height is approached can make the prediction inaccurate. On the other hand, the more time between warning and flood, the better prepared the people.

A flood consists of several phases, each of which can be at least to some extent controlled. The first phase is the land phase and is characterized by flow of water on the surface. This occurs when the ground surface is either impermeable or saturated so the water must flow overland to the streams. Typically this occurs either during very intense storms or after the ground has been saturated by a long-duration storm. Typical damage caused by this phase is gullying. Control measures are revegetation and terracing. A gross example of how this phase leads to floods occurs when forest or brush fires destroy

vegetation. In the following rainy season the water that was previously used by the plants runs off and floods occur downstream. This happens almost every year in southern California.

The second phase of a flood is the channel phase, and this, of course, occurs downstream from the land phase. The objective now is to keep the water within the channel. Dikes, dams, and channel improvements are the main methods of control. Dikes or levees may be temporary or permanent and will prevent flooding only if they are higher than the water level. They are brittle defenses in the sense that where overtopped and broken, the flooding is locally worse than if the water, in the absence of dikes, had spread over a larger area upstream. Channel improvements consist of dredging and straightening the channel to permit more efficient movement of the water. Dams hold back the water and release it at a rate that will not permit flooding downstream. Again, if the dam is too small, or the reservoir is allowed to fill at the wrong time, dams are brittle defenses, and once overtopped, the resulting flood is more destructive.

Figure 8-30. Successive uplifts have formed a terraced valley. A. An uplift caused the river initially in a broad valley to downcut, forming a V-shaped valley. B. Further erosion widened the new valley. C. A new uplift caused renewed downcutting. Ultimately erosion will destroy the older terraces, and it will not be possible to decipher this history.

Dams are expensive and controversial. Many dams are multipurpose and produce electricity and irrigation water as well as flood control. Recreational use is also important. These sometimes conflicting uses lead to controversy. Large dams may affect several states or other countries which, of course, leads to other problems of a non-scientific nature.

The scientific controversy that bears on these social and political problems is which is more efficient, big dams or little dams in flood control? Should floods be controlled by small dams in the small drainage basins in which they form, or is one large dam nearer to the place to be protected more efficient? There seems to be no clear-cut answer to this question. Probably each river system needs a carefully tailored approach using both methods of flood control. A larger question is: What other benefits can or should be gained from dams—power, irrigation, and recreation? This question has social, political, and scientific aspects.

Urbanization, Floods, and Erosion

The main effect of urbanization is to increase the runoff, which, in turn, causes increased erosion and, of course, the eroded material is subsequently deposited. The resulting flooding, erosion, and deposition, as we have seen, may also occur in urbanized areas, compounding the problems. In a given drainage basin, a rainfall will, after some time period, cause an increase in the flow of the main stream. The length of the time lag and the amount of the increase in the flow will depend on the amount and the intensity of the rainfall, as well as the characteristics of the river. Fig. 8-35 shows an example. Urbanization of a drainage basin will increase both the peak runoff and the total runoff because buildings and paving reduce the infiltration.

The increase in discharge causes increased erosion. The first effect is a deepening of the channel and removal of accumulated debris. This increases the capacity of the river so that it can carry more water with-

Figure 8-31. Intrenched meanders on the Grande Ronde River in Washington. Broad valley features like meanders in narrow, V-shaped valleys as shown here strongly suggest a change in the regimen of the river, such as a slow uplift. Photo from Washington State Department of Commerce and Economic Development.

Figure 8-32. A typical river valley, showing the channel and the flood plain.

out flooding. As the main channel deepens, the tributary streams also deepen because their base level, which is the elevation at which they join the main stream, is lowered. Gullies may appear at many places in the drainage basin.

The increased runoff means less infiltration, so less recharge of ground water supplies. The water table will fall, and water wells must be drilled deeper.

A dam causes changes in a stream. Dams have finite lives because the sediment carried by the stream eventually fills the lake created by the dam. This deposition upstream from the dam occurs not only in the lake but all along the stream and its tributaries. This happens because the rising lake is the new base level of the river. This upstream deposition causes problems, especially for farmers whose irrigation ditches are silted up. Downstream the opposite problem, erosion, occurs. The clear water released from the dam no longer has the sediment load that it carried in the natural state, so it uses its excess energy to erode and deepen its channel. The tributaries also deepen to keep pace with the main stream, and gullies form where they can. These events have all happened to a greater or lesser extent at some dams and have resulted in many law suits. Climatic changes can also cause these effects, so some experts deny that dams are the whole cause of these troubles.

Other effects of dams are not always considered until after they are

Figure 8-33. Flooding of the Mississippi River at West Alton, Missouri. The river is generally contained behind the row of trees in the middle of the photograph. World Wide Photos.

Figure 8-34. Changes in the Rio Grande River near El Paso from 1827 to 1936. After Boggs, 1940.

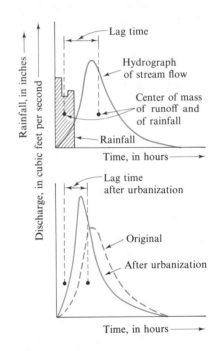

Figure 8-35. The effect of urbanization tion on stream flow. After urbanization, the peak flow is higher and occurs sooner. From U. S. Geological Survey, *Circular* 554, 1968.

built. If the dam was built for water supply or irrigation, the amount of evaporation from the large surface area of a lake may reduce the amount of useful water. Upstream, the rising water table, due to the lake, may waterlog the soil, causing problems for the farmer. This may also cause a change in the natural vegetation that may or may not be good.

Another unexpected man-caused change occurred in Lake Michigan. The navigation channel built to bypass Niagara Falls allowed the alewives to reach Lake Michigan with unfortunate results for the native fish. The dams in the Northwest, especially on the Columbia River, prevented salmon from swimming upstream to their spawning grounds and so reduced the salmon in the Pacific Ocean. Fish ladders were added or built into new dams, reducing this problem.

The increased erosion caused by either urbanization or dams may make the water muddy, a type of pollution. To this may be added domestic sewage and industrial waste. The river is then no longer useful as a water source or for its esthetic beauty; and the long political, social, and economic task of cleaning up the river must be faced. The maximum amount of erosion occurs during the construction phase of urbanization when the natural slope of the ground is altered. More erosion occurs after urbanization than before, in spite of the reduced area where erosion can occur because of building and paving. This is because the increased runoff effectively erodes the uncovered areas.

Southern California, a semiarid area that recently has been urbanized, has many flood and erosion problems. The streams head in mountains, flow across alluvial fans, then across flat valleys into the Pacific Ocean. As described earlier, storms cause intense flows of water or mudflows. The debris is largely deposited on the alluvial fans at the foot of the mountains. As urbanization moved inland, first the valleys and then the fans were built upon. The rivers in many cases only rarely have water in them, so were disregarded in the building. Alluvial fans are built by the rivers changing their position on the fan from time to time. A mud or debris flow is deposited in the channel on the fan, choking it, and the next flow moves on the fan at a different, lower area. In this way, over a long time, the semicircular fan is built. The river valley and the fan are not distinctive features on the ground, and so the tendency to disregard them. They are, however, clear on topographic maps and air photos.

Thus flooding, debris flows, or mudflows cause damage on the fans. In the natural state, much of the water seeps into the ground on the fan. Where building prevents this, the water moves down the valley, flooding the urbanized areas there. Much work has been done to minimize these problems. Debris basins have been constructed behind dams to stop the sediment from reaching the fans. Flood conveyance channels have been constructed to carry the water to the ocean. This work was begun as long ago as 1915, or twenty years before the federal government began any flood control work. These measures have saved countless lives and probably billions of dollars damage, but even so, lives and property are still lost. Zoning laws to control the use of the hazardous areas could prevent some of this loss. This simple measure is difficult to achieve because of political and economic complications. One facet of the problem is the large number of municipalities involved. In Los Angeles County alone, the number of incorporated areas has increased from 44 to 78 between 1935 and 1969.

Many of the ways of preventing

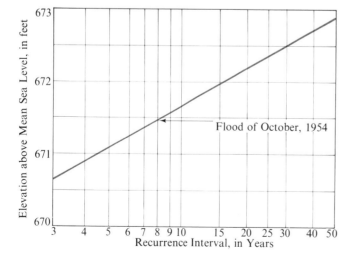

Figure 8-36. Frequency of floods on Salt Creek at Addison near Chicago. This example shows how often a flood of a given height can be expected on the average. From U. S. Geological Survey, *Circular* 601 C, 1970.

flood damage have already been mentioned such as dams, debris basins, and zoning laws. Flood-proofing structures by making the lower floors waterproof, either permanently or during flood danger, can reduce damage. Accurate flood forecasting can make emergency flood-proofing effective. Flood hazard maps, which are available for many areas, can be used for zoning and for predicting how often a given site will be flooded. (See Fig. 8-36.)

Flood-plain zoning is perhaps the best answer in spite of the relocation problems for those now using the flood plains. The resulting green belts could be parks that would improve urban life. The financial institutions that provide the money for buildings could protect themselves and their borrowers by consulting flood hazard maps before lending money. Municipalities should not build streets or sewers in these areas, and utility companies should not provide service. All of these agencies would save themselves and their clients money by these simple measures.

Other environmental aspects of rivers are discussed in Chapter 12.

Drainage Patterns

The development of the drainage pattern is an important part of the erosional history of an area. A few of the ways that streams develop can be observed in nature, but because geologic processes are slow, most of the development must be imagined. Initially, the runoff flows down the irregularities in the slope in many small rills. The flowing water erodes these rills; the ones with greater flow are eroded most. These larger rills become deeper gullies,

Figure 8-37. Rills developed in an excavation in glacial deposits in Iowa. Photo by W. C. Alden, U.S. Geological Survey.

and ultimately one of them will become the master stream. The process of gully development can be seen in many excavations (see Fig. 8-37) and in fields where plowing has removed the natural vegetation that prevented erosion. (See Fig. 8-38.) The rills develop by downcutting and by headward erosion. The rill with the most water will down-cut fastest, and this will accelerate headward erosion by this rill and all of its tributary rills. The lengthening by headward erosion enables this rill to capture the flow of the rills it intersects. This increases the area drained by the main rill and accelerates its erosion. The process of stream capture, called *stream piracy,* continues, and the rill develops first

Figure 8-38. Gullies cut in a sandy slope in a two-year period in Wisconsin. Photo by W. C. Alden, U.S. Geological Survey.

into a gully and finally into a master stream. (See Figs. 8-39 and 8-40.)

Stream patterns develop in many ways. Most stream patterns are probably in part inherited from an earlier development and in part dependent on the underlying rock structure and changes in elevation. Other factors that influence drainage development include climate, runoff, and vegetation. The preceding description tacitly assumes an initially fairly smooth surface that was rapidly uplifted. These conditions, however, are probably not commonly found in nature. One can also envision an uplift slow enough that the initial streams are merely intrenched. Thus uplift may be slow or rapid, or may accelerate or slow down, and each case will affect the erosional development of an area in a different way. The landscapes developed by these differing conditions will be described in the next section.

The effect of bed rock structure on stream pattern can be seen in many places. Areas of uniform bed rock commonly have branching streams in a pattern similar to the veins in a leaf. (See Fig. 8-41.) In other areas, the joints, or cracks in the rocks, may influence the stream pattern, and rectangular patterns commonly develop in this way. At other places, alternating hard and soft rocks may produce a trellis pattern. (See Fig. 8-42.) The initial slope of an area also affects the stream pattern, as shown in Fig. 8-43.

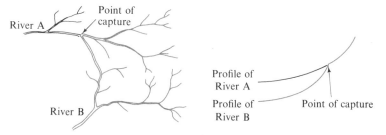

Figure 8-39. A common type of stream piracy. Rapidly downcutting minor tributary of River B captures the headwaters of River A. The gradients of the two rivers near the point of capture are shown in the right diagram.

Figure 8-40. An example of stream capture is shown in this photo of the southern part of the Arabian Peninsula taken from Gemini IV. The Gulf of Aden is in the background. Wadi Hadramaut in the foreground clearly flows toward the right because it is wider there. Many of the tributary steams flow toward the left, and their flow is abruptly reversed where they join Hadramaut. Such reversals are common where rivers have been captured. Photo from NASA.

Development of Landscape

Development of landscapes is even more complex than the development of river valleys and drainage patterns, and all of the factors discussed above are involved. Thus it is no wonder that much controversy exists. We will begin with the simplest case and use it to illustrate other possibilities, which, hopefully, will show why controversy exists. This is an area of active research; and as more data are obtained, much of the present controversy may be resolved.

The simplest viewpoint on the development of landscape was an American idea, and it had a pronounced influence on the history of

Figure 8-41. Branching stream pattern (dendritic pattern) developed in massive rocks. From U.S. Geological Survey, Mount Mitchell quadrangle, North Carolina-Tennessee.

Figure 8-42. Alternating hard and soft rocks produce a trellis drainage pattern.

geology. The cornerstone of modern geology is the doctrine of *uniformitarianism,* which, in the present context, states that the present processes at work on the earth, given enough time, caused the present landforms. This principle, although recognized by a few earlier men, was widely adopted near the end of the eighteenth century. Geology, however, developed mainly in Western Europe during the nineteenth century; and most geologists, especially the island-dwelling English, thought that ocean waves rather than rivers were responsible for most erosion. The opening of the American West following the Civil War provided a vast new area for geologic study. The great rivers and canyons, such as the Colorado, and the many mountain ranges at various stages of dissection, together with the dry climate that bared great vistas, made recognition of huge amounts of erosion simple. (See Fig. 8-44.)

Late in the nineteenth century, William Morris Davis of Harvard codified the earlier studies into what he termed the *cycle of erosion.* Davis visualized every landscape going through a series of changes from initial uplift to complete leveling by weathering and erosion. His emphasis was on the forms produced, and he paid little attention to the processes involved. He clearly recognized that an area rarely goes through the complete cycle, but, rather, the cycle generally is restarted by new uplifts. This accounts for the fact that there are very few examples

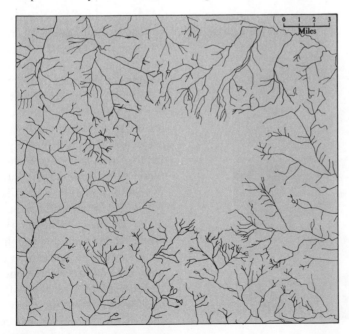

Figure 8-43. Radial stream pattern. From U.S. Geological Survey, Mount Rainier quadrangle, Washington.

Figure 8-44. A typical scene in the arid Southwest. The lack of vegetation and the fact that the flat-lying sedimentary rocks can easily be traced for many miles make recognition of the immense amount of erosion that has occurred easy.

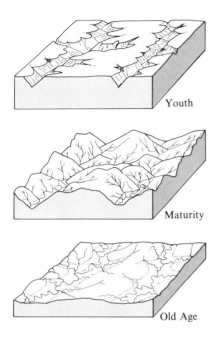

Figure 8-45. The cycle of erosion as envisioned by Davis.

Figure 8-46. An example of youthful topography. Oblique aerial view looking west at the Grand Canyon in Arizona. The flat surface is being attacked by headward erosion of the tributary streams. The inner canyon of the Colorado River is being cut in metamorphic rocks. Photo from U.S. Geological Survey.

of areas that have gone through the complete erosion cycle. In the youthful stage, a newly uplifted area is relatively flat with deeply entrenched streams (Figs. 8-45 and 8-46). As erosion proceeds, the interstream areas are reduced in size as a result of downcutting by the rivers and widening of the valleys by downslope movement of material. When no flat interstream areas exist, the stage has progressed to maturity. (See Figs. 8-45 and 8-47.) This is the stage of maximum relief. As the interstream areas are rounded and lowered, the area passes into old age (Fig. 8-45). In the ultimate stage, when even these gentle interstream areas are worn almost flat, the area is called a *peneplain* ("almost a plain," sometimes spelled *peneplane*). Flat eroded areas such as peneplains are rare features on the earth's surface, and this fact constitutes one of the weaknesses of the cycle-of-erosion hypothesis. However, many events, such as renewed uplift, can occur to prevent the complete development of a peneplain.

Figure 8-47. An example of a mature area in the Sierra Nevada of California. Glacial erosion has modified this area. Photo from U.S. Geological Survey.

Perhaps the main objection to the Davis cycle is that it requires rapid uplift and then a stillstand during which erosion occurs. If the uplift of an area is slow enough, perhaps erosion keeps pace with the uplift and no change in form occurs. However, any type of uplift from rapid to slow, or from accelerating to slowing, is geologically possible, so at best the Davis cycle describes a special case.

Another objection is that the hills might not be reduced in the manner envisioned by Davis. Instead, a balance might result when the hill slopes and stream gradients become adjusted. Such an equilibrium can, in a way, be thought of an an extension of the graded-stream concept discussed earlier. After this stage is reached, the shape of the area remains substantially the same as erosion progresses. Thus there is no great change in shape, as there is in the Davis cycle of erosion, because once an equilibrium shape develops, no new shape can form unless the balance is upset by uplift or some other change. (See Fig. 8-48.)

In an effort to resolve this latter objection, a number of recent studies have been made of hill slopes. The studies have been made in areas where the climate, vegetation, relief, and rock type are uniform. The slope angle of such hillsides is related to both stream gradient and density of streams. Where direct measurement of slope erosion was made, it was found that during retreat of slopes the slope angle remains constant if the debris is removed from the foot of the slope. (See Fig. 8-49.) If the erosional debris is not removed from the foot of the hill, the slope angle is reduced. The latter case is probably more common in humid areas where weathering may be deep and

Figure 8-48. Successive profiles of the erosion of a river valley contrasting two theories of the development of slopes. A. The Davis concept of downwearing of hillsides. B. Backwearing of hillsides, preserving approximately the same slope.

creep may be active. All of this suggests that the Davis cycle is not only a special case, but an over-simplification as well.

The length of time and the size of the area considered in the study of landscape may also influence one's ideas. A single hillside or stretch of river may be in equilibrium during the time of a short study of the type described above; however, if the whole drainage area is considered over a geologically long time interval, erosion must remove appreciable amounts of mass and so change the system. Thus, in the short term, landforms may not seem to change with time; but over a long interval they might actually do so. The dynamic earth may not remain still long enough, however, for these changes to occur, for the earth's surface is the result of the constant struggle between erosion and uplift.

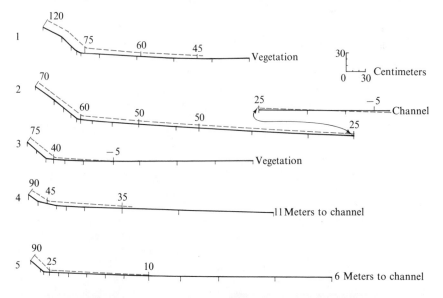

Figure 8-49. Measurement of erosion in the Badlands of South Dakota. The dashed profiles are July, 1953, and the solid lines are May, 1961. The numbers are the erosion depth in millimeters. From S. A. Schumm, *Bulletin,* Geological Society of America, Vol. 73, 1962.

Summary

Rivers are agents of erosion, transportation, and deposition. They are the most important agents of transportation. Weathering, downslope movements, and rivers shape the landscape.

Average velocity of river flow is less than five miles per hour.

Discharge or *flow* is the amount of water passing a given cross-section of the river in a time unit (area of cross-section times the velocity, usually given in cubic feet per second, c.f.s.). The factors controlling the flow vary from place to place on a river, and vary with season.

The main geologic work of rivers is transportation. *Dissolved load* comes from chemical weathering and solution by the river. *Suspended load* is material carried in suspension — clay, and, in turbulent water, silt. *Bed load* is material that rolls or slides along the river bottom, in some cases only during floods.

River erosion is mainly by solution and abrasion. Maximum erosion occurs in areas with about ten inches of rain per year.

River erosion deepens, lengthens, and widens the valley. *Headward erosion* is downcutting at the head of a stream; it lengthens the valley. The valley widens by creep and landsliding, and by lateral cutting by the stream.

Distinctive curves called *meanders* form which distribute the river's loss of energy most uniformly. Developing meanders may intersect, cutting off a stretch of river, forming *oxbow lakes*.

A graded stream is one which has just the right gradient to be able to transport the debris delivered to it by its own erosion, tributaries, downslope movements, and rain wash. The gradient of a given stretch is maintained by the river's eroding or building up its bed in response to climate or other conditions. The concave slope of the long profile of a river provides for the most uniform rate of work along the river, balancing slope with discharge and velocity.

Base level is the limit to which a river can erode, the elevation of its mouth.

Fluvial (river) deposits are *deltas* (formed where rivers enter standing water), *alluvial fans* (formed where rivers flow out of mountains into broad, flat valleys), and *levees* (formed along river banks when flood-waters spill over, reducing velocity). Rivers also build up their beds by deposition. Deposition occurs when velocity is slowed.

Terraces form from changes in river regimen or conditions; they can form by either deposition or erosion.

Braided streams form when load is large and coarse for the slope and discharge. The channel divides and rejoins many times.

Narrow valleys have V-shaped cross-sections and are generally formed by downcutting, non-graded rivers near their headwaters. They commonly have falls, rapids, and lakes, which are temporary features.

Valley widening begins after a stream has reached grade, usually first near the mouth. Meanders are common features.

A V-shaped valley being cut in a wider valley indicates uplift or some change in river's regimen. Slow uplift can cause meanders to be intrenched in narrow V-shaped valleys.

A river valley consists of a channel, a flood plain, and sloping valley sides. The flood plain is the channel of the river during times of high water, and flooding is the normal behavior of the river.

Drainage patterns are affected by climate, runoff, vegetation, and underlying bedrock structure. The rill with the most water downcuts fastest, and by headward erosion captures other rills (*stream piracy*), accelerating its development into a master stream.

Uniformitarianism is the doctrine that the present processes at work, given enough time, caused the present landforms.

William Morris Davis' *cycle of erosion* is a deductive sequence of landforms from initial uplift to complete leveling by weathering and erosion. *Youth* is characterized by deeply incised streams cutting a relatively flat, newly uplifted area. *Maturity*, the stage of maximum relief, exists when no flat interstream areas remain. *Old age* is the stage of rounded, lowered interstream areas and wide valleys. *Peneplain* is the term given to the ultimate flat eroded surface, a rare feature on the earth. Davis' cycle has objections to it, and at best is a special case or an oversimplification.

A balance might result when hill slopes and stream gradients reach equilibrium, and no new shape may develop unless the balance is upset.

Questions

8-1. How is most of the energy of a river used?

8-2. How do you account for the fact that the velocity of a river increases downstream?

8-3. What are the three ways by which a river transports material?
    1. _____ 2. _____ 3. _____

8-4. What are the processes of river erosion?

8-5. Describe the sequence of events in the formation and migration of a curve in a river course. (Use diagrams.)

8-6. A placer deposit is a place where a river has deposited fragments of heavy minerals, such as gold. Where would you look in a river for such deposits?

8-7. Can you think of some reasons why doubling the velocity of a river increases the amount of material transported four to eight times but only increases the maximum size of material transported four times?

8-8. Show with a series of sketches the erosion of a river valley.

8-9. Describe the erosion of an upland area. How does the topography change as erosion proceeds? Your answer depends on several assumptions you must make.

8-10. What is the final surface produced by long erosion called?

8-11. What are some characteristics of narrow river valleys?

8-12. Discuss whether building should be allowed on the flood plains of rivers.

8-13. Under what conditions are mudflows most apt to occur?

8-14. How can flood danger be reduced?

8-15. How does urbanization affect the runoff of streams?

8-16. Describe the evidence showing rapid uplift of an area.

8-17. To what was Twain referring in the quotation at the beginning of this chapter?

Supplementary Reading

General

Bloom, A. L., *The Surface of the Earth*. Englewood Cliffs, New Jersey: Prentice-Hall, Inc., 1969, 152 pp. (paperback).

Curtis, W. F., J. K. Culbertson, and E. B. Chase, *Fluvial-Sediment Discharge to the Oceans from the Conterminous United States*. U. S. Geological Survey Circular 670, Washington D. C.: U. S. Government Printing Office, 1973, 17 pp.

Davis, W. M., *Geographical Essays,* especially Part II, "Physiographic Essays." New York: Dover Publications, Inc., (1909), 1954, 777 pp. (paperback).

Flemal, R. C., "The Attack on the Davisian System of Geomorphology: A Synopsis," *Journal of Geological Education* (January, 1971), Vol. 19, No. 1, pp. 3-13.

Hunt, C. B., *Physiography of the United States*. San Francisco: W. H. Freeman and Co., 1967, 480 pp.

Janssen, R. E., "The History of a River," *Scientific American* (June, 1952), Vol. 186, No. 6, pp. 74-83. Reprint 826, W. H. Freeman and Co., San Francisco.

Judson, Sheldon, "Erosion of the Land," *American Scientist* (Winter, 1968), Vol. 56, No. 4, pp. 356-374.

Leopold, L. B., and W. B. Langbein, "River Meanders," *Scientific American* (June, 1966), Vol. 214, No. 6, pp. 60-70.

Leopold, L. B., and Thomas Maddock, Jr., *The Hydraulic Geometry of Stream Channels and Some Physiographic Implications*. U.S. Geological Survey, *Professional Paper* 252, 1953, 57 pp.

Leopold, L. B., M. G. Wolman, and J. P. Miller, *Fluvial Processes in Geomorphology*. San Francisco: W. H. Freeman and Co., 1964, 504 pp.

Morgan, J. P., "Deltas—A Resume," *Journal of Geological Education* (May, 1970), Vol. 18, No. 3, pp. 107-117.

Morisawa, Marie, *Streams: Their Dynamics and Morphology*. New York: McGraw-Hill Book Co., 1968, 175 pp. (paperback).

Peixoto, J. P., and M. A. Kettani, "The Control of the Water Cycle," *Scientific American* (April, 1973), Vol. 228, No. 4, pp. 46-61.

Schumm, S. A., "The Development and Evolution of Hillslopes," *Journal of Geological Education* (June, 1966), Vol. 14, No. 3, pp. 98-104.

Shimer, J. A., *This Sculptured Earth: The Landscape of America*. New York: Columbia University Press, 1959, 256 pp.

Thornbury, W. D., *Principles of Geomorphology* (2nd ed.), Chapters 5, 6, 7, 8, pp. 99-208. New York: John Wiley & Sons, Inc., 1969, 594 pp.

Environmental Hazards

Bue, C. D., *Flood Information for Flood-Plain Planning*. U.S. Geological Survey Circular 539, Washington, D.C.: U.S. Government Printing Office, 1967, 10 pp.

Chorley, R. J., *Water, Earth, and Man*. London: Methuen and Co., Ltd., 1969, 588 pp.

Glancy, P. A., *A Mudflow in the Second Creek Drainage, Lake Tahoe Basin, Nevada, and Its Relation to Sedimentation and Urbanization*. U.S. Geological Survey *Professional Paper* 650C, Washington, D.C.: U.S. Government Printing Office, 1969, pp. C195-C200.

Guy, H. P., *Sediment Problems in Urban Areas*. U.S. Geological Survey Circular 601-E, Washington, D.C.: U.S. Government Printing Office, 1970, pp. E1-E8.

Hoyt, W. G., and W. B. Langbein, *Floods*. Princeton, N.J.: Princeton University Press, 1955, 469 pp.

Leopold, L. B., *Hydrology for Urban Land Planning — A Guidebook on the Hydrologic Effects of Urban Land Use.* U.S. Geological Survey *Circular 554,* Washington, D.C.: U.S. Government Printing Office, 1968, 18 pp.

Rantz, S. E., *Urban Sprawl and Flooding in Southern California.* U.S. Geological Survey *Circular* 601-B, Washington, D.C.: U.S. Government Printing Office, 1970, pp. B1-B11.

Sheaffer, J. R., and others, *Flood-Hazard Mapping in Metropolitan Chicago.* U.S. Geological Survey *Circular* 601-C, Washington, D.C.: U.S. Government Printing Office, 1970, pp. C1-C14.

Thomas, H. E., and W. J. Schneider, *Water as an Urban Resource and Nuisance.* U.S. Geological Survey *Circular* 601-D, Washington, D.C.: U.S. Government Printing Office, 1970, pp. D1-D9.

U.S. Geological Survey, Hydrologic Atlases Portraying Areas Struck by Hurricane Camille on the Mississippi Gulf Coast. HA395 to HA408, 1969.

Geologic Processes in Arid Regions and the Work of Wind

9

As outlined in the introduction to the preceding chapter, some of the classical studies on erosion by running water were made in the arid American southwest. Other aspects of those same studies will be considered in this chapter. Thus it should be no surprise that the prevailing opinion is that running water, although rare, is the main agent of erosion there. Wind erosion is much less effective. However, in drier areas, such as parts of north Africa, wind is an important geologic agent. (See Fig. 9-1.) Recognition of this has come in part from study of photographs made by the earth-orbiting satellites. Much of our knowledge of wind erosion comes from wind tunnel experiments published in 1942 by British Brigadier R. A. Bagnold, who became interested in deserts when posted to Egypt.

Geologic Processes in Arid Regions

Dry regions form over one-quarter of the total land surface of the earth. Most of these arid areas are hot deserts or near-deserts, but some of the dry areas are cold. The main desert areas are in the subtropics near thirty degrees north and south latitude. They are caused by the primary atmospheric circulation of the earth that causes dry air to descend at these latitudes, and the resulting compression heats the already dry air. Other arid areas occur in the rain shadows of high mountain ranges. Southwestern United States is an example of such an area. The moist winds from the Pacific Ocean are cooled and precipitate their moisture when they are forced to rise over the Sierra Nevada and other mountain ranges in California. Here, too, the dry descending air is heated by compression.

In most arid areas, the lack of moisture changes the emphasis but not the types of geologic processes. Rainfall, although infrequent and irregular, is the main agent of erosion. Rain may fall only once in many years at any one place, but is generally in the form of an intense cloudburst or thunderstorm. The lack of vegetation enables such storms to erode very efficiently by splash and runoff. Flash floods in normally dry stream channels many miles from the storm may trap unwary travelers. The runoff carries much loose surface material to the stream channels, and the resulting flow may be a thick mudflow rather than a normal stream. Such mudflows can carry huge boulders and can do great

Edge of the dune complex of Imperial Valley, California. Photo by W. C. Mendenhall, U. S. Geological Survey.

A.

B.

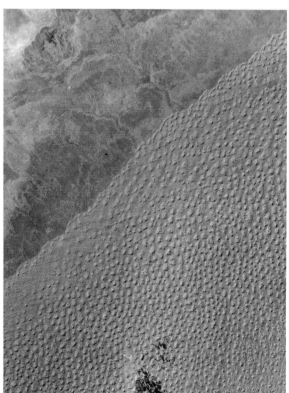

C.

Figure 9-1. Satellite photographs of sandy deserts or sand seas. A. Simpson Desert, Australia. B. Rows of dunes. Algerian desert. C. Dunes similar to B but not in rows. Algerian desert. ERTS photograph.

damage to anything in their path. Weathering is slow, and because mechanical weathering is much more important in arid regions than chemical weathering, little soil is formed. Limestone that is rapidly dissolved in humid climates forms bold cliffs in the desert.

Desert Landscapes

Infrequent rain and the resulting lack of moisture cause desert landscapes to develop slowly. Because of the lack of rain, throughgoing rivers rarely develop. Exceptions are the Colorado and Nile Rivers that begin in high, well-watered mountains and then cross deserts. Most deserts, however, are areas of internal drainage, and this profoundly affects their erosional development.

The development of drainage patterns in deserts is well shown by the Great Basin of southwestern United States. This is an area of many mountain ranges formed by relatively recent faulting. The Basin and Range country is another name sometimes applied to much of this region because basins separate the ranges. The basins are low areas, and lakes may form in them during times of heavy rainfall, such as the ice age described in the next chapter. Lake-deposited sediments are fairly common in these basins, and dry lake beds form the surfaces of many. The lakes generally contain salt and other dissolved materials that come from the rain water and chemical weathering. Thus many of them are what are called alkali lakes, and the dry lake beds are alkali flats. Great Salt Lake and Bonneville Salt Flats are examples. Because these lakes are not connected to the ocean, they

Figure 9-2. Development of drainage systems in the basin and range area. As each basin is filled, the drainage spills over the lowest divide into the next basin. This process continues, slowly integrating the drainage of many basins until the ephemeral streams reach the ocean — a goal rarely reached.

Figure 9-3. A complex drainage system that resulted from integration of basins. The low range in the foreground and middleground is being vigorously eroded, probably because its base level was lowered by the integration of basin drainage. Some possible future captures can be seen. In the far middleground, several basins that have integrated are visible. Note the lack of water in the stream courses and the almost complete lack of vegetation in this view near the southern border of California. Photo from U.S. Geological Survey.

develop in much the same way as do the oceans. They are, in effect, small oceans.

The usual development of drainage patterns in a desert begins with erosion in the mountains and deposition in the basins. The deposition may occur well out in the basin or as alluvial fans near the mountain front. These processes continue until the basin is filled to the level that the ephemeral streams can flow over the lowest divide separating it from an adjoining basin. (See Fig. 9-2.) The processes now continue in the new basin until it too is filled. While the second basin is being filled, the divide between the two basins is generally lowered by erosion; this initiates a stage of erosion in the first basin because of a lowered base level. (See Fig. 9-3.) This complicated series of events will continue until an outlet to the ocean is established; however, a connection to the ocean is rarely developed because lack of rain makes these processes very slow.

The erosional reduction of desert mountains is somewhat different from that of more humid climates. The differences may be basic, but are more likely the result of different emphasis of the various erosional processes. Also, as just noted, local base levels and small drainage areas are involved, rather than large watersheds, so that each desert range or part of a range develops at its own rate, making the developmental sequence easy to see.

The first step is generally the development of fans at the mountain front. After this, the wearing back of mountains begins. The initial steep mountain front retreats, and a fairly smooth surface in the bedrock is eroded. This surface is called a *pediment* and is generally continuous

Figure 9-4. Formation of pediments. A. Shortly after uplift. B. Pediments have begun to form. C. Late stage. The mountain range is mainly pediment.

with the fan. The pediment is commonly covered by a thin layer of gravel, making it very difficult to distinguish from a fan unless the generally present gullies cut through the gravel. The method by which pediments are formed is not clear, but streams issuing from the mountains may change their courses back and forth across the pediment in a process similar to the way alluvial fans are formed. Runoff is probably also important in pediment formation. Pediments have concave-upward slopes, much like river profiles. The slope of most pediments is

Figure 9-5. The faulted eastern face of the Sierra Nevada is only slightly modified by erosion. Photo from U.S. Geological Survey.

Figure 9-6. The main range shown here has been embayed by erosion with the development of pediments and fans. In the right foreground, the mountains have been reduced to low pediments. Photo from U.S. Geological Survey.

between one-half and seven degrees and is related to the particle size of the gravel that covers it, and probably the amount of runoff. The steep mountain front behind the pediment seems to maintain its slope angle during retreat. The various stages in the erosion of a desert range

Figure 9-7. Desert armor or pavement, Greenwater Range, eastern California. Photo by C. S. Denny, U. S. Geological Survey.

are shown in Figs. 9-4, 9-5, and 9-6. The examples shown are faulted mountains similar to those in the arid areas of southwestern United States. Other types of mountains are found in other arid regions. The formation of faulted mountains is described in Chapter 14.

Geologic Work of Wind

Only in dry areas is wind an active agent of erosion, transportation, and deposition. The rapid heating and cooling of deserts leads to strong winds. The behavior of wind is similar to that of water in some ways, but it is much less effective. Wind erosion occurs by two processes, deflation and sandblasting (or abrasion). *Deflation* is simply the removal by wind of sand- and dust-sized particles. Typical places where deflation may take place are dry, unvegetated areas such as deserts, dry lake or stream beds, and actively forming, glacier-outwash plains in dry seasons. Deflation may, in some areas, produce hollows, sometimes called blowouts, that are recognized by their concave shape, which could not have been produced by water erosion. (The only other common way that closed depressions form is by solution of soluble rocks, such as limestone, by ground water.) At other places deflation removes the fine material from the surface, leaving behind pebbles, to produce a surface armored by a pebble layer against further deflation. (See Fig. 9-7.) Alternate wetting and drying can also cause pebbles to rise by a process somewhat like frost heaving discussed in Chapter 7. *Sandblasting* by wind-driven sand grains can cause some erosion near the base of a cliff or on a boulder, but probably is much less effective than weathering. (See Fig. 9-8.) Sandblasting does produce some interestingly shaped pebbles called *ventifacts,* or *dreikanter* (German for "three edges"). Sand hitting a pebble on the side facing the prevailing wind wears a round surface that may be polished and slightly pitted. Most ventifacts have several such surfaces, three being common. (See Fig. 9-9.) The other surfaces may be produced by other common wind directions or, probably more commonly, by the turning of the pebble, perhaps due to the undermining of the pebble by deflation of its supporting sand. (See Fig. 9-10.)

Wind transportation is similar to water transportation in that both a suspended load and a bed load are involved. The suspended load produces dust storms, and the bed load forms sandstorms. Most wind-blown sand is less than one millimeter (mm) in diameter, and the average size is about ¼ mm. Sand begins to move when the wind reaches 11 miles per hour. All material finer than $\frac{1}{16}$ mm (silt) is considered dust, and most wind-blown dust is much smaller than this. The separation into these two size ranges is very sharp, much more so than in the case of water transportation, so that wind-deposited sediments are generally much better sorted by size than water-laid deposits. The reason for this better sorting is that wind can raise sand particles only a few feet above the ground, although it can raise dust thousands of feet. At one place where measurements were made, 50 per cent, by weight, of the sand was raised less than five inches above the ground and 90 per cent was raised less than 25 inches, al-

Figure 9-8. Double Arch in Arches National Monument, Utah. The arches are believed to be partly the result of wind erosion of thick sandstone beds. Water from infrequent rains probably loosens the sand near the surface, and wind removes the loose sand grains and causes some abrasion. Photo from National Park Service.

though some large fragments rose over 10 feet. At this place, maximum abrasion occurred nine inches above the ground. Such abrasive effects (only a few inches above ground) are common on fence posts and power poles in areas of sandstorms. As might be imagined, sandstorms

Figure 9-9. Ventifacts. Photo by M. R. Campbell, U.S. Geological Survey.

Figure 9-10. Formation of ventifacts by wind abrasion. The main facet faces the prevailing wind direction. Other faces are formed by other common wind directions or the pebble is turned, exposing another side to the wind.

Figure 9-11. A dust storm caused by a cold front at Manteer, Kansas, in April, 1935. The dust is in the cold air mass and rises to great heights. Photo from NOAA.

the dust into motion, and the turbulence of the wind will tend to keep it in motion.

The amount of wind-blown material raised by a storm can be very high. For example, during some of the intense dust storms in the United States in the 1930's, the load in the lower mile of the atmosphere in some areas was estimated at over 150,000 tons per cubic mile of air.

Wind deposits are of two types — loess composed of dust, and drifts and dunes composed of sand. *Loess* forms thick, sheet-like deposits that are relatively unstratified because of their rather uniform grain size, generally in the silt size range. (See Fig. 9-13.) Thick loess deposits in China are believed to have originated from the deserts of Asia. Most American loess deposits are associated with glacial deposits and are believed to have formed when glacial retreat left large unvegetated areas on which winds could act. Wind-blown sand, unlike loess, does not generally form blanket deposits. *Dunes* are naturally formed accumulations of wind-blown sand. Most dunes probably start in the lee of an obstacle (Fig. 9-14). Unless they have become fixed by vegetation, most sand dunes move slowly across the desert (Fig. 9-15). As might be expected, many dunes and dune areas are not easily classified; however, several distinct types of dunes are of wide occurrence (Figs. 9-16, 9-17, 9-18, and 9-19). The sand grains in dune deposits are commonly frosted by the abrasion during transport.

are associated with dust storms, but dust storms may occur without sandstorms; so, in general, dust is moved much farther than sand. (See Fig. 9-11.)

The mechanism for moving sand is similar, in part, to the action of water. The process produces a bouncing motion of the sand grains. It begins when a grain rolls into or over another grain. When the first grain hits, its impact transfers its kinetic energy to one or more grains which, in turn, also bounce up and, on falling, continue the process by hitting other grains (Fig. 9-12.) Except in large downdrafts, dust cannot be lifted by wind alone because at ground level there is a thin zone of no wind about one-thirtieth of the diameter of the average particle. The impact of falling sand grains can put

Figure 9-12. Movement of bed load. Sand grain 1 rolls over and hits grain 2 which bounces up and hits grains 3 and 4, both of which also bounce up and hit more sand grains.

Figure 9-14. Accumulation of sand in lee of fence post.

Figure 9-13. Loess near Vicksburg, Mississippi. The upper layer of loess is somewhat different from the lower. Although loess can stand in vertical cliffs, it is very easily eroded where a little water spills over the cliffs. Photo by E. W. Shaw, U.S. Geological Survey.

Figure 9-15. The movement of a sand dune. The wind moves individual sand grains along the surface of the dune until they fall off the steep face.

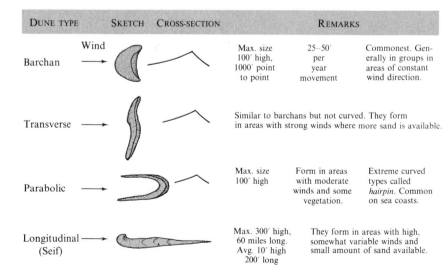

DUNE TYPE	SKETCH	CROSS-SECTION	REMARKS
Barchan	Wind →		Max. size 100′ high, 1000′ point to point — 25–50′ per year movement — Commonest. Generally in groups in areas of constant wind direction.
Transverse	→		Similar to barchans but not curved. They form in areas with strong winds where more sand is available.
Parabolic	→		Max. size 100′ high — Form in areas with moderate winds and some vegetation. — Extreme curved types called *hairpin*. Common on sea coasts.
Longitudinal (Seif)	→		Max. 300′ high, 60 miles long. Avg. 10′ high 200′ long — They form in areas with high, somewhat variable winds and small amount of sand available.

Figure 9-16. Types of sand dunes.

Figure 9-17. Barchan dunes in the Columbia River Valley near Biggs, Oregon. Photo by G. K. Gilbert, U.S. Geological Survey.

Figure 9-18. Asymmetric wind ripples and the steep lee face of a complex barchan dune near Paxton, Montana. Photo by C. E. Erdman, U.S. Geological Survey.

Figure 9-19. Aerial view of Imperial Valley, California, showing a dune complex. Most areas of dunes are complex and do not fall into the common classification of dunes. Photo by J. R. Balsley, U.S. Geological Survey.

Summary

Intermittent rainfall and running water are the main agents of erosion in most arid areas, but wind may be important in very arid areas.

Rainfall erodes very efficiently by splash and runoff. Runoff carries much loose surface material to stream channels, and the resultant flow may be a thick, destructive mudflow.

Mechanical weathering is more important than chemical in arid regions, so little soil is formed.

Limestone may form cliffs rather than be dissolved as in more humid climates.

Throughgoing rivers rarely develop because of lack of rainfall.

Lakes may form in basins during heavy rainfall; they generally contain salt and other dissolved material from rain or chemical weathering. They are called *alkali lakes,* or, when dry, *alkali flats.*

Drainage patterns in deserts develop as the result of erosion in mountains, and deposition in basins or on alluvial fans. When a basin is filled and the mountains reduced sufficiently, the stream flows over the lowest divide into the adjacent basin. This lowers the base level and so initiates new erosion in the first basin and deposition in the adjacent basin.

Pediments are smooth, gravel-covered, concave-upward erosional surfaces cut in the bedrock of desert mountains. They are continuous with the alluvial fans deposited at the mountain fronts.

Wind erosion occurs by two processes. *Deflation* is the removal of sand- and dust-sized particles. It may produce closed depressions called *blowouts.* The removal of fine material may leave a surface armored by pebbles. *Sandblasting* is abrasion by wind-driven sand grains.

Ventifacts or *dreikanter* are pebbles with polished facets or surfaces produced by sandblasting.

Wind transportation involves *suspended load* (dust storms) and *bed load* (sandstorms). All material finer than $\frac{1}{16}$ mm is considered dust; most wind-blown dust is smaller. Most wind-blown sand is less than 1 mm, average about $\frac{1}{4}$ mm. The size separation is very sharp, and wind-deposited sediments are well-sorted by size because dust can be raised much higher and transported longer than sand.

Loess is a wind deposit composed of silt; it forms thick, sheet-like, unstratified deposits.

Dunes are naturally formed accumulations of wind-blown sand. Unless fixed by vegetation, dunes move slowly across the desert. Although many dunes are not easily classified, some widely occurring types are *barchan*, *transverse*, *parabolic*, and *longitudinal* or *seif*. Type depends on strength and variability of wind, amount of sand, vegetation, and other factors.

Questions

9-1. What is the role of climate in erosion and the development of landforms?
9-2. Describe pediments.
9-3. Briefly describe wind erosion (two kinds).
9-4. Wind-deposited silt is called _____.
9-5. Deposits of wind-blown sand are called _____.
9-6. Describe the sorting of wind deposits.
9-7. How might wind-deposited sediments differ from water-deposited sediments?
9-8. Describe the development of desert pavement.
9-9. Why is wind not an important agent of erosion in humid, windy climates?

Supplementary Reading

Bagnold, R. A., *The Physics of Blown Sand and Desert Dunes.* London: Methuen & Co., 1941, 265 pp.

Sharp, R. P., "Wind-driven Sand in Coachella Valley, California," *Bulletin Geological Society of America* (September, 1964), Vol. 75, No. 9, pp. 785-804.

Thornbury, W. D., *Principles of Geomorphology* (2nd ed.), Chapter 12, pp. 288-302. New York: John Wiley & Sons, Inc., 1969, 594 pp.

Geologic Work of Ice — Glaciers

10

The recognition that glaciers in the recent geologic past had covered much of the northern hemisphere was made in 1836 in the Alps. From observations that deposits far down valleys were similar to the materials currently being deposited by glaciers, Charpentier and others had concluded that glaciers were at one time much more extensive. In 1836 Charpentier showed the field evidence to Jean Louis Rodolphe Agassiz (1807–1873). Agassiz, at first skeptical, was quickly convinced. He quickly grasped that glaciation could not occur only in Switzerland, but that a lower temperature would be world wide, and this realization led to the recognition of the Ice Age. He announced his conclusions in 1837, and in a few years the evidence for past continental glaciation was recognized all over Europe and North America. Agassiz traveled extensively in his study of glaciation and spent his last 27 years at Harvard. Prior to Agassiz, glacial deposits were thought to be remnants of the Flood or to have been carried on icebergs.

At the time of Agassiz' discovery, uniformitarianism was being adopted by many as a result of the publication of Lyell's book in 1830. Neptunism had died out, and the other theory, catastrophism, was on the wane. Catastrophism is the idea that geological changes come about by rapid changes, such as earthquakes creating great valleys. The climatic changes necessary to bring about an ice age seemed rapid, so Agassiz' discovery slowed the acceptance of uniformitarianism for a short time.

Although glaciers are much less important than rivers in overall erosion, they have shaped many of the landforms of northern North America and Eurasia. In addition, most of the high mountain ranges of the world have been greatly modified by mountain glaciers.

A glacier is a mass of moving ice. Glaciers form as a result of accumulation of snow in areas where more snow falls than melts in most years. This can occur at sea level at the poles to about 20,000 feet near the equator. The accumulation of snow must, however, become thick enough so that it recrystallizes to ice. The recrystallization process depends on the pressure of the overlying snow, which transforms the light, loosely packed snow into small ice crystals; with increased pressure, the small crystals become larger. (See Fig. 10-1). The process is quite similar to the metamorphism of sandstone into quartzite and of limestone into marble. The tendency of snow to recrystallize and form larger crystals is well known to anyone who skis or who has lived in snowy areas. Falling snowflakes are generally small and light, but after a few sunny days, or in the spring, they are much larger at the surface; digging will reveal even larger crystals. When the amount of ice formed in this way becomes large enough that it flows under its own

Icebergs forming where a glacier meets the ocean. Storströmmen, east Greenland. Reproduced with the permission (A.477/73) of the Geodetic Institute, Denmark.

Figure 10-1. Some of the steps in the transformation from snow to glacial ice. Crystals in glacial ice can be several inches in diameter.

weight, a glacier is born. Ice, as we see it, is a brittle substance, so the fact that a glacier flows under its own weight, much like tar, may seem strange; however, a mass of ice a few hundred feet thick behaves as a very viscous liquid.

Glaciers are of two main types: the alpine or mountain glacier, and the much larger continental or icecap glacier. Present-day continental glaciers occur only near the poles, but mountain glaciers occur at all latitudes where high enough mountains exist. (See Table 10-1.)

Mountain Glaciers

Mountain glaciers (see Fig. 10-2) develop in previously formed stream valleys, which, because they are lower than the surrounding country, become accumulation sites for snow. When enough ice has formed, the glacier begins to move down the val-ley. How far down the valley it extends will depend on how much new ice is formed and how much melting occurs each year. As the glacier extends further down the valley, the amount of melting each year will increase until a point is reached at which it equals the amount of new ice added. Such a glacier is in equilibrium. If the amount of snowfall increases or the summers become cooler, it will advance further; and if the snowfall decreases or the summers become warmer, it will melt back. However, it generally takes a number of years for the more- or less-than-normal snowfall of a single year, or group of years, to become ice and to reach the snout of the glacier. This means that although a glacier is sensitive to climate changes, it tends to average snowfall and temperature changes over a number of years. Hence, the advance or the retreat of glaciers gives some informa-

Figure 10-2. Mountain glaciers. Several mountain glaciers join in this valley. Note the crevasses on the surface of the ice and the moraines both on the valley sides and separating the ice tongues from the different glaciers. Photo from Swissair Photo Ltd.

Figure 10-3. Movement of glaciers and the development of crevasses. The movement is shown by the successive locations of stakes driven into the ice in the upper part of the figure. The opening of crevasses in the brittle ice as a result of the greater movement of the center of the glacier is shown in the lower part of the figure.

Table 10-1. Glaciers in the United States. Data from U.S. Geological Survey.

STATE	APPROXIMATE NUMBER OF GLACIERS	TOTAL GLACIER AREA IN SQUARE MILES
Alaska	?	About 17,000
Washington	800	160
Wyoming	80	18
Montana	106	10
Oregon	38	8
California	80	7
Colorado	10?	1
Idaho	11?	1
Nevada	1	.1
Utah	1?	.1?

Figure 10-4. Oblique aerial view of Mount Hood, Oregon, showing crevasses on the glaciers. Note the U-shaped valleys. Photo from U.S. Geological Survey.

Figure 10-5. Longitudinal section through a mountain glacier showing development of a cirque by plucking, and reshaping of the valley by plucking and abrasion. This sketch shows an early stage in the development of a glacial valley.

tion on long-term climatic changes. Study of old photographs and maps shows, for instance, that the glaciers in Glacier National Park—and elsewhere—have retreated in the last few decades. Glaciers reached maximum advances about 1825, 1855, and 1895. In general, they have been receding since the turn of the century although with some minor advances. One point needs emphasis: the glacial ice is always moving down-valley, even if the snout of the glacier is retreating because of melting.

The movement of glacial ice can be measured by driving a series of accurately located stakes into the glacier and surveying them periodically. As would be expected, because of friction, the sides of the glacier move more slowly than the center. The upper 100 to 200 feet of the glacial ice behaves brittlely so that the different rates of flow at the surface open large cracks, called *crevasses,* in this brittle zone. (See Figs. 10-3 and 10-4.) The movement of mountain glaciers varies from less than an inch a day to over 50 feet a day.

Erosion by mountain glaciers produces such spectacular mountain scenery as Yosemite Valley in the Sierra Nevada of California. Glaciers erode by plucking large blocks of bedrock and by abrasion. Plucking is accomplished by melt-water that flows into joints in the bedrock and later freezes to the main mass of ice which, on advancing, pulls out the blocks of loosened bedrock. These blocks arm the moving ice with rasp-like teeth that grind the sides and bottom of the glacial valley (Fig. 10-5). Glacial erosion greatly modifies the shape of the stream valley occupied by the glacier. Most of the erosion is probably done by plucking, and abrasion generally smooths and

Figure 10-6. Glacial erosion of river valleys. Steps are commonly developed and may contain lakes.

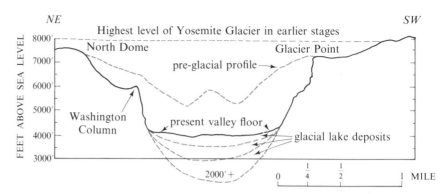

Figure 10-7. Glacial modification of Yosemite Valley, California. From Clyde Wahrhaftig, California Division of Mines and Geology, *Bulletin* 182, 1962.

even polishes the resulting form. A glaciated valley differs from a stream valley in that it is deepened, especially near its head, and the sides are steepened so that its cross-section is changed from V-shaped to U-shaped (Figs. 10-6, 10-7, and 10-8). Glacial erosion is most active near the head of the glacier, and the deepening there flattens the gradient of the valley. This deepening, together with the steepening of the sides and the head of the valley, produces a large amphitheater-like form that closely resembles a teacup cut vertically in half. *Cirque* is the name applied to this form. (See Fig. 10-6.)

A glacially eroded valley is recognized by the U-shape, the cirque at the head, and the grooves, scratches, and polishing due to abrasion. (See Fig. 10-9.) In addition, any small hills or knobs in the valley have been overridden by the glacier, and they are rounded and smoothed by abrasion on the struck side and steepened by plucking on the lee side. (See Fig. 10-10.) Many glaciated valleys are nearly flat with rises or steps where more resistant rocks crop out. (See Figs. 10-11 and 10-12.) A glacier moving down a valley also tends to straighten the valley, because the glacier cannot turn as abruptly as the original river could. This has the effect of removing the spurs or ridges on the insides of curves of the stream valley. The amount of ice in a main-stream valley, whose source is near the crest of the moun-

tain range, is generally much greater than the amount of ice in a tributary-stream valley; hence, the main valley is eroded deeper by the ice than is the tributary. After the ice melts, the tributary stream, in a valley called a *hanging valley,* is higher than the

main stream into which it flows via a waterfall. (See Figs. 10-13 and 10-14.)

Even the high part of a mountain range that is above the level of the glaciers develops characteristic forms. Increased frost action here

Figure 10-8. Oblique aerial photograph of a fiord. A fiord is a U-shaped, glacially eroded valley that is an arm of the sea. Valley glaciers in this fiord are fed by a continental glacier. Evighedfjord, west Greenland. Reproduced with the permission (A.477/73) of the Geodetic Institute, Denmark.

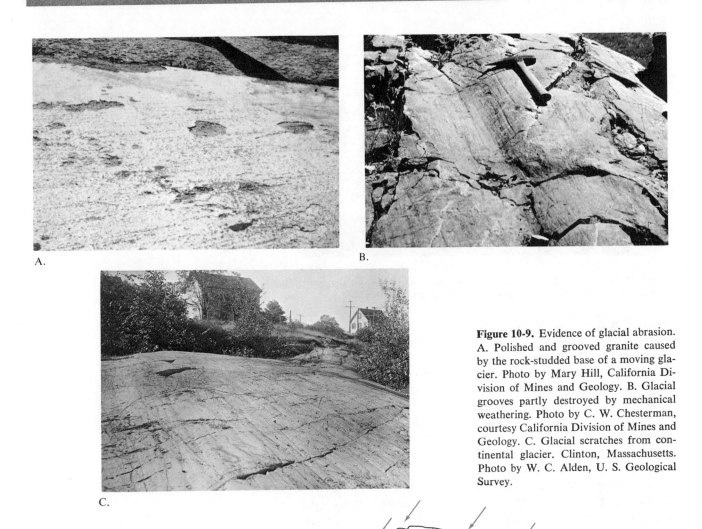

A.

B.

C.

Figure 10-9. Evidence of glacial abrasion. A. Polished and grooved granite caused by the rock-studded base of a moving glacier. Photo by Mary Hill, California Division of Mines and Geology. B. Glacial grooves partly destroyed by mechanical weathering. Photo by C. W. Chesterman, courtesy California Division of Mines and Geology. C. Glacial scratches from continental glacier. Clinton, Massachusetts. Photo by W. C. Alden, U. S. Geological Survey.

Figure 10-10. Small hills of resistant rock are smoothed by abrasion on the struck side and steepened by plucking on the lee side. The arrows show the direction of the main force of the glacier that moved from right to left. The jointing in the rock is indicated. From F. E. Matthes, U.S. Geological Survey, *Professional Paper* 160, 1930.

Figure 10-11. Origin of steps in a glacial valley. AA is the profile of the preglacial valley and BB the post-glacial profile. Closely jointed rocks at c and c_1 are readily plucked by the glacier, but sparsely jointed rocks as at d, d_1, d_2 are reduced by abrasion. From F. E. Matthes, U.S. Geological Survey, *Professional Paper* 160, 1930.

Figure 10-12. A glacial valley showing rock steps with lakes on the steps. Photo by S. A. Davis, courtesy California Division of Mines and Geology.

produces narrow, jagged ridges extending up to the peaks. The peaks develop pointed, pyramidal shapes, called *horns,* that are largely due to headward erosion by cirque development. (See Figs. 10-2 and 10-12.)

Glaciers also deposit the debris that they carry. *Drift* is the name used for all types of glacial deposits, and *till* is used for ice-deposited sediments. The material in the glacier is carried by the ice to the snout, where it is deposited to form the terminal moraine which marks the point of furthest advance of the glacier. (See Fig. 10-15.) Because the glacier carries a great variety of sizes of fragments, moraines are unsorted; huge boulders are mixed with all sizes of fragments down to silt and finer. (See Fig. 10-16.) The pebbles are generally faceted and striated by abrasion. Because the snout of a glacier is curved, the terminal moraines are characteristically curved. They continue along the sides of a glacier as lateral moraines. The lateral moraines develop, in much the same manner as terminal moraines, as a result of melting of the slow-moving ice at the valley sides. (See Figs. 10-2 and 10-15.) Many of the rocks deposited by glaciers are scratched or striated from abrasion during transport. The melt-water leaving the snout of a glacier also transports and deposits glacial material, forming outwash deposits that in some cases form thick sheets. The melt-water from an active glacier is colored white due to fine material, called *rock flour,* from the abrasive

A.

B.

C.

D.

Figure 10-13. The development of Yosemite Valley illustrates both river and glacial erosion. A. The broad valley stage prior to the uplift of the Sierra Nevada. B. After uplift, the rivers downcut about 700 feet. C. After the second uplift, the Merced River cut a canyon about 1300 feet below A. D. Yosemite Valley just after the glacier retreated. The lake was dammed by glacial moraine, and the filled lake is the present valley floor. B, Mount Broderick; BV, Bridalveil Creek; C, Clouds Rest; CC, Cascade Cliffs; CR, Cathedral Rocks; EC, El Capitan; EP, Eagle Peak; HD, Half Dome; LC, Liberty Cap; LY, Little Yosemite Valley; MR, Merced River; MW, Mount Watkins; ND, North Dome; R, Royal Arches; SD, Sentinel Dome; TC, Tenaya Creek; W, Washington Column; YC, Yosemite Creek. From F. E. Matthes, U.S. Geological Survey, *Professional Paper* 160, 1930.

erosion of the glacier. This material resembles clay, but microscopic examination shows that it is composed of fine mineral fragments, not clay. It is the source of the varved "clays" to be discussed later. When a glacier melts, its transported load is dropped wherever it happens to be and forms ground moraine. Upon melting, blocks of ice buried in ground moraine leave depressions, which may become small ponds. (See Fig. 10-17.)

Continental Glaciers

The geologic work of continental glaciers is very similar to that of mountain glaciers. Their erosive

Figure 10-14. Yosemite Valley. Figure 10-13 shows the steps in the development. Note that the present river is very small for the size of the valley. Photo by S. A. Davis, courtesy of California Division of Mines and Geology.

Figure 10-15. Idealized sketch of a glacier leaving successive moraines as it melts back. Both end moraines and lateral moraines are shown. From F. E. Matthes, U.S. Geological Survey, *Professional Paper* 160, 1930.

Figure 10-16. Glacial moraine containing many kinds of rock in all sizes from boulders to silt. Photo by Mary Hill, courtesy California Division of Mines and Geology.

Figure 10-17. Pond created by the melting of a block of ice buried in moraine. Note the variety of sizes of rocks in the moraine. Baird Glacier, southeastern Alaska. Photo by A. F. Buddington, U.S. Geological Survey.

Figure 10-18. This sketch shows many glacial features and the names that have been applied to them.

Figure 10-19. Aerial view of the 6288-foot summit of Mount Washington, New Hampshire, the highest point in the northeast. The continental glaciers overrode and smoothed all of the rounded mountains in the view. Photo by Dick Smith, New Hampshire Division of Parks.

Figure 10-20. Glacially eroded area near Hidden Lake, Northwest Territories, Canada. Continental glaciers have removed the soil and caused many small depressions occupied by lakes. The bedrock structures can be seen. The light-colored areas are granite bodies, and the white lines are pegmatite dikes; both features are discordant to the foliation of the bedrock. Oblique aerial photo from Department of Energy, Mines and Resources, Ottawa, Canadian Government Copyright (A5032-62.R).

effects are much less and are generally limited to rounding the topography and scraping off much of the soil and overburden. (See Figs. 10-19 and 10-20.) Their deposits are like those of mountain glaciers, but are more extensive. One type of deposit that helped lead to the recognition of continental glaciation is the *erratic boulder*. (See Fig. 10-21.) This term is applied to boulders, some of which weigh up to 18,000 tons, that are foreign to their present locations. In some areas, occurrences of distinctive erratic boulders can be traced back to their origin, thus indicating the path of the glacier.

Continental glaciers occupied parts of North America and Europe in relatively recent geologic time, and they occupy Greenland and Antarctica now. (See Fig. 10-22.) During the recent glacial period, there were four main times of glacier development in North America and at least five in Europe. In North America, there were several centers of ice accumulation in Canada. Glaciers moved out from these centers in all directions. Thus, although the ice moved generally southward in the United States, it moved northward in northern Canada. In no sense did the last glacial age result because ice from near the north pole moved south. Recognition of four advances depends largely on field observations. In some places fresh, unweathered moraines overlie weathered moraines which in turn overlie more deeply weathered moraines. Tracing such relationships led to the recognition of four main stages with the extents shown on the map in Fig. 10-23.

Moraines, however, are not the only types of deposits left by continental glaciers. In some places, blocks of ice were buried in the mo-

Figure 10-21. Erratic boulders dropped by a glacier when it melted. Photo by Mary Hill, courtesy California Division of Mines and Geology.

raine and when the ice melted, depressions formed. These depressions are called *kettles,* and lakes form in many of them (Fig. 10-17).

Eskers are long, sinuous ridges, generally 10–50 feet high, that run for fractions of a mile to many miles. They are believed to be the filling of melt-water channels that ran under the glacial ice (Fig. 10-24).

Drumlins are streamlined hills 25–200 feet high and several hundred feet long. Their origin is not known, but the streamlined shape suggests that they formed under the moving glacial ice. They are not symmetrical, and the blunt end is in the direction of the origin of the ice. (See Fig. 10-25.)

The most recent continental glaciation of Pleistocene age consisted of four major advances and began about three million years ago. This date is estimated from the degree of weathering of deposits, from the amount of soil produced between glacial advances, from the study of fossils, and from other information obtained by a variety of methods. The last glacial advance began about 30,000 years ago, reached its maximum advance about 19,000 years ago, and retreated about 10,000 years ago. Another minor advance occurred between 1550 and 1850 and is some-times called the Little Ice Age. These dates are fairly accurately known from carbon-isotope studies. This method is based on the fact that organic carbon has a fixed ratio of carbon isotopes when formed, but this ratio changes after the death of the organism due to radioactive decay of one of the carbon isotopes (carbon 14). The method is not suitable for material more than about 40,000 years old. Radioactive dating methods are discussed elsewhere.

Before radiocarbon was used, other methods, particularly the study of varved clays, were used. This method is much like the study of tree rings, as the varves are assumed to be yearly deposits in glacial lakes. The dark part of each varve is deposited during the summer, and the light part in the winter. (See Fig. 10-26.) Thus the thickness of varves records the climatic conditions, and the sequence of relative thicknesses is correlated from lake to lake until the life span of the glacier is covered; then the total number of varves is counted. Apparently due to error in correlation, this method gave too high an age estimate.

Figure 10-22. The edge of a continental glacier where it meets the ocean. Inglefield-land, northwest Greenland. Reproduced with the permission (A.477/73) of the Geodetic Institute, Denmark.

Figure 10-23. Maximum extent of Pleistocene ice sheets. Glaciers also formed in the mountains in both eastern and western North America.

Figure 10-24. Esker. The long, somewhat curving ridge is an esker and probably originated as a meltwater channel under the continental glacier. Oblique aerial photo from Department of Energy, Mines and Resources, Ottawa, Canadian Government Copyright (A2711-94).

Cause of Glaciation

To discover the cause of glacial periods, we must first see under what conditions glaciers form. Average temperatures must be lower than at present, but not greatly lower; and precipitation must be high. These two conditions will cause the accumulation of snow, which will form glaciers. The temperature must be low enough to insure that precipitation will be in the form of snow, but if it is too low, it will inhibit precipitation. This latter point is illustrated by the lack of glacier formation, because of low precipitation, in the many very cold regions today. Heavy precipitation during at least parts of the glacial age is suggested by the huge lakes that formed in areas south of the glaciers, where it was too warm for snow to accumulate. On the other hand, a humid climate without much evaporation may also have been involved in the formation of these lakes. These lakes formed in the basin areas of western United States, and the present Great Salt Lake is but a small remnant of Lake Bonneville, which covered much of northwestern Utah. Other lakes covered large areas in Nevada, and some of the present saline lakes, such as Carson Sink, are remnants. (See Fig. 10-27.) These lakes are saline because, without outlets to the sea, dissolved material builds up in them while evaporation keeps their levels fairly constant.

A number of theories of glacier origin have been proposed. A successful theory must explain the change in climate just discussed and must account for the four separate advances of the recent ice age as well as for other periods of glaciation at different places on the earth. The other periods of glaciation occurred about 275 and 600 million years ago. The younger of these two, and the better known, occurred in Africa, India, and Australia—far from the present poles. The main theories are as follows:

1. Changes in the amount of energy received from the sun due to changes either in output by the sun, or in the amount of energy that penetrates our atmosphere. There is no evidence to suggest any changes in output by the sun of the type necessary to support this theory. Dust in the atmosphere, perhaps from volcanoes, could cool the earth, but whether this effect is enough to cause glacier development is debatable. Decrease in the amount of carbon dioxide in the atmosphere would cool the earth, as this gas allows radiant energy from the sun to pass through, but prevents radiation from the earth from leaving. Thus the atmosphere acts much like a greenhouse; however, it is doubtful if the necessary changes in the amount of carbon dioxide are possible.

2. Movement of the continents relative to each other (continental drift). This is discussed in Chapter 16.

3. Changes in the location of the earth's rotational poles. This could

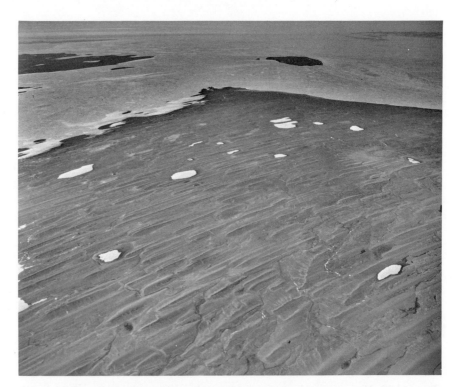

Figure 10-25. Aerial view of drumlins, Stefansson Island, Northwest Territories. The ice flowed toward the upper right. Photo from Geological Survey of Canada.

Figure 10-26. Varves or yearly layers. Photo shows varves in the Green River Formation, Colorado. Diameter of area photographed is 2 inches. Photo by W. H. Bradley, U.S. Geological Survey.

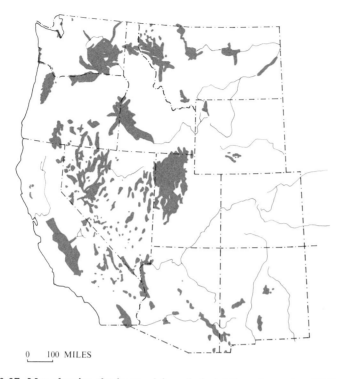

0   100 MILES

Figure 10-27. Map showing the known lakes of glacial times in western United States. From J. H. Feth, U.S. Geological Survey, *Professional Paper* 425B, 1961.

change the climate of some areas by moving them toward the cool poles. The regular periodic movements of the poles, the precession and other movements, which are mentioned later, are well known, but are not great enough to cause glaciers; moreover, the known glacial periods do not coincide with these movements. However, other, larger movements could have occurred in the past, although no reasonable cause for such movements has been suggested yet.

4. Changes in the circulation of the oceans. This is a promising hypothesis that seems adequate for the recent glacial periods. The hypothesis begins with an ice-free Arctic Ocean. With the present pole posi-

tions and our present atmospheric circulation, the ice-free Arctic Ocean would cause the heavy precipitation necessary to initiate a glacial advance. By the middle of the glacial advance the Arctic Ocean would freeze, cutting off much of the supply of moisture. Precipitation would continue until the North Atlantic became too cold to provide enough moisture. Melting would then begin and continue until the Arctic Ocean was again ice-free. The stage is now set for the cycle to begin again.

This hypothesis can account for glacial periods only when the poles are located as they are now. If both poles were over open ocean, then the atmospheric circulation could not develop glacial periods. Thus, to

initiate a glacial period, the poles would have to shift. Thus, this theory accounts for the four glacial advances of the recent glacial period; it can also explain the irregular periods of older glaciation. Recent studies, however, suggest that the present Arctic ice may be much older than the last glacial advance, a finding which casts doubt on this theory.

It should be obvious from this discussion that the climatic conditions that cause glaciation also had far-reaching effects, even far from the areas of actual glacial accumulation. It is estimated that sea level was lowered about 500 feet during the glacial advances, and that if the present-day glaciers melt, sea level will rise between 100 and 150 feet.

Summary

A glacier is a mass of moving ice, which forms as a result of accumulation of more snow than can melt. The thick snow recrystallizes under pressure to ice. When enough ice is formed, it flows downhill as though it were a very viscous fluid.

Mountain or *alpine glaciers* develop in previously formed stream valleys, whose relative deepness makes them accumulation sites for snow.

The snout of the glacier advances or retreats in response to long-term climate changes in snowfall and temperature. The glacial ice is always moving down-valley, even though the snout may be retreating because of melting.

Crevasses, or large cracks, are formed by different rates of flow in the brittle upper portion of the glacier.

Glaciers erode by two methods. *Plucking* is the process in which melt-water flows into joints in bedrock, later freezes to the main ice, which, on advancing, pulls out the loosened block of bedrock. *Abrasion* is performed by the transported blocks of rock grinding, scratching and polishing the sides and bottom of the valley.

Glaciated valleys have U-shaped cross-sections rather than the stream valley's V shape. The sides are steepened; the valley is flattened and straightened.

A *cirque* is the large amphitheater-shaped form cut at the head of a glaciated valley.

A *hanging valley* is the glaciated valley of a tributary stream. The glacier erodes the main valley deeper, steepening its sides, and the water flows from the hanging valley into the main stream via a waterfall.

A *horn* is the pointed pyramidal peak formed above glacier level by headward development of adjacent cirques.

Drift is the name for all types of glacial deposits.

Till is the term used for ice-deposited sediments.

Moraines are unsorted deposits of rock material carried by the glacier. *Terminal moraines* are characteristically curved deposits at the snout. *Lateral moraines* are deposits along the glacier's sides. *Ground moraine* is the deposit left when a glacier melts and drops the load it is transporting.

Rock flour, composed of fine mineral fragments, colors melt-water from an active glacier white; it forms varved clay deposits.

Continental or *icecap glaciers* cover large areas. In relatively recent geologic time there were four advances in parts of North America and at least five in Europe. Continental glaciers occupy Greenland and Antarctica now.

Erratic boulders are boulders foreign to their present locations that were glacially transported.

Kettles are depressions formed by the melting of blocks of ice buried in moraines.

Eskers are long, sinuous ridges, believed to be the filling of melt-water channels that ran under glaciers.

Drumlins are blunt-ended, streamlined hills composed of till.

The Pleistocene glaciation began about three million years ago. The last glacial advance began about 30,000 years ago; maximum, 19,000 years ago; retreat, 10,000 years ago. The Little Ice Age was a minor advance between 1550 and 1850.

Dating of glaciation is by several methods. More recent than 40,000 years is dated by radioactive decay of carbon 14. Before radiocarbon was used, varved clays (yearly layers deposited in glacial lakes, dark in summer and light in winter) were counted and correlated. Other methods are study of weathering of deposits, soil produced between advances, fossil study, and other means.

Glaciers form under special conditions. The temperature must be low enough to insure that the precipitation is snow, but not so low as to inhibit precipitation. The precipitation must also be heavy. Where it is too warm for snow to accumulate, the heavy precipitation may form huge lakes such as Lake Bonneville in Utah, whose remnant is Great Salt Lake, and lakes in Nevada, one of whose remnants is Carson Sink.

Older glaciation occurred 275 million years ago in Africa, India, and Australia, and also 600 million years ago.

Theories of causes of glacier formation are:

1. Changes in amount of energy received from the sun, due to changes either in output by the sun or in the amount that penetrates our atmosphere.
2. Continental drift.
3. Changes in location of the earth's rotational poles.
4. Changes in circulation of the oceans.

Melting and forming of glaciers causes changes in sea level.

Questions

10-1. What is a glacier?

10-2. The two main types of glaciers are _____ and _____.

10-3. What changes does a mountain glacier produce in its valley?

10-4. How does a present-day glacier show climatic changes?

10-5. Why do crevasses form on the surface of glaciers?

10-6. Describe the processes of glacial erosion.

10-7. What are the erosional effects of continental glaciers?

10-8. What is a glacial erratic?

10-9. How many glacial stages are recognized in the Pleistocene?

10-10. Which way did the ice move in northern Canada? What is the evidence?

10-11. What type of climate is most favorable for the development of glaciers?

10-12. What evidence do we have as to the climate during glacial times in southwestern United States?

10-13. List some suggested causes of continental glaciation.

10-14. Outline the new theory of glacier origin that depends on changes in ocean currents between the Arctic and the Atlantic oceans.

10-15. Which of the three agents of transportation — wind, water, or ice — is able to transport the largest boulders; which is most limited as to the largest size it can transport?

10-16. How do mechanical weathering and glacial erosion work together?

10-17. What is the evidence for continental glaciation?

10-18. How do glacially deposited rocks differ from those found in river deposits?

10-19. How does the fine material in a river with active glaciers at its head differ from the material in a river in a non-glacial area?

10-20. Describe several types of glacial deposits.

Supplementary Reading

Davis, W. M., *Geographical Essays*, especially "The Sculpture of Mountains by Glaciers," "Glacial Erosion in France, Switzerland, and Norway," and "The Outline of Cape Cod." New York: Dover Publications, Inc. (1909), 1954, 777 pp. (paperback).

Damuth, J. E. and R. W. Fairbridge, "Equatorial Atlantic Deep-Sea Arkosic Sands and Ice-Age Aridity in Tropical South America," *Bulletin* Geological Society of America, (January, 1970), Vol. 81, No. 1, pp. 189-206. Describes climatic changes far from the areas of glaciation.

Denton, G. II. and S. C. Potter, "Neoglaciation," *Scientific American* (June, 1970), Vol. 222, No. 6, pp. 101-110.

Donn, W. L., "Causes of the Ice Ages," *Sky and Telescope* (April, 1967), Vol. 33, No. 4, pp. 221-225.

Dyson, J. L., *The World of Ice*. New York: Alfred A. Knopf, Inc., 1962, 292 pp.

Flint, R. F., *Glacial and Quaternary Geology*. New York: John Wiley & Sons, Inc., 1971, 892 pp.

Hough, J. L., *Geology of the Great Lakes*. Urbana: University of Illinois Press, 1958, 313 pp.

Thornbury, W. D., *Principles of Geomorphology* (2nd ed.), Chapters 14, 15, 16, pp. 345-444. New York: John Wiley & Sons, Inc., 1969, 594 pp.

Shorelines

11

Waves breaking on the shore are also agents of erosion, transportation, and deposition. Breaking waves have a great deal of energy, and anyone who watches them cannot fail to be impressed by this. (See Fig. 11-1.) It is so obvious that they are agents of erosion that for many years they were considered more important than the much less obvious, less spectacular rivers. Before discussing shoreline processes, we will consider the origin of the various movements of ocean waters.

Circulation of the Oceans

The movement of the surface waters of the oceans is caused by wind and the earth's rotation. The wind blowing over the water exerts a drag and sets the water in motion. In this process, first ripples and then waves are generated. The circulation developed in this manner is shallow, at the most only a few hundred feet

deep; and the water moves slowly, generally between a fraction of a mile per hour and a few miles per hour. The resulting circulation is rotational because the earth's rotation and the continents deflect the moving water. The earth's rotation causes a deflection of the water's forward motion to the right in the northern hemisphere, and to the left in the southern hemisphere. In the southern part of the southern hemisphere, where no continents intervene, the circulation is latitudinal. (See Fig. 11-2.)

The deep circulation is caused largely by density differences in the ocean water. The cold, heavy Arctic, and especially the Antarctic, waters flow near the bottom toward the equator and in some cases to points beyond. In contrast, the waters of the warm Mediterranean, which have a high salinity because of evaporation, flow into the Atlantic at middle depths.

Much more study is needed before the circulation of the oceans can be

described in detail, but the circulation is much less important geologically than waves. Oceanic circulation is, however, a very important means of heat transfer, greatly affecting the climates of the world and so affecting geologic processes.

Tides

Tides are associated with gravitational attraction, mainly due to the moon. Newton's law of gravity shows that there is a mutual attractive force between two bodies. The oceans are free to move and so are deformed by this force. Thus, the water on the side of the earth toward the moon moves a little closer to the moon, causing a high tide at that point. It is not quite so apparent that a high tide forms at the point on the opposite side of the earth at the same time. The reason for the two tidal bulges is that centrifugal forces caused by the earth's rotation are

Barrier beaches, North Carolina. Cape Hatteras juts farthest into the Atlantic Ocean. Cape Lookout is near the bottom of the picture. Photo from NASA.

Figure 11-1. Breaking waves are agents of erosion. Photo from California Division of Beaches and Parks.

Figure 11-2. Circulation of surface waters of the oceans.

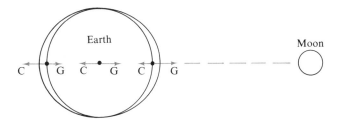

Figure 11-3. Origin of tides. On the side of the earth toward the moon, the gravitational attraction (G) between the moon and the earth is greater than the centrifugal (C) force. On the other side of the earth, the centrifugal force is greater. At the earth's center, the two forces balance. The tidal bulge is greatly exaggerated, and the earth-moon distance is greatly shortened.

also involved in producing tides. In an equilibrium system such as the earth-moon system, the gravitational force between the two bodies is exactly balanced by the centrifugal force. This balance is exact at the centers of the two bodies, but at the earth's surface the two forces are not equal. On the side near the moon, the gravitational force toward the moon is greater; and on the opposite side, the centrifugal force acting in the opposite direction is greater. (See Fig. 11-3.) Because the moon orbits the earth, it rises about 51 minutes later each day. Thus, successive high tides occur about 12 hours and 25 minutes apart. The sun also affects tides because of its gravitational attraction, and all of these effects change somewhat with the season. This is the origin of tides, but this theoretical approach cannot be used to predict either the height or the time of actual tides at a coastline because tides are greatly affected by the shape of the coastline and the configuration of the ocean bottom.

Tidal currents may move sediments at some places; but, except at these places, tides probably are not important agents of transportation or erosion. A notable exception occurs where tides move through narrow inlets, constantly scouring these entrances to many good harbors that would otherwise be blocked. (See Fig. 11-4.) The vertical movement of the tides is in the range of a few feet to a few tens of feet, and this constantly changing datum complicates the near-shore processes.

Do not confuse tidal currents with the misnamed *tidal waves* (seismic sea waves) that are caused by earthquakes in the oceans, discussed in Chapter 13.

The solid rock earth is, like the oceans, also deformed by the tidal forces. Although difficult to measure, the movement is three to six inches. Other aspects of tides are discussed in Chapter 18.

Waves

Waves can be described by their height and wavelength. These terms are defined in Fig. 11-5. The wavelength and period are related to the velocity at which the wave travels by the equation,

Wavelength = period × velocity

The velocity, however, depends on the wavelength; and longer wavelength waves (longer period waves) travel at higher velocities. Period is much easier to measure than wavelength, so most relations are expressed in terms of period. In deep water, the velocity in miles per hour is about 3.5 times the period in seconds. The height of a wave has almost no effect on its period or velocity.

Waves, like surface circulation, are caused by wind. Wind blowing over water causes ripples to form; and, once formed, the ripples provide surfaces on which the wind can act more efficiently. The buildup of ripples into waves is a complex process. Turbulence in the wind forms ripples of differing size and direction that combine to form other waves. The energy of the shorter wavelength waves is largely absorbed by longer waves, increasing the height of the longer waves. The height of the waves produced depends on the strength of the wind, how long the wind blows, and how large an area the wind blows over. Most large waves originate in storms; but, once formed, the waves can travel great distances with little loss of energy.

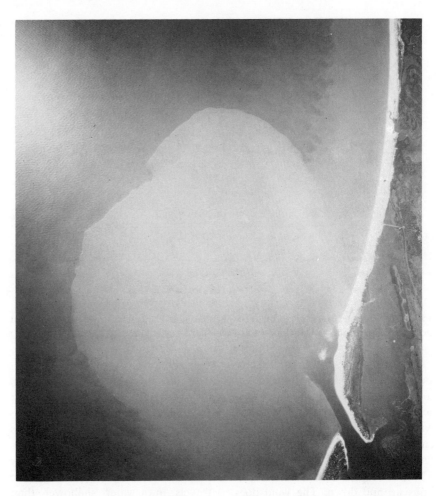

Figure 11-4. Vertical aerial photograph of inlet to Apalachicola Bay, Florida. The outgoing tide has carried out into the ocean the sediment brought into the bay by the Apalachicola River. At the time the photo was taken, clear water was moving through the inlet and the cloudy water forms a turbidity current in the ocean. In the upper right of the photo, clouds of fine sediment are being moved by wave action.

Thus, a storm's energy may be transmitted by waves to a distant shore.

Typical waves in a storm area have periods near 10 seconds. As the waves move out of the storm area, their periods (and wavelengths) increase. At distances of a few thousand miles from the area of their formation, waves have periods between 15 and 20 seconds and may travel faster than the winds that generated them.

Figure 11-5. Characteristics of a wave.

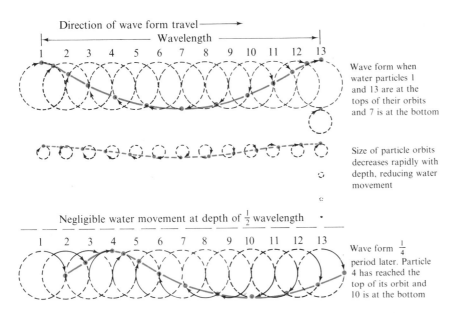

Direction of wave form travel ⟶

Wavelength

1 2 3 4 5 6 7 8 9 10 11 12 13

Wave form when water particles 1 and 13 are at the tops of their orbits and 7 is at the bottom

Size of particle orbits decreases rapidly with depth, reducing water movement

Negligible water movement at depth of $\frac{1}{2}$ wavelength

1 2 3 4 5 6 7 8 9 10 11 12 13

Wave form $\frac{1}{4}$ period later. Particle 4 has reached the top of its orbit and 10 is at the bottom

Figure 11-6. Water particles move through a circular orbit as a wave passes. Almost no water motion occurs below a depth of one-half wavelength.

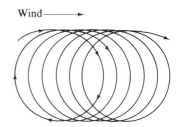

Wind ⟶

Figure 11-7. The wind moves the water slowly forward, causing the surface circulation of the oceans. The circles show the actual movement of water particles.

In deep water, waves are surface shapes; and though the shape moves forward with the wind, the water essentially does not. An individual water particle in a wave moves in a circular path and returns to almost its original position. (See Fig. 11-6.) Thus, in deep water a floating object bobs up and down. The wind does drag the water slowly forward, causing the slow surface circulation (see Fig. 11-7); but in the case of waves, it is the form, not the water, that moves. The motion is much like a wind wave through tall grass or wheat.

In shallow water the motion is very different. (See Fig. 11-8.) Because waves form at the surface, their energy decreases downward. At a depth of one-half the deep water wavelength, the water motion is negligible. As long as the water is this deep, a wave is unaffected by the depth of the water. If a wave travels into water shallower than this, however, it feels the bottom and changes. The wave travels slower, and some of its energy is used in moving small grains on the bottom back and forth. The height of a wave increases as the water becomes shallower; and on a sloping shore, the front of a wave is slowed

more than the rear. These effects cause waves to break, creating surf. In the surf zone, the breaking waves run up on the shore and are active agents of erosion and transportation. (See Fig. 11-9.) Their return to the ocean creates the "undertow" feared by many swimmers. On the landward side of breaking waves, the surface water moves on shore and the bottom water moves outward, thus tending to upset a bather. At times the returning water is concentrated at a few places, creating what are called *rips,* or *rip currents.* (See Fig. 11-10.)

Along many coasts, after breaking, the water moving onshore forms new waves that in turn break closer to the shore. This process may be repeated several times, especially if the waves are large and the bottom slopes gently.

Wave erosion, then, is concentrated in shallow water. Waves generally break at depths of about one and one-half times their height. Waves over 20 feet high are rare, although waves up to 100 feet high have been reported. At depths less than 30 feet, therefore, the majority of wave energy is expended and most of the geologic work of waves, or surf, is done. Because waves first begin to act on the bottom at about one-half their wavelengths, however,

Wavelength

$\frac{1}{2}$ wavelength

No water motion on bottom

Bottom motion increases

Breaking waves

Higher, steeper front, shorter wavelength

Water runs onto beach

Figure 11-8. The changes that occur when a wave moves onto the shore.

some wave erosion can occur as deep as 1000 feet. Waves with 10-second periods have wavelengths of about 500 feet, and 20-second waves have about 2000 foot wavelengths.

Another effect of shallow water on waves is to cause their bending, or refraction. When a wave feels the bottom, it slows. If parts of the wave do not feel the bottom, these parts are not slowed. In this case, because different parts of the wave are traveling at different speeds, the wave bends or refracts. The effect of refraction is to concentrate the waves on headlands, as shown in Fig. 11-11.

Wave Erosion

Waves erode mainly by hydraulic action and abrasion, and to a much smaller extent by solution.

Hydraulic action is the impact and pressure caused by waves. Storm waves, in particular, are able to move huge blocks of bedrock. Breaking waves can cause great

Figure 11-9. Aerial view of breaking waves and surf zone. Photo from California Division of Beaches and Parks.

pressure in cracks by the water's force alone or by compressing air in cracks. A breakwater weighing 2,600 tons has been moved in this manner. Windows in lighthouses hundreds of feet above sea level have been broken by rocks thrown up by storm waves. The effects of storm waves in moving large fragments may be likened to river floods, during which the largest fragments are moved. Newspapers occasionally show pictures of large and small boats thrown long distances inland in this way. Waves breaking on a coastline very commonly develop cliffs by undercutting the bedrock. (See Fig. 11-12.)

Abrasion is the grinding done by material moved by the waves. It is especially effective in the surf zone, but can occur, though only to a much smaller extent, in the entire

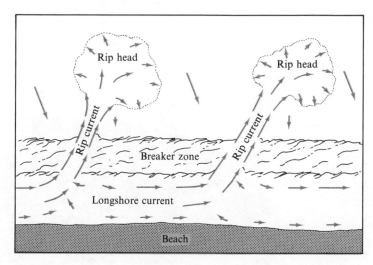

Figure 11-10. Rip currents are formed by the return of the water that is moved on shore by the waves. The location of the rip currents is variable. They can carry an unwary swimmer into deep water.

Figure 11-11. Wave refraction along the California coast. Note the changes in wavelength that occur in the bay on the right. Photo from U.S. Department of Agriculture.

area where the waves feel bottom and where some of their energy is dissipated on the bottom. This action rounds and reduces the size of the pebbles. Waves running up on the shore carry pebbles and sand grains up with them and then back as they return to the sea. This rolling, churning action is very similar to the movement of the bed load of a river. Acting in the surf zone, such abrasion cuts a smooth surface, generally at a depth of 30 feet or less. The surface is called a surf-cut bench. (See Fig. 11-12.) The presence of such a bench may cause the waves to break far from the shore and so limit the effect of the waves on the actual shore.

Wave erosion is less effective than generally thought. It is estimated

that only about one per cent of the sediments in the oceans came directly from wave erosion. Most beaches are formed of material mov-

ing onshore. At some places wave erosion is very active, and rates as high as 30 feet per year of recession are known. Most wave erosion occurs during large storms.

Wave Transportation

The movement of sediment by waves is a selective process that effectively sorts the sediment by size, especially in the near-shore surf zone. Only sand and coarser material are found in the surf zone, and the main movement of this material is along the shore. In this active zone the waves run up on the beach, carrying sand and coarser material as both bed load and suspended load. The slope of this area is such that the receding wave, aided by gravity, carries this material back with it, even though it loses some of its water by percolation into the beach. Any finer material in the receding wave is carried to deeper water by the "undertow." Thus the sand and coarser materials move back and forth in this zone. This material moves parallel to the beach because refraction of the

Figure 11-12. Surf-cut terrace and undercut cliff near Arecibo, Puerto Rico. Photo by C. A. Kaye, U.S. Geological Survey.

waves generally causes them to move diagonally toward the shore. The waves moving onshore carry the sand forward in this diagonal direction. The receding water, however, moves the sand back down the slope under the influence of gravity. Thus, in each back and forth movement the sand is carried further down the beach. (See Fig. 11-13.) The magnitude of this longshore transportation has been measured with marked sand grains and at places amounts to a few thousand tons (dry weight) per day under normal conditions and several times this much during storms. Another measure of the longshore movement is the effect of barriers sometimes constructed to protect beaches or provide harbors for small boats, as shown in Fig. 11-14.

In the zone where the waves feel bottom, sand moves shoreward when the bottom is disturbed by the waves. The returning water flows offshore with a more constant, slower movement that can only move finer material. Thus, in this zone sand moves shoreward and finer materials may tend to move offshore. (See Figs. 11-4 and 11-15.)

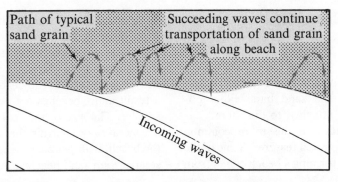

Figure 11-13. Movement of sand along a beach. The incoming waves are not parallel to the shoreline and move sand grains forward perpendicular to the wave. On retreat, the water flows perpendicular to the shore. Thus each wave moves material a little further along the beach.

In deeper water, however, only the long wavelength waves disturb the bottom materials, and the movement is symmetrically back and forth as the wave passes. The bottom generally slopes seaward so gravity may help to move some of the disturbed sediments seaward.

During storms beaches lose sand. This typically occurs during winters. The sand returns to the beach soon after the waves become normal again. On some shores the high storm waves move some of this sand inland beyond the beach. Wind may also move sand inland. In general, storm waves remove sand from the beach and deposit it as a bar at the edge of the surf zone, probably because the higher wave crests of the storm waves need a shallower slope to maintain the equilibrium slope on

which they move the sand back and forth. (See Fig. 11-15.) On some coasts some of the sand is carried to deeper water, and it continues to move seaward down the gentle slope of the shelf.

Rip currents transport fine sand offshore, especially during times when onshore winds as well as waves move surface water onshore.

Sand also moves offshore in submarine canyons. This occurs where sand moving along the shore meets a canyon. For this reason beaches may end abruptly near submarine canyons.

Storm waves or storm winds may stir up the bottom sediments, and the resulting turbid water may move offshore. Turbid water a few tens of miles offshore has been noted from satellites.

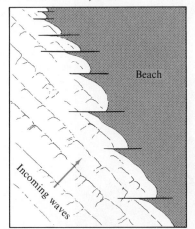

Figure 11-14. The effect of barriers on longshore movement of sand is shown in this oblique view of a beach.

Figure 11-15. Transportation in the near-shore area.

Wave Deposition

Wave deposits are of two main types: beaches and sand bars, either offshore or attached to the shore.

Beaches are the more common wave-deposited features. Some of the material forming a beach comes from wave erosion, and some of it comes from other agents of erosion, especially rivers. In recent years many rivers have been dammed so they deliver little sand to the beaches. As a result many beaches are becoming smaller. The size of the sediment in a general way controls the slope of the beach, with pebble beaches being steeper than sand beaches. As noted before, beaches are in delicate balance and change drastically when wave and current conditions change. (See Fig. 11-16.)

The sediment carried along the shore is deposited as sand or gravel bars at headlands and across bays, as shown in Fig. 11-17. These bars and offshore bars are important in the development of shorelines.

A.

B.

C.

Figure 11-16. Shoreline changes at Little Egg Inlet, New Jersey. A. 1940. B. 1957. C. 1963. These three photographs taken in 1940, 1957, and 1963 show changes in Island Beach south of the inlet. The tidal channels in the salt marsh serve as reference marks for comparison of shoreline changes from year to year. In 1940, a long, narrow, north-trending beach and spit separated the marsh from the ocean. By 1957, the north-trending beach and spit had been replaced by a northwest-trending beach and spit, shorter and slightly west of those of 1940. In addition a new northwest-curving spit about a quarter of a mile wide had been built east of the position of the 1940 spit and extending northward for about a mile. By 1963, the curving spit had grown northwestward an additional 1,000 feet and changed shape to some extent. Photos and caption from C. S. Denny and others, U.S. Geological Survey, *Professional Paper 590,* 1968.

Development of Shorelines

Understanding the development of shorelines is much like understanding the erosional development of rivers or land areas in that an inductive approach is necessary. The great variety of present-day coastlines suggests that shorelines are complex areas; many factors must be considered. Only a few of the better understood cases will be considered here. Any actual shoreline must be studied individually, taking into account such factors as rock type, rock structure, size of waves, direction of waves, number of storms, tidal range, and submarine profile. All coastlines, like all other parts of the crust, are constantly changing, and the type of changes depends partly on the recent changes in regimen. As with rivers, major changes are initiated mainly by changes in elevation.

Two types of sea level changes are possible. The water level itself may rise or fall, or the land may be uplifted or depressed. Sea level may change as a result of uplift or depression of the ocean basins. Evidence that such changes have taken place will be presented in later chapters. Another way that sea level is changed is by the formation and melting of glaciers. This has happened in the recent past, and it is very difficult to separate the changes caused by the recent withdrawal and return of the sea from the normal shoreline processes. Uplifts and depressions of the continents are also common and will also be described later. Changes in sea level from this second cause are more localized, but cannot always be distinguished from

Figure 11-17. Oblique aerial photo of Morro Bay, California, showing a sand bar enclosing a bay. Photo from California Division of Beaches and Parks.

rise or fall of the ocean itself. Many coastlines show evidence of repeated relative changes of land and sea. The melting of the last glacial advance caused a rise in sea level, but the effects of this are not apparent on many coasts.

A suitable classification of coastlines has not yet been devised because of the difficulty of recognizing the many possibilities. Some possibilities can be discussed, but examples are hard to find. One reason for this may be that most coasts are

A. Neutral coastline

B. Submergent coastline. Wave action may develop cliffs on the headlands, and the eroded material may form sand bars across the bays.

C. Emergent coastline

D. Profile of a coastline that has been uplifted (or the sea level fallen) several times, producing a series of wave-cut terraces and cliffs.

Figure 11-18. Possible types of shorelines. A. A neutral coastline. B. The same coastline after the land has sunk or sea level has risen. Note the drowning of the river valleys, producing bays and estuaries. Such irregular coastlines are typical of submergence. C. The same coastline after the land has risen or sea level has sunk. Note the wave-cut cliff produced by breaking waves on the newly exposed land. The rivers form falls when they meet the ocean. Offshore sandbars may form. D. The profile of a coastline that has emerged in several stages similar to the coastline in C. The recent rise of sea level caused by the melting of glaciers has resulted in submergence of most coastlines; however, examples similar to D are common on the West Coast.

Figure 11-19. A deeply embayed submergent or drowned coastline. Marlborough Sound, New Zealand. Photo from New Zealand National Publicity.

unstable and the relation between land and sea is constantly changing, making classification difficult. The factors involved are probably whether the coast is being eroded or built-up, on the one hand, and whether it is emergent or submergent, on the other.

One way to classify coasts is to determine whether they are emergent or submergent. A submergent coastline would develop if either the land went down or the ocean rose. A submerged coastline should have deep embayments and estuaries caused by the filling of coastal valleys with sea water. (See Figs. 11-18 and 11-19.) This is, of course, not the only way that such a coast can develop. Few coastlines are of this type, although the recent melting of the glaciers should have made almost all coasts submergent.

Some types of emergent coastlines are easily recognized. Rising land or falling water level will bring beaches and surf-cut benches above

sea level. In some cases, several benches are visible above a present

beach. This is common at many places along the California coast where recent faulting and uplift are responsible. (See Figs. 11-18, 11-20, and 11-21.) At some places warping of the wave-cut benches reveals differential uplift.

Assuming stability, it is possible to describe the development of some types of shorelines. An initially irregular coastline with headlands and embayments will be greatly modified by wave action. Refraction of the waves will concentrate the wave erosion on the headlands (see Fig. 11-22), and a wave-cut bench and sea cliffs will develop there. The waves will carry some of the eroded material into the embayments, which, together with river-eroded material, will tend to fill them. Sand bars of various types will also develop across the embayments. Ultimately, a

Figure 11-20. Sea cliffs undergoing active erosion by waves and uplifted surf-cut benches on the skyline. Photo from Oregon State Highway Department.

smooth coast will develop. A coast-line of this type tends to retreat landward.

Shallow, smooth coastlines, such as much of the Atlantic and Gulf coasts, have offshore bars or barrier islands with shallow lagoons behind them. The origin of these islands is a problem. Recent studies have suggested that they may be former beaches above which onshore winds had caused the development of sand dunes such as are common at many of our present beaches. These beaches are believed to have formed during glaciation when sea level was lower and winds may have been more active. Melting of the glaciers would cause sea level to rise and flood the area behind the beach-dune complex. The resulting lagoon may be partially filled by material carried into it by rivers. (See Fig. 11-23.)

Figure 11-21. Wave-cut beaches showing the many former levels of Great Bear Lake, Northwest Territories, Canada. In the right foreground, frost polygons and an old land-slide can be seen. Photo from Department of Energy, Mines and Resouces, Ottawa, Canadian Government Copyright (A4116.37).

Summary

The circulation of the surface waters of the earth is caused by winds and the earth's rotation. The deep circulation is caused by density differences in ocean water.

Tides on the opposite sides of the earth are caused by the gravitational and centrifugal forces of the earth-moon system. Tides are not important agents of transportation or erosion except where tidal currents move through and scour narrow inlets.

For waves, wavelength = period × velocity. In deep water, the velocity in miles per hour is about 3.5 times the period in seconds. Wave height has almost no effect on period or velocity.

Waves are caused by wind. In deep water the wave shape moves forward with the wind, but the individual water particles move in circular paths almost returning to their original positions. At a depth of ½ wavelength, water motion is negligible.

In water shallower than ½ wavelength the wave is slowed. The height of the wave increases and the front of the wave is slowed more than the rear, causing the waves to break, creating surf. Breaking waves are active agents of erosion and transportation.

Rips or *rip currents* are caused when the water returning to the ocean from waves breaking on the beach is concentrated at a few places.

Waves generally break at depths of 1½ times their height. The majority of the geologic work of waves is done at depths less than 30 feet.

The refraction or bending of waves is caused if parts of the wave do not feel bottom and are not slowed; refraction concentrates the waves on headlands.

Waves erode by hydraulic action (impact and pressure of waves) and by abrasion (grinding by material moved by waves). Only about one per cent of ocean sediments comes directly from wave erosion; most wave erosion occurs during large storms.

Surf-cut benches are cut by abrasion in the surf zone, generally at a depth of 30 feet or less. The presence of the bench may cause the waves to break far from the actual shore.

Sand and coarser materials are transported back and forth in the surf zone. Wave refraction generally causes the waves to move diagonally toward shore, so each onshore wave moves the sand along the beach, and the returning water moves it back down the slope. In this way sand moves along the beach.

Between the surf zone and the depth of ½ wavelength, in general, sand moves shoreward and finer material

Figure 11-22. Development of an initially irregular coastline.

Figure 11-23. Development of a gently sloping, smooth coastline. A. Beach-dune complex forms at a lower sea level during glaciation. B. Melting of the glaciers floods the area behind the beach-dune complex forming a lagoon. C. River deposition begins to fill the lagoon.

moves outward.

During storms beaches lose sand. In general, the storm waves deposit the removed sand as a bar at the edge of the surf zone, and it returns to the beach when the waves are normal again.

Wave deposits are of two types: *beaches* and *sand bars*.

Some of the material forming a beach comes from wave erosion, some from rivers and other agents. Pebble beaches are steeper than sand beaches. Beaches change drastically with changing wave and current conditions.

Sand or gravel bars, at headlands and across bays, are formed of sediment carried along the shore and are important in the development of shorelines.

Two types of sea level changes are possible. Water level may rise or fall, or the land be uplifted or depressed. Formation and melting of glaciers affects the rise and fall of sea level.

A *submergent* coastline would result if land went down or ocean rose; it would have deep embayments and estuaries.

An *emergent* coastline would result from rising land or falling water level, and may result in beaches or surf-cut benches being above sea level.

Questions

11-1. What causes the surface circulation of the oceans?

11-2. Explain why high tides occur simultaneously on both sides of the earth.

11-3. What are the relationships among wavelength, period, and velocity of water waves?

11-4. Describe the motion of water in waves in both deep and shallow water.

11-5. Describe the processes and the results of wave erosion.

11-6. Discuss the transportation of both sand and finer material by waves.

11-7. Do sea cliffs always mean uplift of land (or lowering of sea level)?

11-8. How might one distinguish between uplift of the continent and depression of sea level at a given place?

11-9. Discuss whether sea level is a horizon to which absolute movements of the continents can be referred?

Supplementary Reading

Bascom, Willard, "Ocean Waves," *Scientific American* (August, 1959), Vol. 201, No. 2, pp. 74-84. Reprint 828, W. H. Freeman and Co., San Francisco.

Bascom, Willard, "Beaches," *Scientific American* (August, 1960), Vol. 203, No. 2, pp. 80-94.

Bascom, Willard, *Waves and Beaches.* Garden City, New York: Doubleday and Co., Inc., 1964, 267 pp. (paperback).

Carson, Rachel, *The Sea Around Us* (revised edition). New York: Oxford University Press, 1961, 230 pp. (also available in paperback).

Gross, M. G., *Oceanography.* 2nd ed.

Charles E. Merrill Publishing Co., Inc., 1971, 150 pp. (paperback).

Higgins, C. G., "Causes of Relative Sea-level Changes," *American Scientist* (December, 1965), Vol. 53, No. 4, pp. 464-476.

Hoyt, J. H., "Field Guide to Beaches," *ESCP Pamphlet Series*, PS-7. Boston: Houghton Mifflin Co., 1971, 45 pp.

Hoyt, J. H., "Shoreline Processes," *Journal of Geological Education* (January, 1972), Vol. 20, No. 1, pp. 16-22.

Inman, D. L., and B. M. Brush, "The Coastal Challenge," *Science* (July 6, 1973), Vol. 180, No. 4094, pp. 20-32.

King, C. A. M., *Beaches and Coasts.* London: Edward Arnold, 1959, 403 pp.

Kort, V. G., "The Antarctic Ocean," *Scientific American* (September, 1962), Vol. 207, No. 3, pp. 113-128. Reprint 860, W. H. Freeman and Co., San Francisco.

Munk, Walter, "The Circulation of the Oceans," *Scientific American* (September, 1955), Vol. 193, No. 3, pp. 96-104. Reprint 813, W. H. Freeman and Co., San Francisco.

Schubel, J. R., and D. W. Pritchard, "The Estuarine Environment," *Journal of Geological Education*, Part I (March, 1972), Vol. 20, No. 2, pp. 60-68. Part II (September, 1972), Vol. 20, No. 4, pp. 179-188.

Ground Water and
Water Resources
12

All the rivers run into the sea;
Yet the sea is not full:
Unto the place from whence the rivers come,
Thither they return again.

Ecclesiastes 1:7

Ground Water

Ground water is an important source of water, especially in dry areas. In many of the countries of north Africa and the Near East, ancient peoples, hundreds of years before Christ, dug tunnels many miles long to collect and transport ground water. In Europe in Medieval times, a technique of well drilling was invented, and in a few areas the wells flowed naturally.

It was not until the latter part of the seventeenth century that it was shown that the origin of river water is rainfall. When this measurement was undertaken, however, it was discovered that only a fraction of the amount of water that fell as rain on a drainage basin flowed out as the river. The fate of the rest of the precipitation is not hard to imagine.

Some of it evaporates, some is used by plants, and some seeps into the ground. The latter is the origin of ground water.

In the middle 1850's, Henry Darcy, a French engineer, worked out the law of fluid flow through a permeable medium. In the last half of the nineteenth century, a parallel development of ground water geology occurred in Europe and the United States, and the same discoveries were made in some cases in several countries. In the United States in 1877, T. C. Chamberlin wrote the first report on artesian wells (which will be described below). In 1888, J. W. Powell, who was then the Director of the United States Geological Survey, stimulated much work on ground water. He was deeply concerned that the Great Plains might be opened to home-

steading on the same basis as the better watered parts of the country were. He felt that much larger homesteads would be necessary because of the dryness of that area. In the early 1900's, the first quantitative studies of ground water movement were made possible by the development of new instrumentation. A classic study of ground water in the United States was made in 1923 by Oscar Meinzer, who directed the ground water branch of the United States Geological Survey for many years.

Ground water is a very important part of the water resources of many areas. Since World War II the problem of water resources has received much attention, and all aspects, including ground water, are being studied vigorously. Water studies have important legal, social, and eco-

Vertical aerial photograph of an area of sinkholes near Park City, Kentucky. Photo from U. S. Geological Survey.

nomic aspects as well as the mainly scientific aspects considered here.

Movement of Ground Water

The surface below which rocks are saturated with water is called the *water table.* (See Fig. 12-1.) Some water is retained above the water table by the surface tension of water. (See Fig. 12-2.) In general, the water table is a reflection of the surface topography, but is more subdued, that is, has less relief, than the surface topography. (See Fig. 12-3.) Lakes and swamps are areas where the land surface is either below or at the water table. Springs occur where the water table is exposed as on a valley side. The position of the

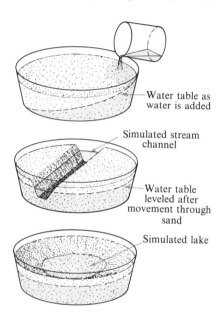

Figure 12-1. An illustration of ground water. The dish is filled with sand, and the water poured in percolates into the sand, filling the pore spaces. A hole in the sand can represent a well, or a slot a stream channel. This experiment works best in a transparent dish with water colored by a few drops of ink.

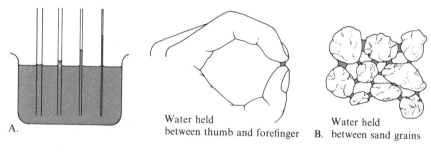

Figure 12-2. Capillary action is caused by surface tension and can cause water to rise above the water table. A. In a very narrow tube, water rises well above the water level. B. In the ground, the narrow spaces between particles take the place of the tubes. For this reason, the top of the water table is a thin zone rather than a sharp line. This is also the way that soil retains its moisture.

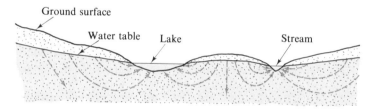

Figure 12-3. The water table has less relief than the topography. The arrows show the movement of the ground water.

Well-sorted sandstone. Large amount of pore space gives high porosity and permeability.

Poorly sorted sandstone. Has much lower porosity and permeability.

Figure 12-5. The sorting of sandstone affects its porosity and permeability.

Figure 12-4. Spring formed by a perched water table created by impervious shale bed. Such springs tend to dry up in late summer.

Figure 12-6. Springs and wells can also occur in fractured rocks such as granite.

water table changes seasonally, which explains why some springs are dry in summer. Springs may form where the water table is *perched*. This situation results where the downward percolation of rain water is stopped by a relatively impermeable rock, such as a shale. Springs of this type may flow only during the wet season. (See Fig. 12-4.)

The movement of ground water is controlled by the physical properties of the rocks. The amount of water that can be stored in the rocks is determined by the amount of pore, or open, space. The availability of the water is determined by the interconnections of the pore space, which is, of course, the permeability (Fig. 12-5). The most common reservoir rock is sandstone, although fractured granite or limestone, as well as many other rock types, can serve equally well. (See Fig. 12-6.)

A very important type of well is the artesian well. The reservoir of an artesian well is confined above and below by impermeable rocks. The reservoir is charged at a place where it is exposed, and the water is forced to move through the reservoir. A well drilled into an artesian reservoir may flow without pumping because the released water seeks its own level—that of the recharge area. However, the resistance to flowage of the water through the reservoir prevents the water from flowing to the same elevation as its recharging area; therefore, flowing wells must be lower in elevation than the recharge area. (See Fig. 12-7.)

Ground water is a very valuable economic commodity. It is recovered from wells for domestic, industrial, and agricultural use. (See Fig. 12-8.) In areas where ground water is used extensively, care must be taken that, on the average, no more water is

Figure 12-7. Artesian water results when the flow of ground water is restricted by impermeable layers.

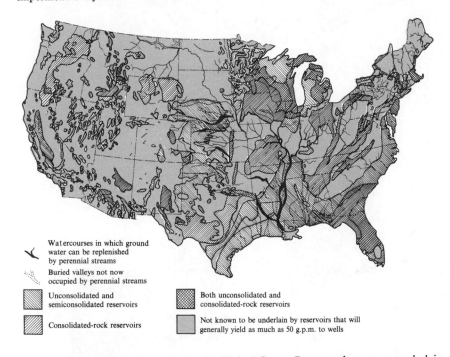

Figure 12-8. Ground-water areas in the United States. Patterns show areas underlain by reservoirs capable of yielding to a single well 50 gallons per minute (g.p.m.) of water containing less than 0.2 per cent dissolved solids (includes some areas where more mineralized water is actually used). From U.S. Department of Agriculture, 1955 Yearbook, *Water.*

Figure 12-9. A cone of depression results when water is pumped too rapidly from a well.

withdrawn in a year than is replaced by natural processes or artificial infiltration. This can be determined by seeing that the water table is not lowered. The conservation of ground water is very important because it moves very slowly and many years may be required to replace hastily pumped water. The average rate of movement of ground water through rocks is only about 50 feet per year, although at some places the movement is much faster.

This slow movement of the water requires another caution. If water is withdrawn from a well by pumping too rapidly, the water is removed from the immediate vicinity of the well faster than it can be replaced. The result is called a *cone of depression* (Fig. 12-9). If cones of depression are developed around several closely spaced wells, the water table can be drastically lowered. Thus, the spacing and the rate of pumping of wells need to be regulated in order to insure the most efficient use of the water. Figure 12-10 shows the changes that have occurred as ground water use has increased on Long Island. Note that as fresh water was removed faster than it was replaced, the heavier salt water from the ocean encroached. At some places the water has been withdrawn so rapidly that hundreds of years may be required to replace it. Such usage is similar to mining, in the sense that a natural resource is used up.

Another effect of the withdrawal of ground water at a more rapid rate than it is replaced is the gradual subsidence of the surface of the ground. Subsidence has also occurred at some places because of the withdrawal of oil from the ground. (See Fig. 12-11.) Up to 20 feet of subsidence caused by withdrawal of ground water has occurred in parts of the Central Valley of California.

Pollution is a problem in some areas where well water is used. Drainage from cesspools can contaminate nearby wells. This danger is always present but is most acute where the ground water flows freely as in wide joints or cavernous limestones. If the polluted water seeps slowly enough through sand or gravel, especially above the water table, it is commonly purified in only a few tens of feet of travel. The purification is accomplished by filtering

A.

B.

Figure 12-10. The encroachment of salt water from the ocean into ground-water reservoirs in Long Island, N.Y. This was caused by removal of ground water faster than it was recharged. From U.S. Geological Survey *Circular* 524, 1966.

Figure 12-11. Flooding at Long Beach, California, caused by subsidence resulting from withdrawal of oil. Photo courtesy of City of Long Beach.

and oxidation. (See Fig. 12-12.) Advantage of this means of purifying water has been taken lately by spraying sewage and industrial waste on the ground surface, especially in wooded areas. In this way the ground water is recharged and the contaminants removed. However, surface disposal or burial both have the danger of contaminating water, especially underground water. In times of flood or heavy precipitation, surface waters can also be polluted. Decomposing garbage in a landfill produces methane, carbon dioxide, ammonia, and hydrogen sulfide. The carbon dioxide combines with water, forming carbonic acid that increases the amount of many materials that can be dissolved in the water. When such water seeps into surface or ground water, it may pollute it. It is possible to avoid pollution from disposal sites by paying attention to the ground water conditions as shown in Fig. 12-13.

One special type of underground disposal is the pumping of fluids into deep rocks. This is a common way to dispose of oil-field brines. The earthquake problems at the Denver Arsenal, described in the next chapter, suggest caution.

Figure 12-12. Pollution of wells. In this example, because of the cavernous limestone, the shallow well is more likely to produce pure water than is the deep well that is further away from the septic tank.

Geologic Work of Ground Water

Much of the geologic work of ground water was covered in the discussion of weathering in Chapter 4. Water is a very good solvent, especially if it contains carbon dioxide; and the role of ground water in solution and cementation has been described. It is not surprising that most ground water contains dissolved material and so is what is termed hard water. This is the origin of the taste of most spring and well water. In some cases ground water contains so much dissolved material that it is not suitable for domestic or industrial use. Some ground water used for irrigation deposits dissolved salts in the soil and so can ruin soil unless enough water is used to flush the salts continually from the soil. Some of the water encountered in drilling is salty, and some of these waters were once sea water that was buried with the sediments when the bed was deposited in the ocean.

Ground water forms caves, but

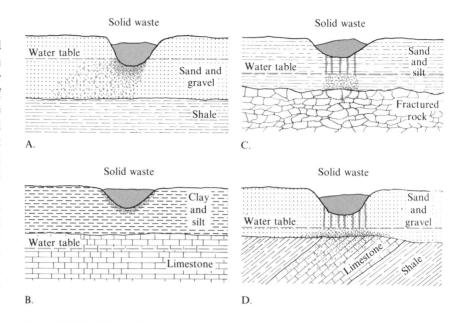

Figure 12-13. Effects of waste disposal on ground water. Only in B is contamination prevented by impermeable rocks. In all the other examples, contaminants from the waste move through permeable rock to the water table. From U. S. Geological Survey, *Circular* 601 F, 1970.

exactly how is not known. They are formed mainly in limestone, but can form in any soluble rock. (See Fig. 12-14.) Calcite is soluble in water, especially in water that has carbon dioxide dissolved in it, as noted in the discussion of chemical weathering. Probably caves form in the zone of seasonal water table fluctuation by solution of limestone along joint planes. Lowering of the water table may make the cave accessible. In times of flood caves may be enlarged by erosion by running water. A process going on in caves (Fig. 12-15) is the deposition of calcite in the form of stalactites, stalagmites, and other features, all of which are caused by the evaporation of carbonate-charged water.

In limestone areas, caves or other channels may carry most of the water. The surface water may sink underground and flow through caves. In such an area there are only a few short surface streams, and they end in closed depressions called *sinks*.

Figure 12-14. Development of caves. The caves begin by solution of limestone in the zone of seasonal water table fluctuation. Lowering of the water table makes the caves accessible. An area of cavernous limestone generally has few, if any, surface streams.

(See Fig. 12-16.) This, as might be expected, produces an uncommon type of surface topography. It is worth noting that only in the case where caves are developed is anything like an underground stream developed; many people have the erroneous impression that all ground water flows in underground streams similar to surface streams. In the same sense, there are no underground lakes, but only reservoir rocks whose pore space is filled with water.

Hot Springs and Geysers

Geysers are periodic discharges of hot ground water, in contrast to hot springs which flow more or less continuously. (See Fig. 12-17.) Studies of the isotopes in hot spring and geyser waters have shown that most, if not all, of it is ordinary ground water that has been heated. This heating is due to deep circulation of the water or to contact with hot igneous rock bodies. (See Fig. 12-18.) Geysers result when water accumulates in underground chambers where it is heated. The pressure of the overlying water causes the boiling point near the bottom to rise. The heating also causes the column of water to expand and spill over near the top. This reduces the pressure, and the superheated water at the bottom flashes into steam and causes the geyser to erupt. (See Fig. 12-19.) This process is repeated more or less periodically in a geyser. All ground water contains dissolved material, and the hot waters contain, in general, more. Thus, hot springs and geysers commonly deposit calcite and other minerals. (See Fig. 12-20.)

Figure 12-15. Stalactites and stalagmites in Carlsbad Caverns, New Mexico. The stalactites form by dripping water that evaporates, depositing its dissolved calcite. Stalagmites build up when drops of water with dissolved calcite fall on the floor of the cave and evaporate. Photo by National Park Service.

Petroleum

Much of the foregoing information on ground water is applicable to the other natural fluids—petroleum and natural gas. These materials originate mainly from organic material in marine sedimentary rocks, although some do form in some lake-deposited sedimentary rocks. The beds that contain the oil-forming organic material are generally shale. It takes millions of years and increased temperature and pressure, due to burial, to form petroleum and natural gas. Petroleum and natural gas move very slowly in the shale source rocks; they are not economically recoverable by drilling until they have migrated to more permeable reservoir rocks. Thus, to have an oil field, it is necessary to have source rocks, reservoir rocks, and a trap, generally an impermeable cover. Because petroleum is lighter than water, it rises to the top of a reservoir. Several types of traps are shown in Fig. 12-21. The same cautions concerning well-spacing and rate of pumping that applied to ground water apply to oil fields.

Figure 12-16. A sink hole caused by solution of limestone by ground water. This sink probably formed by collapse into a cavern. Photo by N. H. Darton, U.S. Geological Survey.

Water Resources

Water is one of the cheapest and most useful and important commodities that we have. Most people take water for granted, but more and more often water shortages in different areas are reported in the newspapers. Are these shortages ominous signs that some regions have reached their limit of development, or are we mismanaging our water resources? Water is a world-wide problem, but only the United States, excluding Alaska and Hawaii, will be discussed here. The same natural problems occur everywhere, and they are aggravated to a greater or lesser extent by the stage of industrial and technological development of the country.

The United States receives an average of 30 inches of rain per year over the entire country. Of this 30 inches, about 21.5 are lost by evaporation or used by plants. The remaining 8.5 inches become the runoff of rivers and the ground water; this is the water available for use by man. (See Fig. 12-22.) This available water amounts to about 1,200,000 million gallons per day, and in 1970 total withdrawals of water were about 370,000 million gallons per day. It thus appears that we have more than three times as much available water as we need. The true situation is not this simple, unfortunately, and several other factors must be considered. They are the natural distribution of water, the need for runoff to carry away refuse, and the water used by plants that are the base of the food chain on which all life depends.

Natural Distribution of Water

Precipitation is not uniform, and neither is the distribution of population. Precipitation ranges from about 85 inches per year in northwestern United States to about 4 inches in the southwest. (See Fig. 12-23.) If individual stations are considered, the range is even greater. These are average figures and may vary from year to year by large amounts, especially in the drier regions. Climate, too, varies; and as shown in Fig. 12-24, the potential or possible evaporation and consumption of water by plants exceeds the rainfall in most of western United States. These are the arid regions where runoff occurs only during infrequent heavy storms. The season at which rain occurs also affects the use of the water. In most of eastern United States, much of the rain falls during the growing season, but in the west, most of the precipitation comes during the winter months. For all of these reasons, some places have sufficient water, a few places have a surplus, and many areas have a deficiency, even though the overall available water is more than four times the amount withdrawn.

In an effort to overcome these problems of natural distribution, dams and other structures have been built to control and divert rivers. Many of these projects have multiple purposes and may provide such benefits as municipal water supplies, flood control, irrigation, electricity, and recreation areas. Deciding which rivers to control, and for which benefit, is always a problem because of the competition for use of the water and use of reservoir sites. Also, any

Figure 12-17. Old Faithful at Yellowstone National Park in eruption. Photo by National Park Service.

reservoir or lake that is formed increases overall evaporation, and thus some water is lost. In arid regions this loss can be high, but recent studies suggest that much of it can be eliminated by placing a film of fatty alcohol on the water surface.

Like precipitation, river flow is seasonal; and during times of heaviest flow, floods can occur and much runoff is lost to the ocean. The maximum flow of a river is, in general, five to ten times the minimum flow. The most regular major river in the United States is the St. Lawrence, whose average maximum flow is less

than twice its minimum flow. This is because its main flow comes from the Great Lakes. (See Table 12-1.) In the same way, dams that create lakes can make the flow of rivers more uniform and so increase the use of the water. At places such dams may eliminate most floods; but if, in an exceptional year, a flood does occur, much damage may result because the previously avoided flood area is now in use.

Use and Consumption of Water

Distinction must be made between

the water *used* and the water *consumed*. Only the water lost to the atmosphere by evaporation or plants is consumed. The rest of the water withdrawn for use is returned either to the ground or to the runoff, and thus is available for other use. In some cases the water returned is unaffected by the use, but in most cases its quality is impaired. The water withdrawn is not the only water used by man; the runoff itself is used for some purposes in most rivers. Much of the water used to grow crops comes from soil moisture and so is not withdrawn. All of these compli-

Table 12-1. Large rivers. Data from U. S. Geological Survey, Water Resources Division, 1961.

A. World

RANK	RIVER	COUNTRY	DRAINAGE AREA (THOUSANDS OF SQUARE MILES)	AVERAGE DISCHARGE AT MOUTH (THOUSANDS OF CUBIC FEET PER SECOND)
1	Amazon	Brazil	2,231	3,000 to 4,000
2	Congo	Zaire	1,550	1,400
3	Yangtze	China (mainland)	750	770
4	Bramaputra	Bangladesh	361	700
5	Ganges	India	409	660
6	Yenisei	USSR	1,000	614
7	Mississippi	USA	1,244	611
8	Lena	USSR	936	547
9	Paraná	Argentina	890	526
10	St. Lawrence	USA & Canada	498	500
26	Danube	Romania	315	218 (Largest river in Europe)

Figure 12-18. Locations of hot springs and geysers in the United States. They are mainly found in the geologically young, active areas. From G. A. Waring, U.S. Geological Survey, *Professional Paper* 492, 1965.

B. United States

RANK	RIVER	LENGTH (MILES)	DRAINAGE AREA (THOUSANDS OF SQUARE MILES)	AVERAGE DISCHARGE AT MOUTH (THOUSANDS OF CUBIC FEET PER SECOND)	DISCHARGE AT GAGING STATION NEAREST MOUTH (THOUSANDS OF CUBIC FEET PER SECOND)	
					AVERAGE	MAXIMUM RECORDED
1	Mississippi	3,892	1,243.7	611[a]	553	2,080
2	Ohio	1,306	203.9	259	257	1,850
3	Columbia	1,214	258.2	256	183	1,240
4	St. Lawrence	—	302.0[b]	241[b]	236	314[c]
5	Missouri	2,714	529.4	69	69	892
6	Tennessee	900	40.6	64	61	460
7	Mobile	758	42.3	58	—	—
8	Red	1,300	91.4[d]	57	51[d]	233
9	Snake	1,038	109.0	49	49	409
10	Arkansas	1,450	160.5	42	41	536

a—About 25% of flow occurs in the Atchafalaya River; b—At international boundary, lat. 45°; c—Maximum monthly discharge; d—Flow of Ouachita River has been added.

cations must be taken into account to understand water needs.

The uses and consumption of the withdrawn water are shown in Figs. 12-25, 12-26, and 12-27. Note that only about one-quarter of the withdrawn water is actually consumed, and that this is a small per cent of the total available water. The reuse of water in public water supplies is well illustrated by communities along a river. An upstream community withdraws its water supply from the river and discharges its wastes at a lower point on the river. The downstream community also draws its water from the river and so reuses some of the water. It is estimated that 60 per cent of the public water supplies withdraw some used water.

This reuse works well only if upstream communities adequately treat their sewage before discharging it into the river. Detergent types have recently been changed so that sewage treatment plants can prevent foaming in downstream water supplies. Unfortunately, only about two-thirds of the population of the United States have sewers, and only about two-thirds of these sewer systems treat the sewage before discharging it. Much of the water withdrawn for industrial use is also returned. In

Figure 12-19. Idealized diagrams showing hot spring and geyser. A hot spring in which the water temperature no place exceeds the boiling point may have almost any configuration. If, however, high temperatures are involved, the configuration must be such that the heat is easily distributed by convection as in the figure shown. A geyser can form if the heat is not distributed by convection. In the figure above, the water near the bottom is heated to near its boiling point. The boiling point is higher there than at the surface because the weight of the water above increases the pressure. The water higher in the geyser system is also heated and so expands and flows out at the top. This reduces the weight of the water on the bottom and, at the reduced pressure on the bottom, boiling occurs. The bottom water flashes into steam, and the expanding steam causes an eruption.

some cases this industrial return, like the domestic return, must be purified either naturally or by man before it can be reused. In other cases the industrial return is only warmer, but even this can drastically alter the ecology of a stream.

The runoff is not wasted water, but is used by man. It is used to produce hydroelectric power, for recreation, for transportation, and, as described above, to carry away wastes. The widespread pollution of rivers and streams suggests that we are nearing maximum usage for waste disposal in many areas, although many improvements in treating waste before discharge into rivers are possible. The magnitude of the waste disposal problem is shown by the fact that the city of St. Louis, on the Mississippi River, produces 200,000 gallons of urine and 400 tons of solid body waste every day, in addition to industrial wastes.

Rivers are used to carry away the excess heat produced in many industrial processes and especially at power plants. (See Fig. 12-28.) It may seem strange that heat energy

Figure 12-20. Terraces built by hot spring deposits at Mammoth Hot Springs, Yellowstone National Park. Photo by National Park Service.

must be disposed of after the discussion of energy resources. The reason is that in any engine using heat energy, not all of the energy is available to do work. This is a well-known law of thermodynamics and is the reason why a perpetual-motion machine cannot be invented. Steam boilers are less than 40 per cent efficient, nuclear burner reactors are about 30 per cent, and breeder re-

actors will be nearly 40 per cent efficient. The rest of the energy must go to the environment as heat. The current nuclear reactors must use water cooling. In other cases, cooling towers can be used so that the heat escapes into the atmosphere. The amount of water needed for cooling will increase. Raising the temperature of a river, lake, or the shallow nearshore ocean can have

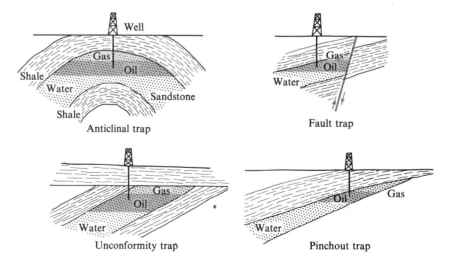

Figure 12-21. Several types of oil and gas traps. The terms anticline, unconformity, and fault are discussed in later sections.

a deleterious effect on the aquatic life. Heating the atmosphere can change the climate, and the climate near cities is changing.

A current problem is that the runoff water from fertilized fields carries some of the fertilizer to rivers. In rivers and lakes the fertilizer provides nutrients that increase the growth of algae. The algae use the oxygen dissolved in the water, and the lack of oxygen causes the death of fish and other aquatic life. Phosphates in laundry detergents have the same effect. Pesticides used on crops get into rivers in this way, too. Long-lived insecticides such as DDT have gotten into most types of life in this way, and from animals such as birds eating poisoned insects. Thus, keeping soil productive without causing harm to the environment is a problem that requires much study.

The main consumption of withdrawn water is by irrigation. (See Fig. 12-25.) It was noted earlier that over 70 per cent of total precipitation either evaporates or is used by plants. Plants, then, consume most of our water and must be considered in any study of water resources.

All life on the earth ultimately depends on vegetation. Plants are the base on which the food chain, or pyramid, is built. The amount of water needed to grow plants is very large. An acre of corn gives off about 4,000 gallons of water per day to the atmosphere. Five hundred pounds of water are required to grow one pound of wheat (dry weight). Most of this water is lost to the atmosphere through the leaves. About one-half of the wheat is discarded in milling, so about 1,000 pounds of water are required to produce one pound (dry weight) of bread. Thus the produc-

tion of the 2½ pounds of bread that could be a daily diet requires 2,500 pounds of water, or 300 gallons. Meat requires even more water. About 30 pounds of alfalfa and 12 gallons of drinking water are required to produce one pound of beef. The 30 pounds of alfalfa require 2,300 gallons of water to grow. Thus the minimum water needed to maintain life, on bread alone, is 300 gallons per day, and if meat is included, is about 2,500 gallons per day. Water is also required to produce clothing, and in the same way, cotton requires much less water than wool.

Water Needs Present and Future

Water resources are two types: the available water composed of runoff and ground water, and the soil moisture that either is used by plants or evaporates. Both types of water are necessary to man, but most estimates of water needs are concerned with

withdrawn water alone. The average rainfall on the United States is 30 inches per year or 4,200,000 million gallons per day. Of this, 8.5 inches or 1,200,000 million gallons per day is the available water and 21.5 inches or 3,000,000 million gallons per day either evaporates or is used by plants. The water needs of plants on which all life ultimately depends will be considered first.

The minimum amounts of water necessary to produce food can best be determined by multiplying the number of people times the water consumed by plants to produce food. The 1960 population of the United States was about 180 million people. To feed these people on bread alone would require 300 gallons per day (2½ pounds of bread) times 180 million or 54,000 million gallons of water per day. Our present diet is closer to a pound of meat plus some vegetables each day and consumes about 2,500 gallons per day per per-

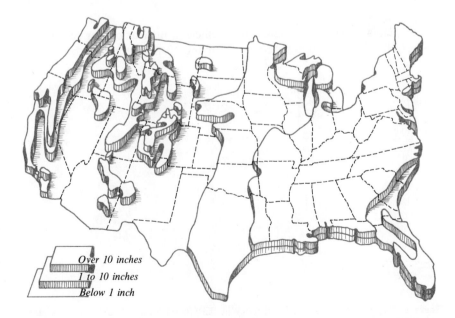

Figure 12-22. Average annual runoff of rivers in the United States. From U. S. Department of Agriculture, 1955 Yearbook, *Water.*

son, or about 450,000 million gallons per day for the total population. Only about 51,000 million gallons of this is withdrawn for irrigation. Thus about 400,000 million gallons per day of soil water is used to produce food. This soil water comes from the 3,000,000 million gallons of water per day that either evaporate or are used by plants. It is estimated that about half of this 3,000,000 million gallons per day evaporates and half is used by plants. However, it is also estimated that about one-quarter of the total rainfall falls in arid areas and so is not used productively for useful crops. Although these are very rough estimates, they suggest that at most 1,200,000 million gallons per day are available for crops and that 400,000 million gallons of this are currently being used to feed us. It would be overly optimistic to conclude that we can feed three times as many people as we do now, because this would mean replacing all

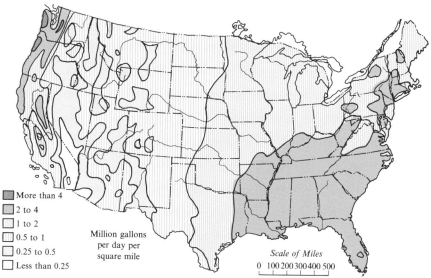

Figure 12-23. Average annual precipitation in the United States. Units are millions of gallons per day per square mile. From A. M. Piper, U. S. Geological Survey, *Water Supply Paper* 1797, 1965.

our natural vegetation, including forests, with productive crops. How much more food can be grown and where the water will come from is a problem that must be faced in the relatively near future.

The withdrawn water that will be needed in the future has been estimated. These estimates show that we will have to withdraw more than twice as much water in 1980 as we now do and over three times as much by the year 2000. The amounts of water actually consumed will increase even more, but all of these figures are less than the total available water. If these estimates are accurate, we will withdraw almost three-quarters of the total available water in 2000. To be able to do this, our water resources will have to be much better controlled and managed than they are today. This is an area where much thought, research, and money must be expended soon if we hope to have enough water for the future.

It may be possible to increase the available water for the future. Controlling the evaporation of water bodies and soil moisture is one possibility. Reducing the need of run-

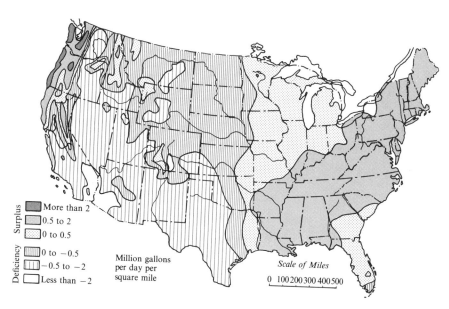

Figure 12-24. Water surplus and deficiency in the United States. Average precipitation less potential evaporation and consumption by plants is plotted on this map. From A. M. Piper, U. S. Geological Survey, *Water Supply Paper* 1797, 1965.

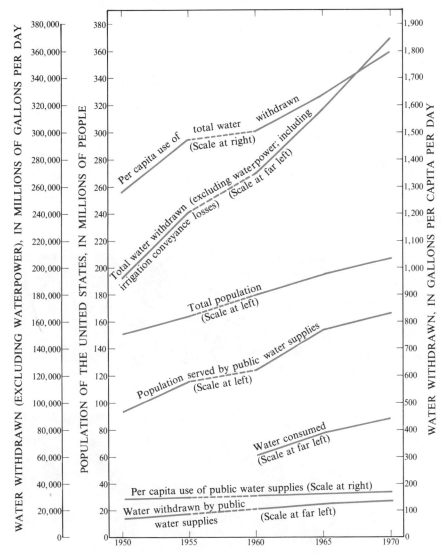

Figure 12-25. Graph showing trends in population, withdrawals of water, and consumption of water in the United States, 1950-1970. From U. S. Geological Survey, *Circular* 676, 1972.

off by better waste disposal is another possibility. Regulating rivers so that more of the runoff can be used will be necessary, and this will entail moving surplus water from one drainage area to another. Desalinization of sea water, probably by atomic power, will help in some areas but may be expensive. More efficient use of ground water will help in some areas, but it must be recognized that ground water is really part of the runoff and is the source of the flowing water in most perennial rivers. To use ground water faster than it is replaced will only reduce the average runoff and will increase the available water only temporarily.

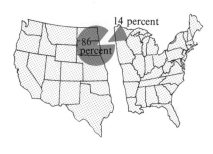

Figure 12-26. Map showing fresh-water consumption in the 17 western states compared with that in the 31 eastern states, 1970. From U. S. Geological Survey, *Circular* 676, 1972.

With an appreciation that he may be oversimplifying, the writer ventures that the United States can be assured of sufficient water of acceptable quality for essential needs within the early foreseeable future, provided that it (1) informs itself, much more searchingly than it has thus far, in preparation for the decisions that can lead to prudent and rational management of all its natural water supplies; (2) is not deluded into expecting a simple panacea for water supply stringencies that are emerging; (3) finds courage for compromise among potentially competitive uses for water; and (4) accepts and can absorb a considerable cost for new water-management works, of which a substantial part will need be bold in scale and novel in purpose.

U.S. Geological Survey Water Supply Paper No. 1797, "Has The United States Enough Water?" by A. M. Piper.

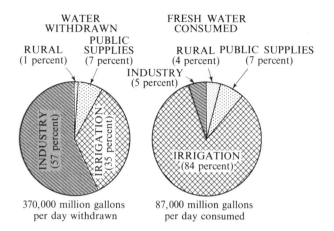

WATER
WITHDRAWN
FRESH WATER
CONSUMED
PUBLIC
SUPPLIES
RURAL
(1 percent)
(7 percent)
RURAL PUBLIC SUPPLIES
(4 percent)   (7 percent)
INDUSTRY
(5 percent)
INDUSTRY
(57 percent)
IRRIGATION
(35 percent)
IRRIGATION
(84 percent)

370,000 million gallons
per day withdrawn
87,000 million gallons
per day consumed

Figure 12-27. Diagrams showing off-channel water withdrawals and fresh-water consumption in 1970, by category. From U. S. Geological Survey, *Circular* 676, 1972.

A.

B.

Figure 12-28. Infrared images showing water temperature at Connecticut Yankee nuclear power plant on the Connecticut River near Haddam. A. Plant is not operating. B. Plant in operation. Light-colored areas are warm water from the plant. Both images were made at flood tide. Photos from Environmental Analysis Department, HRB-Singer, Inc.

Summary

The origin of ground water is precipitation that seeps into the ground.

The *water table* is the surface below which rocks are saturated with water. The water table is a reflection of the surface topography with less relief. Lakes and swamps are areas where the land surface is below or at the water table. The position of the water table changes seasonally.

Springs occur where the water table is exposed, as on a valley side.

A *perched* water table is caused by a relatively impermeable rock (such as shale) stopping the downward percolation of rain water. It may result in a spring.

The *porosity* of a rock is a measure of its open or pore space.

The *permeability* of a rock is a measure of the interconnections of the pore space, affecting the ability of water or petroleum to move through the rock. Poorly sorted or fine-grained sedimentary rocks have low permeability, although they may have high porosity.

The most common reservoir rock is sandstone, although fractured granite or limestone or other rocks can serve as well.

An *artesian well* has its reservoir confined above and below by impermeable rocks. A well drilled into such a confined reservoir will flow nearly to the height of the charging area (the area where the water enters the reservoir rock) because the released water seeks its own level.

The conservation of ground water is important as its average rate of movement through rocks is only about 50 feet per year, and many years may be required to replace hastily pumped water.

A *cone of depression* is the result of pumping water from a well faster than it can be replaced. Cones of depression around a number of closely spaced wells drastically lower the water table. Rapid removal of ground water or petroleum may also cause subsidence of the ground surface.

Pollution is a major problem in use of ground water and in waste disposal.

Water is a good solvent and ground water plays an important role in weathering. Most ground water contains dissolved material and is, therefore, hard or occasionally unsuitable for use.

Caves probably form in the zone of seasonal water table fluctuation by solution of limestone by ground water.

Sinks are closed depressions in areas of cavernous limestone.

Geysers are periodic discharges of hot ground water; *hot springs* flow more or less continuously.

Petroleum and natural gas originate from organic material in sedimentary rocks, generally shale. They are not economically recoverable by drilling until they have migrated to permeable reservoir rocks. As with ground water, well spacing and rate of pumping are important in oil recovery.

The United States receives an average of 30 inches of rain per year. Of this 21.5 inches are lost by evaporation or used by plants, leaving 8.5 inches available for man's use in the form of runoff and ground water. In 1970 we used 370,000 million gallons per day, less than one-third of the available water. However, distribution of population and of precipitation is not uniform, so many areas have a deficiency of available water.

Water that is *consumed* is the water lost to the atmosphere by evaporation or plants; the main consumption of withdrawn water is by irrigation.

Water that is *used* is returned to the ground or runoff and is available for reuse. Its quality may not be impaired by its use, although it commonly is polluted by heat, fertilizer, phosphates, insecticides, and so forth.

All life on earth ultimately depends on vegetation, as plants are the base of the food chain or pyramid. A diet of vegetable matter and clothing of cotton requires much less water than the use of meat and wool.

Our water resources will have to be very carefully managed to have enough water in the future. Our water withdrawals have been estimated to double by the year 1980 and triple by 2000.

Questions

12-1. What is the general relationship between the water table and the topography?

12-2. What is porosity?

12-3. What is permeability?

12-4. What happens near a well that is pumped too fast?

12-5. The average rate of movement of ground water is _____.

12-6. Sketch an artesian well. How long ago did the water in such a well fall as rain?

12-7. Describe capillary action as it applies to geology.

12-8. Sketch several types of springs.

12-9. Sketch a perched water table.

12-10. In what type of rock is pollution most likely to become a problem?

12-11. Describe one way a geyser eruption may occur.

12-12. How are caves formed?

12-13. Describe how the topography is different in an area underlain by soluble limestone compared to an area of sandstone. Assume humid climate.

12-14. Sketch several types of oil traps.

12-15. What is the nature of the United States' water problem?

12-16. What is the main use of runoff at present?

Supplementary Reading

General

Baldwin, H. L., and C. L. McGuiness, *A Primer on Ground Water*. Washington, D. C.: U. S. Government Printing Office, 1963, 26 pp.

Bradley, C. C., "Human Water Needs and Water Use in America," *Science* (October 26, 1962), Vol. 138, No. 3539, pp. 489-491.

Hackett, O. M., *Ground-Water Research in the United States*. U.S. Geological Survey *Circular* 527, Washington, D.C.: U.S. Government Printing Office, 1966, 8 pp.

Keefer, W. R., *The Geologic Story of Yellowstone National Park*. U.S. Geological Survey, *Bulletin* 1347, Washington, D.C.: U.S. Government Printing Office, 1971, 92 pp.

Leopold, L. B., and K. S. Davis, *Water*. Life Science Series, New York: Time, Inc., 1966, 200 pp.

Leopold, L. B., and W. B. Langbein, *A Primer on Water*. Washington, D.C.: U.S. Government Printing Office, 1960, 50 pp.

Meinzer, O. E., *Ground Water in the United States*. U.S. Geological Survey Water Supply *Paper* 836-D, Washington, D.C.: U.S. Government Printing Office, 1939, pp. 157-229.

Piper, A. M., *Has the United States Enough Water?* U.S. Geological Survey Water Supply *Paper* 1797, Washington, D.C.: U.S. Government Printing Office, 1965, 27 pp.

Swenson, H. A., and H. L. Baldwin, *A Primer on Water Quality*. Washington, D.C.: U.S. Government Printing Office, 1965, 27 pp.

Thomas, H. E., and L. B. Leopold, "Groundwater in North America," *Science* (March 5, 1964), Vol. 143, No. 3610, pp. 1001-1006.

U.S. Department of Agriculture, *Water*. U.S.D.A. Yearbook, Washington, D.C.: U.S. Government Printing Office, 1955, 218 pp.

Vogt, E. Z., and Ray Hyman, *Water Witching, U.S.A.* Chicago: University of Chicago Press, 1959.

Caves

Mohr, C. E., and H. N. Sloane, *Celebrated American Caves*. New Brunswick, N.J.: Rutgers University Press, 1955, 339 pp.

Moore, G. W., "Origin of Limestone Caves: A Symposium," *Bulletin*, National Speleological Society, 1960, Vol. 22, Part 1, 84 pp.

Moore, G. W., and G. Nicholas, *Speleology: The Study of Caves*. Boston, Mass.: D.C. Heath and Co., 1964, 128 pp. (paperback).

Waste Disposal

Clark, J. R., "Thermal Pollution and Aquatic Life," *Scientific American* (March, 1969), Vol. 220, No. 3, pp. 18-27.

Merriman, Daniel, "The Calefaction [warming] of a River," *Scientific American* (May, 1970), Vol. 222, No. 5, pp. 42-52.

Schneider, W. J., *Hydrologic Implications of Solid-Waste Disposal*. U.S. Geological Survey *Circular* 601-F, Washington, D.C.: U.S. Government Printing Office. 1970, 10 pp.

Earthquakes and the Earth's Interior

13

Knowledge of the earth's interior cannot be obtained by direct observation and, as a result, comes from very diverse kinds of data. Astronomical observations reveal the size and shape of the earth and indicate that its interior is much different from the surface. Study of earthquake vibrations reveals much of the internal structure, and meteorites provide clues to the possible internal composition. The use of data from such diverse fields shows why the geologist's interest and training must be so broad.

Shape, Size and Weight of the Earth

Shape and Size of the Earth

The Greeks and other ancient peoples were excellent observers. They observed the earth's round shadow during eclipses of the moon. From this, and from noting that ships sailing away appear to sail downhill, they concluded that the earth is a sphere. They felt that this inference must be correct because they considered the sphere to be the most perfect of shapes.

About 235 B.C., Eratosthenes went a step further and measured the diameter of the earth. His method was simple (Figure 13-1) and involved only two assumptions:

1. The earth is a sphere.
2. The sun's rays are all parallel.

His method required the measurement, at the same time at two places, of the angle between the sun's rays and a plumb bob in the plane defined by the sun's rays and the plumb bob. The two places had to be widely separated to obtain a measurable angle, and he had to know the distance between the two cities. He chose noon, the time when the shadow is shortest and points due north (he had no other way to synchronize the time), and two cities that he believed to be on a north-south line so that the line between the two cities would be in the same plane as the sun's rays and the shadows. He chose to make the measurement on the day that the sun was directly overhead at noon at Syene (Assuan); he measured 7½ degrees at Alexandria. He then used simple geometry to compute the diameter.

He knew that for a circle,

$$C = \pi D$$

and for any arc of a circle,

$$C' = \frac{\theta}{360}\pi D \quad \text{or} \quad D = \frac{C'}{\pi}\frac{360}{\theta}$$

His measurement was about 15 per cent too high (we are not sure about

Damage in downtown Anchorage, Alaska, caused by the Good Friday Earthquake of 1964. The differences in amount of damage in the various buildings suggest that changes in building codes or building practices are necessary to protect lives. Photo from National Ocean Survey — NOAA.

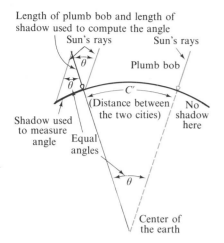

Figure 13-1. Measurement of the earth's diameter.

the length of his linear unit), which was remarkably accurate for the time. Later, less accurate measurements indicated a much smaller earth and resulted in Columbus' thinking that he had reached Asia.

More precise repetition of this measurement at different places in the last few hundred years has shown that the diameter of the earth is less if the measurement is made nearer the poles. From this we reach the conclusion that the earth is ellipsoidal, not truly spherical. The equatorial bulge is apparently due to the rotation of the earth. The difference is not great:

Equatorial radius:
  6,378,388 meters = 3,963.5 miles
Polar radius:
  6,356,912 meters = 3,950.2 miles
Difference:

21,476 meters =	13.3 miles

The most precise knowledge we have on the shape of the earth comes from observations made on satellites. The earth's gravitational pull on a satellite determines its orbit; therefore, the shape of the earth's mass controls the exact orbit. The orbit can be measured by radar, and the shape of the earth can be computed. As a result, we now know that the cross-section at the equator is not an exact circle and that the southern hemisphere is slightly larger than the northern. These discrepancies are, however, very small.

Weight of the Earth

To weigh the earth is a more difficult problem than to determine its general shape. Work on this problem was not possible until after Newton, in 1687, deduced the behavior of bodies at rest and in motion. One of the results of the three laws of motion that bear his name was his law of universal gravitation. With this law, which states that the gravitational attraction between two bodies is proportional to their masses and inversely proportional to the square of their distance apart, he was able to explain the motion of the solar system. Stated mathematically,

$$F_{\text{gravitational}} = G \frac{m_1 m_2}{D^2}$$

where G is the universal gravitational constant.

The universal gravitational constant G should not be confused with the acceleration of gravity (generally called g and equal to about 32 ft/sec²). One of Newton's important contributions was to differentiate between the *weight,* or force acting on an object, and its *mass,* an inherent, unvarying property. The background to this is that Aristotle stated that a heavy object falls faster than a light one; Galileo showed experimentally that both fall at the same speed; and Newton showed why this is true. Newton's second law states that the acceleration (rate at which speed changes) of an object is proportional to the force acting on it and inversely proportional to its mass:

$$a = \frac{F}{m}$$

(more commonly written $F = ma$)

The force F acting on a falling object is its weight. Because weight and mass are related, the quantity F/m is a constant for all falling bodies at any one place; hence, they all accelerate at the same rate. In the case of a falling body, a is the acceleration due to gravity, g. Note that because the earth is not a sphere, all points on the surface are not equidistant from the center of mass of the earth, and the weight F of an object, which is due to its gravitational attraction, is not a constant. Thus g, the acceleration of gravity, varies with latitude and elevation. Perhaps this should be taken into account in determining world's records in athletic events.

Note that although it is relatively easy to measure g at any point, this is not the universal constant of gravitation G, and there is no relationship between them that will enable calculation of G unless the mass of the earth is known.

In the case of any object on the earth, its weight is equal to its mass times the acceleration of gravity, or

$$F_{\text{weight}} = mg$$

This weight is also the gravitational force between the earth and the object:

$$F = G \frac{m m_{\text{earth}}}{R^2_{\text{earth}}}$$

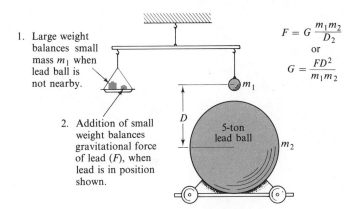

1. Large weight balances small mass m_1 when lead ball is not nearby.

2. Addition of small weight balances gravitational force of lead (F), when lead is in position shown.

D

5-ton lead ball

m_1

m_2

$$F = G \frac{m_1 m_2}{D_2}$$
or
$$G = \frac{FD^2}{m_1 m_2}$$

Figure 13-2. Determination of the gravitational constant. The figure is diagrammatic, and the balance used was of a different, more sensitive design.

Combining, we get

$$F = mg = G \frac{mm_e}{R^2_e}$$

Cancelling and rewriting, we obtain

$$m_e = \frac{gR^2_e}{G}$$

This equation cannot be solved until either m_e (mass of earth) or G (gravitational constant) is known.

The earth was accurately weighed in 1798 by directly measuring the force between two masses so that G could be calculated. (See Fig. 13-2.)

The mass of the earth is computed using the relation derived above:

$$m_e = \frac{gR^2_e}{G}$$

All quantities on the right side of the equation are known, and substitution leads to the result that the earth has a mass of about 7×10^{21} tons (or 7 followed by 21 zeros). Note that in English units the pound is considered by some a force, and by others, a mass unit. It is considered a unit of mass here.

Distribution of Mass in the Earth

A surprising result (predicted by Newton on the basis of planetary motion) is that this mass requires the average specific gravity of the earth to be about 5.5; that is, on the average, a given volume of it weighs 5.5 times as much as does the same volume of water. This is surprising because the rocks at the surface have average specific gravities less than three. Thus the rocks at depth must have much greater specific gravities than those at the surface. This suggests that the composition of the deeper parts of the earth is very different from that of the crust.

The distribution of mass in the earth determines its motion in space. One can imagine two spheres of equal size and weight; one has a uniform composition, but the second has its weight concentrated near the center with the rest hollow, except for spokes that separate the shell from the core. These two spheres will have different mechanical behavior. If we roll them down an inclined plane, they will travel at different rates, the uniform sphere moving

more slowly.

The earth's distribution of mass is determined by study of the earth-moon system. The moon's gravitational attraction on the earth's equatorial bulge causes the earth's axis to wobble very slowly. This motion is called the *precession of the equinoxes* and is quite similar to the wobbling motion of a top as it slows down. Precession is a term used to describe this type of wobble, and equinox is a term that designates a relationship between the earth's position and the "fixed" stars on certain days, occurring twice a year when day and night are of equal duration. This wobble requires 25,735 years to return to the same point, and accounts for the fact that the Egyptian astronomers noted a different star in the position of Polaris, which is our present north star.

Because the amount of wobble (precession) caused by the moon's attraction depends on the mass distribution in the earth, it can be used to calculate the distribution of mass in the earth. This method also reveals that the earth is much denser at deep, than at shallow, levels.

Earthquakes and Seismology

Although earthquakes cause great damage and are greatly feared, they do provide much useful information about the earth's interior. Study of waves generated by earthquakes deep in the earth, for example, reveals that the earth has a layered structure. Combining this information with the distribution of mass deduced from astronomical observations leads to a model of the earth. First, the cause of earthquakes will be examined, and then the types of vibrations produced by earthquakes. The effects of these vibrations at the surface, which include the damage they cause, will be considered next; and finally, how their travel times through the earth reveal the earth's deep structures will be described.

Cause of Earthquakes

The utimate cause of faulting, folding, volcanism, earthquakes, and other geologic processes is not known. The immediate cause of at least some shallow earthquakes is known, however. These earthquakes are caused by faults, breaks in crustal rocks along which movement has occurred. This is known from the formation of fault scarps at the time of earthquakes. The amount of displacement of surface features is not generally an accurate indicator of the amount of movement at depth, as shown in Fig. 13-3. (See also Figs. 13-4 and 13-5.) These movements

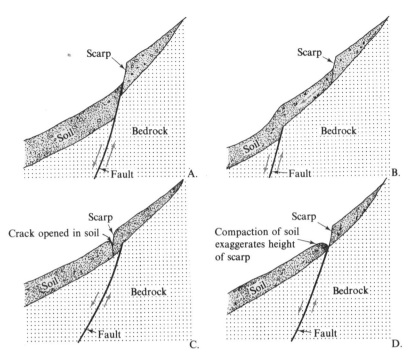

Figure 13-3. Types of fault scarps. Parts A and B after I. J. Witkind, U. S. Geological Survey. *Professional Paper 435, 1964.*

cause vibrations; seismology is the study of these vibrations. The speed at which such vibrations, caused

Figure 13-4. Scarp produced by the Hebgen Lake, Montana, earthquake in 1959. Note the subsidence below the scarp. Photo by J. R. Stacy, U. S. Geological Survey.

by either earthquakes or man-made blasts, travel through the earth provides much information about rock types and structure. Fault types will be described more completely in Chapter 14.

Faulting is one of the few geologic processes that is rapid or catastrophic. Earthquakes result from sudden movement by faulting; but apparently the actual movements within the earth are slow, and rupture only occurs when the forces have built up to the point that they exceed the strength of the rocks. This is suggested by the slow movements that have been detected on some active faults. This was first noted after the 1906 earthquake in San Francisco on the San Andreas fault. Accurate surveys, which had been made on both sides of the fault before the earthquake, showed that a small amount of movement had occurred between the time of the surveys. Re-survey after the earthquake showed that about 20 feet of absolute horizontal movement occurred in the earthquake, in agreement with the 20 feet of relative movement measured by surface features. This lead to the theory that earthquakes occur when the energy stored by elastic deformation in the rocks on both sides of the fault is enough to rupture the rocks or to overcome the friction on an existing fault plane. This elastic rebound process is shown in Fig. 13-6. This theory explains the surface deformation of most earthquakes. Friction on the fault plane may cause sticking after some movement has occurred, and so the total strain may not be relieved in a single earthquake.

Slow creep has been detected along several faults in California in agreement with the elastic rebound theory. (See Fig. 13-7.) At several places buildings, roads, pipelines, and railroads cross active faults, and this slow movement has caused damage. Recording instruments have

Figure 13-5. Tension crack in volcanic rocks on the eastern side of the Sierra Nevada. It is possible to descend 55 feet below the surface in this crack that is believed to have been formed by a prehistoric earthquake. Scarps over 50 feet high are found in this vicinity. Photo from U. S. Forest Service.

been installed at some of these places. Some examples of this movement are:

Warehouse at Irvington on Hayward fault: 6″ between 1921 and 1966 (Fig. 13-8)

Railroad tracks at Niles on Hayward fault: 8″ between 1910 and 1966

University of California stadium at Berkeley on Hayward fault: 2¾″ since 1948; 1¼″ between 1954 and 1966

Winery near Hollister on San Andreas fault: 3¼″ in 7½ years

The movement appears to occur in spasms of a small fraction of an inch followed by weeks or months of quiet.

Deformation of rocks

Deformation of a limber stick

A.

B.

Figure 13-6. Origin of earthquakes. A portion of the earth's crust is shown on the left and a limber stick on the right. A. Slow deformation of the crust is caused by internal forces. B. When the strength of the rocks is exceeded, they rupture or fault, producing earthquake vibrations. Earthquakes on old faults result when the friction along the plane of the old break is exceeded.

Wave Types

Earthquakes are recorded by seismographs. (See Figs. 13-9 and 13-10.)

Three types of waves are sent out by an earthquake. (See Fig. 13-11.) They are

1. P waves—primary or push (compression) waves. "Primary" because they travel fastest and, therefore, arrive first at a seismic station. "Push" because they vibrate as compressions and rarefaction.

2. S waves — secondary or shake (shear) waves. "Secondary" because they are the second arrivals at a seismic station. "Shake" because they vibrate from side to side.

3. Surface waves — slow-moving waves with motion similar to the waves caused by a pebble tossed into a pond.

The propagation, or movement, of these waves will show more of the differences among them. The energy of an earthquake travels in all directions away from the source. We are interested only in the particular rays that are received at one point. The travel of the P waves is easily demonstrated. If a part of a spring is compressed and then quickly released, the resulting wave moves along the spring by successively compressing and spreading the spring. The S-wave movement is shown by displacing, and then releasing, one end of the spring. The wave moves along the spring by successive displacements to one side and the other. (See Fig. 13-12.) This same motion can be demonstrated by shaking a rope. (See Fig. 13-13.) Surface waves are ground motion caused by P and S waves reaching the surface. The transmission of energy by the surface wave is more complex, with two types of waves involved, and cannot be discussed here beyond the previous mention of the water-wave

Total horizontal movement between 1951 and 1963

Figure 13-7. Recent horizontal movements in the San Francisco Bay area revealed by surveying. After A. J. Pope, J. L. Stearn, and C. A. Whitten, *Bulletin*, Seismological Society of America, Vol. 56, 1966.

Figure 13-8. Damage to warehouse built across the Hayward fault at Irvington, California. The originally straight wall is bent several inches near the far end of the second window.

analogy. Typical seismic records show more than three waves because these main waves are commonly reflected by the layers within the earth.

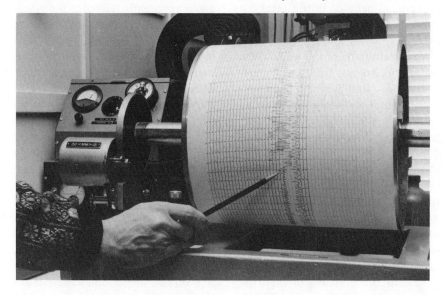

Figure 13-9. Long-period seismograph showing record of earthquake near Formosa on March 12, 1966. Photo from NOAA.

Weight does not move, because of inertia

Flexible wire

Pen

Drum rotated by clockwork

Rigid support is anchored to bedrock and moves with bedrock during an earthquake

A.

B.

Figure 13-10. The operation of a seismograph. A. A weight on a string illustrates the principle of the seismograph. If the arm moves rapidly, the weight remains still. B. In this diagrammatic model of a seismograph, the support is attached to bedrock and moves when the earth quakes. The weight tends to remain still, so the pen records the relative movement between the chart that moves with the bedrock and the weight that does not move. Actual seismographs are designed to respond to movement in a single direction. Instead of using a pendulum, the weights are suspended like doors and so respond only to horizontal movements at a right angle to the plane of the weight.

1 minute   P wave   S wave Surface wave

Figure 13-11. A seismograph record showing the three waves.

Damage

Earthquakes cause damage in a number of ways. They set off landslides, they produce seismic sea waves, and they cause uplift or subsidence of large areas. However, the principal damage done by earthquakes is due to differential movements of buildings. These movements are caused mainly by the surface waves. The P and S waves vibrate fast but with small movement and, therefore, cause very little damage to structures. The surface waves, on the other hand, have much larger amplitude (movement) in lower frequen-

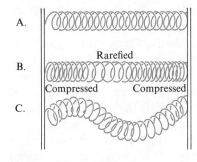

A.

B.

Rarefied

Compressed   Compressed

C.

Figure 13-12. Types of waves. A. A spring stretched between two supports. B. A push wave showing compression and rarefaction. C. A shake wave.

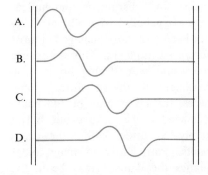

A.

B.

C.

D.

Figure 13-13. Propagation of an S wave shown by displacement of a rope. Part A shows initial displacement; B, C, and D show displacements at later times.

cies and are the cause of most direct damage done by earthquakes. This is easy to understand, for if one part of a building is moved a few inches in one direction, while another part moves in another direction, the building will be damaged. Most earthquake-resistant buildings are designed to be flexible, so that such movements will not damage them. The building site will also affect the amount of movement that a structure experiences. In general, buildings on bedrock will be damaged less than those built on less consolidated, easily deformed material such as natural or artificial fills, because the softer material slows the waves, causing an increase in the amplitude just as occurs with the ocean waves described in Chapter 11. These statements apply only to earthquake-wave damage and do not include the sometimes total destruction that occurs at places where there is an actual break or displacement at the surface.

In urban areas, another major cause of property damage due to earthquakes is the fires started by crossed electric wires and broken gas lines. These fires cannot be controlled because of broken water mains and disrupted communications. In general, private homes have not been too badly damaged in most recent earthquakes. This is because the usual wood-frame house is a reasonably flexible structure. In some cases, the home may be too badly damaged to be salvageable, but it rarely collapses, killing the occupants. Particularly in many older homes the damage may be in the form of the building being pushed off its foundation. Most modern building codes require a tight bond between the foundation and the building. Typically the part of the average home which is damaged in an earthquake is the chimney. Because it is made of brittle bricks, a chimney tends to be cracked or fall down.

In certain parts of the world these statements are not true, particularly in the Near East where most dwellings are built of adobe brick. In an earthquake the brick, being a strong, rigid material, withstands the earthquake until its strength is overcome, at which time the building collapses, commonly killing all of the occupants. In the Near East, after such an experience, a city will be completely rebuilt right on top of the old ruins. This is one cause of the layered cities that archeologists excavate in these areas. In high-rise buildings the damage pattern is quite different. Most modern high-rise buildings are built of steel and concrete and are designed to be flexible. Much of the major structural damage to such buildings occurs where horizontal members are pulled away from vertical members. The tying of such joints is a critical part of such buildings. In general, an attempt is made in high-rise buildings to tie the foundation to bedrock. Generally this means a very deep foundation. In some areas this is not possible and other types of foundations are designed.

Some recent earthquakes, particularly the January, 1968, earthquake in Caracas, Venezuela, the earthquake of March, 1970, in Santa Rosa, California, and the February, 1971, earthquake in San Fernando, California, caused much damage to well-designed buildings. Engineers and engineering geologists are closely studying the damage that occurred at these places in hopes of finding ways to design buildings that are more earthquake-resistant. Part of the problem of this unexpected damage is apparently related to the vibration frequency that the building experiences. If the building is shaken in such a way that it begins to resonate at its natural frequency, much damage can result. The amount of movement is, in this way, magnified. The vibration frequencies of most high-rise buildings are quite low. If such buildings are built on relatively soft foundation material, even with well-designed foundations, apparently the high-frequency earthquake waves are filtered out by the soft foundation material, and the low frequencies move the foundation. Thus, in some cases, well-designed buildings on soft foundation materials are shaken at the natural frequency of the building. The length of time that the earthquake vibrations last is probably also very important in determining the amount of damage. The longer the earthquake, the more opportunity for the high-rise to begin moving at its natural frequency. Although eye-witness observers are not too reliable, earthquake vibrations in major earthquakes seem to last anywhere from about a minute to three or four minutes. In general the amount of damage caused by a nearby earthquake is more related to the local geology, that is, the underlying rock, than it is to distance from the fault or from the epicenter of the earthquake. This applies to those places within a few miles to perhaps a few tens of miles from the actual site of the earthquake.

Many people have a great fear of falling into a fissure created by an earthquake. Fig. 13-3 shows how small fissures can form in earthquakes, but this figure also shows that such cracks are generally shallow. The only recorded cases of loss of life in this way are a woman in a Japanese earthquake and a cow in the 1906 San Francisco earthquake. In both cases the walls of the cracks failed, and the victims were buried before they could climb out.

The likelihood of being injured or killed in a major earthquake in the United States is exceedingly small, but when a major earthquake does strike, the results can be severe, as witnessed by the 116 deaths in 1964 in Alaska; 28 in 1959 at Hebgen Lake, Montana; 115 in 1933 at Long Beach, California; and 64 in 1971 at San Fernando, California. In 1972 over 10,000 were killed at Managua, Nicaragua. If you feel the earth shaking under your feet and realize that you are experiencing an earthquake, you should dive under a table or desk, get in a doorway, or, in short, get into some position where you will be protected against material falling from the ceiling. The worst possible thing to do is to run out into the street, where you may be struck by falling cornices, chimneys and other flying objects. The elimination of buttresses and cornices and other masonry structures that may fall in an earthquake is one of the major considerations in designing earthquake-resistant buildings.

Most earthquakes are followed by aftershocks, some of which can be as severe as the original earthquake itself. For this reason, it is quite dangerous to reoccupy a building after an earthquake, even though the building has withstood the earthquake, because it may now, in its weakened condition, fail in the less severe aftershocks.

Seismic Sea Waves cause much earthquake damage. They are com-

Figure 13-14. This map shows the travel times in hours for seismic sea waves to reach most of the Pacific Ocean for an earthquake off the Aleutian Islands. Courtesy of NOAA.

monly called tidal waves, but this is a very poor term because tides have nothing to do with this type of wave. For this reason the Japanese word for these waves, *tsunami,* is in common use for them; but the translation of tsunami means tidal wave, so it, too, is not really suitable. *Seismic sea wave* is a much better, more descriptive term.

Seismic sea waves are generated by faults or earthquakes under the sea. These undersea movements can cause waves that travel across oceans as small waves imperceptible to ships at sea. In shallow water at shorelines, these waves become huge and cause great destruction. The initial movement of the water is commonly a recession followed in a few minutes by huge waves, some of which have been more than 200 feet high, although they are generally a few tens

of feet high. Because they are so destructive to life and property, warnings are now sent by radio when seismographs detect an earthquake under the sea. These waves travel at velocities of 400 to 450 miles per hour and can cause damage on the opposite side of an ocean. (See Fig. 13-14.)

Size of Earthquakes

The size of an earthquake can be measured in many different ways. The amount of surface displacement is one measure; however, as shown in Fig. 13-3, the surface displacement may be different from the displacement at depth. Surveying shows the absolute displacement, which may differ from the relative displacement shown at the surface. Moreover, most earthquakes have no surface displacement, although in some it is appreciable. The maximum surface displacement so far recorded is 47 feet 4 inches at Yakutat Bay, Alaska, in 1899, and in the Good Friday earthquake of 1964 parts of the sea floor off Alaska were raised over 50 feet.

Another measure of the size of an earthquake is the size of the area in which the displacement occurs. In the Good Friday earthquake in Alaska in 1964, the observable crustal deformation covered an area of between 66,000 and 77,000 square miles. (See Fig. 13-15.) This is the largest of any one historical earthquake, although the 1906 San Francisco earthquake had the longest length of surface movement in a single earthquake. A 205 mile stretch of the nearly 600 mile long San Andreas fault moved in that earthquake.

The duration of shaking is another measure of an earthquake.

None of these ways of measuring an earthquake enables direct comparison of earthquakes, and none is useful for the many earthquakes without surface displacement. Seismologists have devised *intensity* scales to overcome these difficulties. Intensity scales rate the damage that occurs to substantial buildings. (See Table 13-1.) The estimation of intensity on the basis of damage and eyewitness reports is somewhat subjective, but works reasonably well. The main problem is that the damage depends somewhat on the local surface conditions, and so the intensity grades of many earthquakes follow the surface geology. (See Fig. 13-16.) Intensity maps are made by sending postcards to numerous people near the earthquake, asking them to describe the surface effects. In this way the epicenter of an earthquake

Figure 13-15. Map of Alaska showing the area where elevations changed in the Good Friday earthquake of 1964. From George Plafker, *Science,* Vol. 148, p. 1676, June 25, 1965.

can be located reasonably well. Notice that intensity enables direct comparison of surface effects only and that a deep earthquake will, in general, have a lower intensity than a shallow earthquake, even though both release the same amount of energy.

The *magnitude* of an earthquake is a measure of the amount of energy released. This figure is what is reported in newspapers and on the radio soon after an earthquake occurs. The magnitude of an earthquake is determined from the amount of movement on a standard seismometer. This amount of movement is corrected for the distance between the instrument and the earthquake. The early reports of an earthquake commonly give slightly different magnitudes from different seismic stations. This occurs because the movement of the standard seismom-

eters is somewhat affected by the foundations on which they rest. Another reason for this slight uncertainty is that magnitude determinations are based on the wave with the largest amplitude. For most earthquakes this is the surface wave, but the very deep earthquakes have no surface waves. Therefore, a magnitude scale based on P and S waves had to be devised for deep earthquakes. The relationship between these two scales has slight discrepancies.

The magnitude scale in common use is the Richter Scale. The largest magnitude yet recorded on this scale is 8.9, and the smallest felt by humans is 2. The San Francisco earthquake of 1906 was 8.3. For each increase of one unit on the scale, for example, from 6.5 to 7.5, the energy released increases about 31 times and the recorded vibrations at a dis-

Table 13-1. Earthquake intensity scale. This scale indicates the amount of surface damage caused by an earthquake. (Modified Mercalli Intensity Scale of 1931)

I. Not felt except by a very few under especially favorable circumstances.

II. Felt only by a few persons at rest, especially on upper floors of buildings. Delicately suspended objects may swing.

III. Felt quite noticeably indoors, especially on upper floors of buildings, but many people do not recognize it as an earthquake. Standing motorcars may rock slightly. Vibration like passing of truck. Duration estimated.

IV. During the day, felt indoors by many, outdoors by few. At night, some awakened. Dishes, windows, doors disturbed; walls make creaking sound. Sensation like heavy truck striking building. Standing motorcars rocked noticeably.

V. Felt by nearly everyone, many awakened. Some dishes, windows, etc., broken; a few instances of cracked plaster; unstable objects overturned. Disturbances of trees, poles, and other tall objects sometimes noticed. Pendulum clocks may stop.

VI. Felt by all, many frightened and run outdoors. Some heavy furniture moved; a few instances of fallen plaster or damaged chimneys. Damage slight.

VII. Everybody runs outdoors. Damage negligible in buildings of good design and construction; slight to moderate in well-built ordinary structures; considerable in poorly built or badly designed structures; some chimneys broken. Noticed by persons driving motorcars.

VIII. Damage slight in specially designed structures; considerable in ordinary substantial buildings with partial collapse; great in poorly built structures. Panel walls thrown out of frame structures. Fall of chimneys, factory stacks, columns, monuments, walls. Heavy furniture overturned. Sand and mud ejected in small amounts. Changes in well water. Persons driving motorcars disturbed.

IX. Damage considerable in specially designed structures; well-designed frame structures thrown out of plumb; great in substantial buildings, with partial collapse. Buildings shifted off foundations. Ground cracked conspicuously.

X. Some well-built wooden structures destroyed; most masonry and frame structures destroyed with foundations; ground badly cracked. Rails bent. Landslides considerable from riverbanks and steep slopes. Shifted sand and mud. Water splashed (slopped) over banks.

XI. Few, if any, (masonry) structures remain standing. Bridges destroyed. Broad fissures in ground. Underground pipelines completely out of service. Earth slumps and land slips in soft ground. Rails bent greatly.

XII. Damage total. Waves seen on ground surfaces. Lines of sight and level distorted. Objects thrown upward into air.

tant station increase about 10 times. Table 13-2 shows the relationship between intensity and magnitude.

Many more earthquakes occur than cause damage great enough to be reported in the newspapers. Although the average annual number of earthquakes strong enough to be felt at least locally is estimated at about a million, most of the earth's seismic energy is released by the few large earthquakes. It is estimated that about 16 earthquakes of magnitude 7, 5 of magnitude 7.5, and 1.6 of magnitude 8 occur each year.

Examples

Alaska's Good Friday earthquake occurred at 5:36 P.M. local time on March 27, 1964. It was centered in south-central Alaska and had a magnitude of 8.4 to 8.6. It thus released about twice as much energy as the San Francisco earthquake of 1906 that had a magnitude of 8.3. Figures 13-15 and 13-17 show the large area in which the land level changed. Figure 13-15 shows the epicenter of the main shock and also the zone of aftershocks that presumably outlines the area where fault movement occurred at depth. Aftershocks are common in most big earthquakes and are believed to be caused by readjustments made necessary by the main movement. In many earthquakes, the aftershocks cause almost as much damage as the main earthquake. The seismic sea waves caused

Table 13-2. The relationship between the Richter Magnitude and the Mercalli Intensity Scales at the epicenter of an earthquake.

Modified Mercalli Intensity (at the epicenter)	I	II	III	IV	V	VI	VII	VIII	IX	X	XI	XII
Richter Magnitude	–2–		3	4	–5–––6––			–7–––8–				

Figure 13-16. Map of the reported intensity of the March 22, 1957, earthquake. Movement apparently occurred on the San Andreas fault that is also shown. After W. K. Cloud, California Division of Mines and Geology. *Special Report 57, 1959.*

by this earthquake caused much damage in Alaska and as far away as California, and were recorded in Japan. The vibrations caused many landslides, which were especially damaging to coastal installations built on unconsolidated rocks. Examples are shown in Figs. 7-17, 7-18, and 7-19. In this case, a previous geologic report had suggested that the particular clay of the area was susceptible to landsliding.

The Hebgen Lake, Montana, earthquake of August 17, 1959, was a shallow earthquake of magnitude 7.1. It produced two more or less parallel fault scarps seven miles long and up to 20 feet high. (See Fig. 13-4.) Tilting movements are clearly shown by the displaced shorelines of Hebgen Lake. This area is close to Yellowstone National Park, and the shocks affected some of the geysers there, mainly by increasing their activity at least temporarily. The main damage in this remote area was the triggering of a huge rockslide. Unfortunately, the slide fell on a camp-

ground, burying a number of people. (See Fig. 7-13.) It also dammed the Madison River, creating a problem. It was feared that when the lake created by the dam overtopped the dam, the flowing water might rapidly erode the rockslide debris and cause a flood. A channel was later bulldozed through the rockslide to control the water flow.

Earthquake Prediction and Control

The immediate goal of seismologists is the prediction of the time, place, and magnitude of earthquakes so that property damage and casualties can be reduced. This goal may be reached soon. Recently and inadvertently man has triggered some earthquakes, suggesting the possibility that earthquakes can be controlled by man. Such control would enable man to cause several small, harmless quakes rather than allow one natural destructive quake.

"Earthquake weather" is believed by many to be a precursor of earthquakes. Studies of many destructive earthquakes over the last few hundred years have revealed no correlation between weather and earthquakes, but the folklore persists, probably because it is so longstanding. Many large earthquakes have occurred at new or full moon when the highest tides occur because the earth, moon, and sun are in line.

The places where earthquakes are most apt to occur are easily predicted on the basis of past experience. (See Fig. 13-18). In Chapter 16 the reasons why earthquakes occur at these places will be described. As the maps show (Figs. 13-19 A and B), most earthquakes in the United States occur in California and in a belt through part of the Rocky

A. B.

Figure 13-17. Aerial photographs showing the changes at Macleod Harbor, Montague Island, Alaska, caused by the 1964 earthquake. A. 1959. B. 1964. A from U. S. Forest Service, B from National Ocean Survey — NOAA.

Mountains. One obvious way to reduce earthquake damage would be to avoid these areas. Such a solution is not practical because it would mean avoiding most of California, the most populous state; however, anyone moving into an earthquake area should be aware of the potential danger.

Within these broad earthquake areas, the epicenters are located mainly along known faults so it is possible to avoid these high-danger areas. The faults to avoid are those that are active. Such faults are those that are creeping or are the sites of previous earthquakes. Some other faults, generally much older, are no longer active and so are much less dangerous. Just because no historical earthquake has occurred on a fault and there is no evidence of recent movement, the fault is not proven inactive. Faults are generally weak areas, and movement on one could be caused by a nearby large earthquake on an apparently unrelated fault. Also, as noted earlier,

the amount of damage is controlled by the nature of the building and the rocks on which it is built.

The San Andreas fault in California is one of the longest faults; and because it has been the site of several large earthquakes, it has been studied closely. It extends about 700 miles through California from the northern Coast Range to the Mexican border. It is a fracture zone a few hundred yards to over a mile wide, composed of many more or less parallel faults. The total amount of movement is difficult to measure, and may be up to 350 miles. The kinds of evidence used to determine the amount of movement are shown in Fig. 14-22. The behavior of the fault varies along its length. Fig. 13-20 shows that there are three active areas characterized by creep and/or many small, frequent earthquakes. In marked contrast are the areas where large movements (up to 20 feet) occurred in recent large earthquakes. These two parts of the fault are apparently locked, and few

earthquakes occur at those places. These large segments of the earth's crust are capable of storing large amounts of strain and may sometime release their strain in a large earthquake. Thus these are the areas where large destructive earthquakes may occur. Most quakes on the San Andreas fault are shallow and so are capable of causing much damage.

The detailed prediction of earthquakes is based on observation of events that occur just before an earthquake, such as creep, tilting, increase in micro-earthquakes, changes in strain, and changes in such physical properties of rocks as electrical conductivity and magnetic susceptibility. Instruments set up near faults are necessary to make such observations, and such data gathering is one of the main types of current research into earthquake prediction. These observations have only been made on a few earthquake faults to date, and, so far, each area seems to behave differently. The Japanese have been issuing general earthquake pre-

EARTHQUAKES 1966
Located provisionally by the U.S. Coast and Geodetic Survey
from data furnished by many cooperating foreign and
domestic seismological stations.
● Epicenter, normal focus, magnitude 7 or greater.
. Epicenter, normal focus, magnitude less than 7.
▲ Epicenter, focus 71 km. to 300 km., magnitude less than 7.
◢ Epicenter, focus 300 km. or greater, magnitude less than 7.

Numbers Represent Approximate Earthquakes in Area
National Earthquake Information Center

Figure 13-18. Earthquake zones of the earth shown by the locations of 1966 earthquakes. The significance of these zones is described in Chapter 17. Courtesy of NOAA.

dictions for some areas since 1966. Their predictions are based on slight tilting and swarms of micro-earthquakes that precede potentially damaging earthquakes by a number of weeks. Such events do not occur near the San Andreas fault in California, our best instrumented fault. There, deformation of deep rocks changes their magnetic properties and so causes slight changes in the local magnetic field a few tens of hours before creep of a few millimeters

occurs. So far these magnetic changes have occurred only before such movements. In a few cases earthquakes have followed these movements by a few days. In other places along the San Andreas, changes in rate of creep or even reversals in direction of creep have preceded earthquakes. Creep also occurs periodically after some earthquakes on the San Andreas fault, with the period between successive creep episodes increasing. Interestingly, at

such places creep does not seem to occur at locations where aftershocks do occur.

A promising discovery by Russians is that days or weeks before an earthquake, the velocities of P and S waves change so that the interval between their arrivals is reduced. This change is related to the magnitude of the earthquake, and the velocities become normal just before the earthquake. This velocity change could be the result of ground water

Figure 13-19. A. Destructive and near-destructive earthquakes in the United States through 1964. The smallest circles are intensity VII to VIII, the large solid circles are intensity VIII to IX, the smaller ringed circles are intensity IX to X, and the large ringed circles are intensity X to XII. The numbers are the number of earthquakes. Courtesy of Environmental Science Services Administration, U. S. Coast and Geodetic Survey.

Figure 13-20. Earthquake activity on the San Andreas fault in California. The three stippled regions are areas of creep and many small to severe earthquakes. Few earthquakes occur in the areas of the 1857 and 1906 breaks, but because the strain energy may build up to high levels in these "locked" areas, a few very severe earthquakes may be expected. In the area south of San Francisco, the San Andreas fault may be locked because the activity there occurs along other faults. After C.R. Allen, 1968, and L.C. Pakiser and others, 1969.

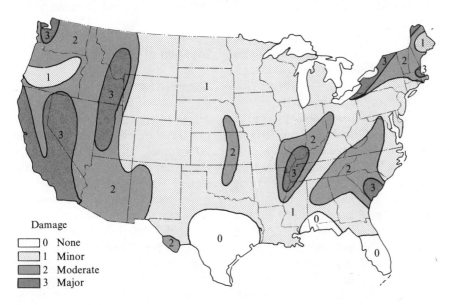

Figure 13-19. B. Seismic risk map for conterminous U. S. The map divides the U. S. into four zones: Zone O, areas with no reasonable expectancy of earthquake damage; Zone 1, expected minor damage; Zone 2, expected moderate damage; and Zone 3, where major destructive earthquakes may occur. Both A and B courtesy of Environmental Science Services Administration, 1969.

entering the many microcracks that develop in a rock when it is stressed to near rupture.

An approach to earthquake safety may be control of quakes. Our knowledge of this came from accidentally triggering earthquakes. The best publicized of these events were the many small earthquakes near Denver that were caused by pumping waste fluids into a disposal well at the Rocky Mountain Arsenal. These earthquakes, some of which had magnitudes up to 5.5 and caused some damage, began in 1962. Until that time there had been few earthquakes in the Denver area, and it was noted that the earthquakes followed periods of pumping waste fluids into the well. A series of seismic stations were set up to monitor the well, and they revealed that the earthquakes occurred in a zone about five miles long and up to a mile deeper than the 12,000 foot well. Analysis of the data suggested that the movements resulted from a reduction in friction caused by the increased pore pressure due to pumping. Soon after this discovery similar micro-earthquakes were observed at the Rangely oil field in northwestern Colorado. Here water was being pumped into the ground to increase the recovery of oil.

These discoveries stimulated the search for other man-made earthquakes. In 1945 a paper had appeared that documented hundreds of small earthquakes in the area of Lake Mead since the building of Hoover Dam. One of these quakes had a magnitude of almost 5, and two had magnitudes of about 4. It was suggested that the weight of the water had caused movements on old faults. More recently in Koyna, India, a lake created in 1962 and 1963 by construction of a dam apparently caused a number of earthquakes that culminated in 1967 with a magnitude 6.5 earthquake that caused about 200 deaths. The geologic environment at Koyna is similar to that at Grand Coulee, Washington, where no earthquakes associated with the lake have been observed. Other occurrences of earthquakes caused by dam-impounded lakes are a magnitude 5.8 earthquake at Lake Kariba on the Zambezi River, Zambia, and a 6.3 magnitude quake at Kremasta Dam, Greece.

The exact cause of any of these earthquakes is not clear. At the Denver well, they could be due to reduced friction as suggested above, or to temperature changes caused by the injection of cool fluids. Withdrawal of fluids at Denver suggests that very little mixing of waste fluids

and ground water occurred. At the dams, increased weight could be the trigger, or the lakes could cause inseeping of fluids and so the situation could be much like that at the injection wells. No earthquakes are associated with many high dams. Much more study is necessary, and hopefully such studies will be included in the construction costs of new large dams. The subsurface geology of oil fields is generally well studied, making them good places for earthquake study if fluid injection is used for secondary recovery late in the production history of the field.

Earthquakes also have been triggered by underground nuclear blasts. The earthquakes produced in this way have all been within about 15 miles of the actual blast and of lower magnitude than the nuclear blast. At the Benham test of 1968, the nuclear blast had a magnitude of 6.3, and the largest associated aftershock had a magnitude of 5.

In the deep gold mines of South Africa, mining at depths of two miles has caused earthquakes of magnitude 4.

Thus, at the moment, we are on the verge of being able to predict earthquakes, and we can trigger them. This combination may enable us to reduce the dangers of earthquakes, but there are many serious problems that must be overcome before this goal is reached. Prediction alone will reduce the likelihood of casualties but not all property damage. Control by triggering many small earthquakes instead of allowing a large natural one could reduce both casualties and property damage, especially if earthquake-resistant construction is practiced more extensively. Fluid injection wells currently appear best for such control. For instance, a series of such wells along the San Andreas fault could be used

to dissipate the strain in a number of small harmless quakes or in creeping movement. The problem is that the San Andreas fault, which is our best studied fault because it causes so many earthquakes, crosses some densely populated areas. Any experimental program triggering earthquakes along it is likely to be blamed for any earthquake damage that occurs. The resulting lawsuits would become a serious problem because the results of such a program are difficult to predict and it is almost impossible to distinguish between natural earthquakes unassociated with fluid injection and those caused by man's interference. In spite of such hypothetical problems, the benefits appear so great that a program of earthquake control should be started as soon as practical. Until such time, the protection of people and structures must depend on zoning laws, building codes, and foundation and structure designs.

Location of Epicenters

The different speeds of the P and S waves can be used to locate the epicenter of an earthquake. A P wave can travel across the earth in about 20 minutes, an S wave travels at about one-half this speed, and the surface wave would require several hours to make the trip. Because the S wave travels more slowly than the P wave, the difference in arrival time between the P and S waves is proportional to the distance away from the station that the earthquake occurred. As shown in Fig. 13-21, reports from three stations are needed to locate an epicenter.

To use this simple principle, it was first necessary to determine the velocities of the P and S waves. This was done by finding the approximate location of an earthquake from in-

Figure 13-21. A, B, and C are seismic stations. The P-S times at each station indicate that the earthquake occurred somewhere on the circle drawn around each station. The point where all three circles intersect is the epicenter.

tensity reports. Then the P and S times at many seismic stations were compared with the distance between the stations and the earthquake. The simplest way to do this is to plot a graph of distance *versus* arrival time for each wave. Generally, it was necessary to refine such a determination by adjusting the location of the earthquake, which was known only approximately from the intensity, until the best velocity determination was found. This was repeated for many earthquakes until the travel-time curves for the whole earth were de-termined. In this process the inter-nal structure of the earth was determined.

Internal Structure of the Earth

Travel times of earthquake waves have revealed a layered structure of the earth. The details of the analysis are best seen in Fig. 13-22. Three main layers have been distinguished: the crust, a very thin shell between 3 and 35 miles thick; the mantle, the next layer, about 1800 miles thick; and the core, with a radius of

about 2123 miles. The velocity of earthquake waves changes abruptly at the contacts between these layers.

Seismic study of the earth's in-terior began in earnest near the beginning of the twentieth century. Earthquake detectors had been built as early as 132 A.D. in China, but it was not until late in the nineteenth century that effective seismographs became available. It was suggested as early as 1906 that the earth had a core, and shortly thereafter its ex-istence was confirmed.

The evidence for the core is that at distances between about 7100 and 9900 miles from an earthquake, no direct P or S waves are received. Beyond 9900 miles, P waves are re-ceived, but no direct S waves are received past 7100 miles. (See Fig. 13-22.) This last observation shows that the core is liquid because shake or shear waves cannot pass through a liquid although push waves can. The P waves that go through the core are slowed, and those that pass through the inner part of the core travel somewhat faster than those that go through the outer core. This suggests that the inner part of the core is solid. The P-wave shadow zone is caused by the bending or refraction of the waves as they go into layers of different velocity.

At the boundaries of each of these layers, and even at the lesser velocity changes within the layers, the waves are also reflected. These reflected waves complicate the simple explan-ation given above. It was not until the mid-1930's that the depths and velocities of each of the layers were calculated. Since about 1960 it has been possible to refine some of this work because extremely sensitive seismograph systems have been built to detect atomic bomb explosions.

The crust is the thin shell that overlies the mantle. A very marked

Figure 13-22. Behavior of P and S waves in the mantle and the core reveals the layered structure of the earth.

change in seismic velocity separates the crust from the mantle. This change is the Mohorovičić discontinuity, commonly called the M-discontinuity or Moho. In 1909, Andrija Mohorovičić, a Yugoslavian seismologist, noted that the velocity of P waves changed sharply at distances more than about 125 miles from the earthquake. (See Fig. 13-23.) He interpreted this to mean that at a depth of about 30 miles, an abrupt change in velocity occurs. Later work shows that this change is worldwide and occurs at depths of about 30 to 35 miles under the continents and about six miles under the oceans. (See Figure 13-24.)

Other smaller differences in travel times (velocity) reveal a layering within the crust itself, as shown in Fig. 13-25. The continental crust is believed to be granitic because intrusive and metamorphic rocks of this composition underlie the sedimentary rocks that cover much of the surface, and because granitic rocks have the observed seismic velocities. In the same way, the oceans are believed to be underlain by basalt, and the lower layer under the continents is believed to be basalt. Strengthening this inference is the fact that basalt volcanoes occur in both oceanic and continental areas. Although this layer has the composition of basalt, seismic data suggest that it is in the form of gabbro or amphibolite, both of which have the same chemical composition as basalt.

The nature of the Mohorovičić discontinuity is not known, except that there is a marked change in seismic velocity with a thickness of 0.1 to 2 miles. For many years this was assumed to be caused by a change in composition from peridotite or dunite in the mantle to basalt

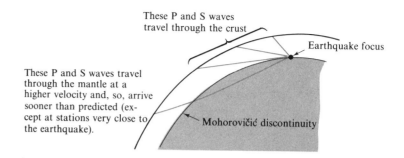

Figure 13-23. Determination of the depth of the Mohorovičić discontinuity.

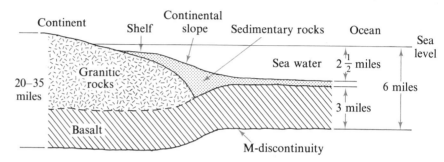

Figure 13-24. Generalized cross-section of continent-ocean boundary at a geologically stable area. The sedimentary rocks on the continental shelf and slope generally have more complex structures than shown. From seismic data.

in the lower layer of the crust. Recently, however, it has been suggested that the change may not be due to a difference in chemical composition, as between dunite and basalt, but due to a change in density or in mineralogy in a rock of the same chemical composition. Such changes are known to take place as a result of increased pressure and temperature, and might occur in a zone narrow enough to account for the rapid change in velocity. A change in mineralogy occurs when basalt is subjected to high temperature and pressure. Under these conditions, the augite and plagioclase of basalt become unstable, and the ele-

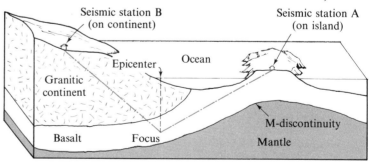

Figure 13-25. Block diagram showing paths of seismic waves in oceanic basalt and continental granite. Both stations A and B are equidistant from the focus of the earthquake, but station A receives waves first because they travel faster in basalt than in granitic rock.

Figure 13-26. Section showing the crustal and upper-mantle structure and the inferred basalt-eclogite transformation zone from the Basin and Range area to the Great Plains. From L. C. Pakiser, U. S. Geological Survey, *Professional Paper 525B, 1965.*

ments recombine to form a type of garnet and a green pyroxene, omphacite. Such rocks are called *eclogite,* and their rare appearance at the surface is at places where uplift and erosion have exposed deep-seated rocks or where intrusive bodies have carried them to the surface. The weight of the overlying rocks supplies the pressure needed to change the mineralogy. Therefore, if the Moho is in fact caused by changes in mineralogy rather than in composition, it should follow the surface topography; however, it does not. Figure 13-26 shows the depth of the Moho from seismic data and the calculated depth of the transition zone in part of the United States. Figure 13-27 shows other lateral variations in the structure of the United States.

Although the Moho is seismically the most obvious discontinuity in the upper part of the earth, it is probably not the most important zone. A low seismic velocity zone in the upper mantle is probably more involved in the formation of surface structures. This zone will be described in the next chapter, and its role in the formation of surface structures is discussed in later chapters.

It is possible to determine the depth of an earthquake from the arrival times of vibrations at nearby stations, as shown in Fig. 13-28. The point at which the vibrations originate is called the *focus,* and the point on the surface directly above is called the *epicenter.* Most earthquakes are shallow, less than 35 miles, but some are as deep as 435 miles. Deep earthquakes were first suspected from the lack of surface waves from some earthquakes. These deep-focus earthquakes occur only at the continent-ocean boundary. We will discuss them later.

The velocity of waves is related to the physical properties of the medium through which they pass; hence,

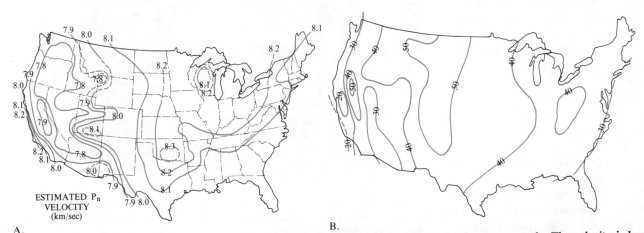

A.   B.

Figure 13-27. Lateral variation in the structure of the United States. A. Seismic velocity in the upper mantle. The velocity is less than 8 kilometers per second in much of western United States. After Eugene Herrin and James Taggart, Seismological Society of America *Bulletin,* Vol. 52, 1962. B. Crustal thickness. The contours are in kilometers. After L. C. Pakiser and J. S. Steinhart, 1964.

it is possible to make some inferences about the material that composes the deep parts of the earth. Remember also the density distribution of the deeper parts of the earth, which was learned from the earlier discussion of the weight of the earth. Both methods of investigation lead to the specific gravity of the mantle, varying from about 3 near the top to about 6 at its base. Similarly, the core has a specific gravity of about 9 at its top and about 12 at the center.

Composition of the Layers of the Earth

Because the earth's interior is not accessible to direct sampling, all knowledge of its composition is based on speculation. Our only evidence concerning the composition comes from meteorites. In Chapter 18 it will be shown that the earth and the meteorites are believed to have formed at the same time and from the same material. Meteorites are the objects that fall into the earth from space. Many of them are burned by frictional heat in the atmosphere to form what are called "shooting stars." Some meteorites are quite large and form impact craters, such as Meteor Crater in Arizona. The craters on the moon, which is not protected by an atmosphere, are believed to have formed in this way.

Meteorites are of three types: stony meteorites, composed mainly of olivine and pyroxene (peridotite is the rock name for this composition); iron meteorites, composed mainly of iron and nickel (see Fig.

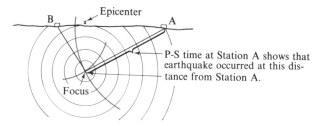

Figure 13-28. Determination of the focus (depth) of an earthquake.

13-29); and stony-iron meteorites, which are mixtures.

The specific gravity and other physical properties of peridotite or dunite fit those inferred for the upper mantle from seismic and other indirect methods of study. The lower mantle (below the low velocity zone described in the next chapter) is probably composed of simple oxides of iron, magnesium, and silicon. In the same way, the core is generally believed to be mainly molten iron with some nickel. Recently on both

seismic and chemical evidence, it has been suggested that iron and sulfur are the main elements in the core. In reaching these assumptions, allowance must be made for the increased temperature and pressure deep within the earth, for these will greatly modify the properties of rocks within the earth. Because these temperatures and pressures are too high to reach in the laboratory, some uncertainties exist.

The pressure deep in the earth is reasonably well known, but the tem-

Figure 13-29. Cross-section of iron-nickel meteorite from Henbury, Australia. The crystal pattern has been shown by acid etching after cutting. Photo from Ward's Natural Science Establishment, Rochester, N. Y.

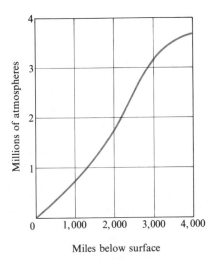

Figure 13-30. Pressures within the earth. Adapted from John Verhoogen, *American Scientist,* Vol. 48, 1960.

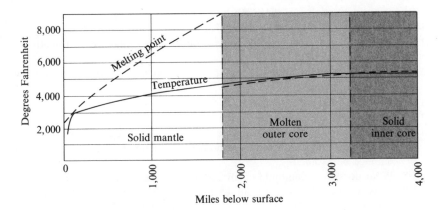

Figure 13-31. Estimated temperatures in the earth. Adapted from John Verhoogen, *American Scientist,* Vol. 48, 1960.

perature is less well known. The pressures are due to the weight of the overlying material, and, with our knowledge of specific gravity gained from seismic studies, can be calculated from the law of gravity. (See Fig. 13-30.) These pressures are hydrostatic; that is, they are equal in all directions because the rocks in the mantle and crust are not, as we will see, completely rigid, even though they are solid. A simple analogy will illustrate hydrostatic pressure. Bricks piled one on the other exert a pressure only on their base, because they are rigid. A pile of sand, however, unless confined by a wall, will assume a cone shape because the sand is not rigid; hence, the pressure due to its weight is exerted in all directions.

Data from drill holes and deep mines reveal that the temperature rises with depth in the earth at the approximate rate of 1°F per 50 feet. If this rise continues to the center of the earth, the temperature there would be 400,000°F, an unreasonably high temperature—hotter than the sun. The upper limit of temperatures in the earth is estimated from knowledge of the melting points of the materials assumed to form each layer, taking into account the rise in melting point at high pressures. The lower limit is estimated from calculations of the lowest temperature which will permit heat transfer by convection. Convection (see Fig. 16-25) is believed to be an important process in the mantle and is discussed later. Although convection requires a fluid, the material in the mantle could flow slowly (and still transmit S waves as a solid) in much the same way that tar behaves at room temperature. Under such conditions tar acts as a solid to rapid deformation, such as the passage of S waves, but over a long period it will flow under its own weight. From this reasoning, the temperature distribution shown in Fig. 13-31 is derived.

Summary

The diameter of the earth is easily measured using the method shown in Fig. 13-1.

The polar radius of the earth is 3950.2 miles, and the equatorial radius is 3963.5 miles.

The mass of the earth was calculated from Newton's law of gravity as soon as the constants in the equation were determined by experiment.

The average specific gravity of the earth is 5.5, but the surface rocks are less than 3, indicating that the deep rocks have high specific gravities.

Earthquakes are vibrations caused by rapid movement on a fault.

A few faults move by slow *creep*.

P waves are primary (first to arrive) or push (compression) waves.

S waves are secondary (second to arrive) or shake (shear) waves.

Surface waves are slow moving, low frequency, large amplitude waves that cause most earthquake damage.

Earthquake damage can be reduced by constructing flexible buildings with good foundations in bedrock.

Seismic sea waves or *tsunami* are destructive waves caused by undersea earthquakes.

Earthquake *intensity* is a measure of the damage to buildings.

Earthquake *magnitude* is a measure of the energy released by an earthquake.

Earthquake prediction may be possible based on creep, micro-earthquakes, and especially local changes in the velocity of seismic waves.

Earthquakes can be triggered by pumping fluid into deep wells, formation of large lakes by dam construction, and underground nuclear blasts.

The *focus* is the place where the earthquake occurs.

The *epicenter* is the place on the surface above the focus.

The difference in arrival time between the P and S waves is proportional to the distance to the epicenter. Such measurements at three stations are necessary to locate the epicenter.

The *crust* is 3 to 35 miles thick, and is separated from the *mantle* by the *Mohorovičić* (*M* or *Moho*) discontinuity.

The mantle is about 1800 miles thick.

The *core* has a radius of 2123 miles, and S waves cannot pass through the outer core, indicating that it is liquid.

The continental crust is 20–35 miles thick and is composed of granitic rocks.

The oceanic crust is about 3 miles thick and is composed of basalt.

The Mohorovičić discontinuity is probably caused by a difference in composition, but might be caused by a change in mineralogy as from basalt to eclogite.

Meteorites are our best clues as to the composition of the deep interior of the earth.

Stony meteorites are composed of olivine and pyroxene (peridotite), and the mantle may have a similar composition.

Iron meteorites are composed of iron and some nickel, and this may be the composition of the core.

Questions

13-1. How was the size of the earth first measured? (Use a diagram.)

13-2. How was the weight of the earth determined?

13-3. What is the average specific gravity of the earth, and what problem does this pose?

13-4. What is the immediate cause of an earthquake? What is the ultimate cause of an earthquake?

13-5. How does a seismograph work?

13-6. Briefly describe each of the three seismic waves.

13-7. Which of the earthquake waves causes the most damage? How is most of the property damage due to earthquakes caused? How might some of this damage be avoided?

13-8. What does the magnitude of an earthquake mean?

13-9. How is the epicenter of an earthquake determined? (Diagram might help.)

13-10. Describe the behavior of S waves passing through the earth.

13-11. What effect does the moon have on earthquakes?

13-12. How can earthquakes be predicted?

13-13. How can man trigger earthquakes?

13-14. Describe the behavior of P waves passing through the earth.

13-15. How is the focus of an earthquake determined?

13-16. The deepest earthquakes are _____ miles.

13-17. The core has a radius of _____ miles and is believed to be composed of _____. What is the evidence for this composition?

13-18. The mantle of the earth is _____ miles in thickness and is believed to be composed of _____. What is the evidence for this composition?

13-19. The crust is separated from the mantle by the _____.

Supplementary Reading

Earth's Size, Shape, and Weight

Bullen, K. E., "The Deep Interior," in *The Earth and Its Atmosphere* (D. R. Bates, ed.), pp. 31-47 (also covers much seismology). New York: Science Editions, Inc., 1961, 324 pp. (paperback). This book was also published under the title *The Planet Earth*. London: Pergamon Press, 1957.

King-Hele, Desmond, "The Shape of the Earth," *Scientific American* (October, 1967), Vol. 217, No. 4, pp. 67-77.

Seismology

Anderson, D. L., "The Plastic Layer of the Earth," *Scientific American* (July, 1962), Vol. 207, No. 1, pp. 52-59.

Anderson, D. L., "The San Andreas Fault," *Scientific American* (November, 1971), Vol. 225, No. 5, pp. 52-68.

Benioff, Hugo, "Earthquake Source Mechanisms," *Science* (March 28, 1964), Vol. 143, No. 3613, pp. 1399-1406.

Bolt, B. A., "The Fine Structure of the Earth's Interior," *Scientific American* (March, 1973), Vol. 228, No. 3, pp. 24-33.

Bullen, K. E., "The Interior of the Earth," *Scientific American* (September, 1955), Vol. 193, No. 3, pp. 56-61. Reprint 804, W. H. Freeman and Co., San Francisco.

Christensen, M. N., and G. A. Bolt, "Earth Movements: Alaskan Earthquake, 1964," *Science* (September 11, 1964), Vol. 145, No. 3627, pp. 1207-1216.

Grantz, Arthur, and others, "Alaska's Good Friday Earthquake, March 27, 1964," U.S. Geological Survey *Circular* 491, 1964, 35 pp.

Hammond, A. L., "Earthquake Predictions: Breakthrough in Theoretical Insight," *Science* (May 25, 1973), Vol. 180, No. 4088, pp. 851-853.

Hansen, W. R., and others, "The Alaska earthquake, March 27, 1964: Field investigations and reconstruction effort," U.S. Geological Survey, *Professional Paper* 541, 1966, 111 pp.

Hodgson, J. H., *Earthquakes and Earth Structure*. Englewood Cliffs, N.J.: Prentice-Hall, Inc., 1964, 166 pp. (paperback).

Iacopi, Robert, *Earthquake Country*. Menlo Park, Calif.: Lane Book Co., 1964, 192 pp. (paperback).

Oliver, Jack, "Long Earthquake Waves," *Scientific American* (March, 1959), Vol. 200, No. 3, pp. 131-143. Reprint 827, W. H. Freeman and Co., San Francisco.

Pakiser, L. C., and others, "Earthquake Prediction and Control," *Science* (December 19, 1969), Vol. 166, No. 3912, pp. 1467-1474.

Plafker, George, "Tectonic Deformation Associated with the 1964 Alaska Earthquake," *Science* (June 25, 1965), Vol. 148, No. 3678, pp. 1675-1687.

Scholz, C. H., L. R. Sykes, and Y. P. Aggarwal, "Earthquake Prediction: A Physical Basis," *Science* (August 31, 1973), Vol. 181, No. 4102, pp. 803-809.

Smylie, D. E., and L. Mansinha, "The Rotation of the Earth," *Scientific American* (December, 1971), Vol. 225, No. 6, pp. 80-88.

Tocher, Don, "Earthquakes and Rating Scales," *Geotimes* (May–June, 1964), Vol. 8, No. 8, pp. 15-18.

Wallace, R. E., "Earthquake Recurrence Intervals on the San Andreas Fault," *Bulletin*, Geological Society of America (October, 1970), Vol. 81, No. 10, pp. 2875-2890.

Witkind, I. J., and others, "The Hebgen Lake, Montana, Earthquake of August 17, 1959," U.S. Geological Survey, *Professional Paper* 435, 1964, 242 pp.

Composition of Layers (See Also Seismology)

Hales, A. L., "A Look at the Mantle," *Geotimes* (July-August, 1964), Vol. 9, No. 1, pp. 9-13.

Takahashi, Taro, and W. A. Basset, "The Composition of the Earth's Interior," *Scientific American* (June, 1956), Vol. 212, No. 6, pp. 100-108.

Verhoogen, John, "Temperatures within the Earth," *American Scientist* (June, 1960), Vol. 48, No. 2, pp. 134-159.

The Continents

14

Topographic Features of the Earth's Surface

The relief features of the earth are summarized in Fig. 14-1. This diagram shows that the very high and very low portions of the earth are very small in area, and that two levels, the continental and deep ocean platforms, make up most of the surface. Note that the difference between the highest and the lowest points is about 12 miles, which is very small compared with the radius of the earth (12/3957 = 0.3%), so that the earth is smoother than a billiard ball. This comparison shows that the features we observe on the surface are very slight compared with the earth as a whole.

The topographic features shown by the diagram are not randomly distributed. In general, the rock types and structures of these features are also distinct. Analysis leads to the following main features of the earth's surface:

Continental features
    Continental platforms (including continental shelves which are submerged parts of continental platforms)
    Mountain ranges
Oceanic features
    Deep ocean platforms
    Submarine mountains*
    Mid-ocean ridges*
Continent-ocean border features
    Continental slopes
    Volcanic island arcs*
    Ocean deeps or trenches

One interesting aspect of the earth's surface is that, for the most part, ocean occurs on the opposite side of the earth (antipodal) from continents. Only 6 per cent of the earth is not this way, or, said another way, 82.6 per cent of the continental areas have ocean antipodal. The reason for this is not known, but the odds against this being a chance relationship have been calculated at 14 to 1.

Differences between the Oceans and the Continents

Certainly the differences between ocean and continent are the most pronounced contrast on the earth's surface. One important contrast is in rock type — the ocean basins are composed mainly of basalt; the continents, of igneous and metamorphic rocks of granitic composition. To understand how these compositional differences might account for the two main levels in the elevation diagram (Fig. 14-1), we must turn to a different type of study.

Surprisingly, an important clue came from a surveying error made about the middle of the nineteenth century when the English were mapping India. Their methods of surveying were much the same as our modern ways; only the instruments were different. Because it is easier and more accurate to measure angles than distances, most surveys use triangulation. To use this method, a

*These features do not show on the diagram and are described later.

Folded mountains in Sulaiman Range, Pakistan. Photo from NASA.

Figure 14-1. Elevations of the earth's crust.

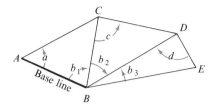

Figure 14-2. Location of points by triangulation. *AB* is base line. Angles *a* and b_1 are measured so that *C* is located relative to *A* and *B*. Angles *c* and b_2 are now used to locate *D*. Angles b_3 and *d* will now locate *E*, and the process can be continued.

base line is accurately measured, generally by actually taping it. From each end of this measured base line, the angle is measured to a third station that must be visible from both stations on the base line. The third station can now be located, either by calculating its position by trigonometry or by making a scale plot. The triangle is checked by measuring the angle at the third station. Now, two of these stations can be used to locate another station, and so forth, until all stations are located. (See Fig. 14-2.)

The location of each station in the Indian survey was determined also by astronomical observations in the same way that a navigator at sea locates his position. It was discovered that the northern stations near the Himalaya Mountains were located too far south by astronomical

methods, according to the locations obtained by the method of triangulation. Although the differences in location by these two methods were small, they were greater than could be due to errors in measurements, especially as corrections for the curvature of the earth had been applied. Apparently, the great mass of the Himalaya Range attracted the plumb bob that was used to determine the zenith in the astronomical determination of location. A level bubble and a plumb bob are basically the same, and one or the other must be used to establish the zenith from which astronomical measurements are made. Knowing the size of the mountain range and using Newton's law of gravitation, it is easy to calculate the expected amount of attraction of the plumb bob by the range; in this case, the observed de-

flection was much less than the expected deflection (Fig. 14-3). This implies either that the mountains are composed of rock lighter than the plain, or, if both mountains and plains are composed of the same rock type, that the rock under the mountains is lighter than the deeper rocks. The latter case is supported by seismic data and geologic mapping and is called the *roots-of-mountains hypothesis.* (See Fig. 14-4.)

This suggests that the granitic continents, like the mountains, may stand higher because they are less dense and thicker than the basaltic oceans, granite being less dense than basalt. A possible layer in which the crustal rocks may float is now sought, and we will begin by considering the strength of crustal rocks.

Laboratory tests show that granite, the main rock composing the continents, has a compressive strength of about 22,000 pounds per square inch. The specific gravity of granite is about 2.7; that is, it weighs 2.7 times as much as an equal volume of water. From this we can calculate how high a column of granite can support itself; that is, at what height the granite will weigh more than its compressive strength.

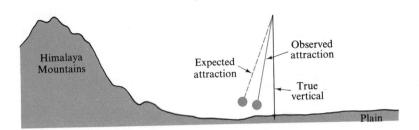

Figure 14-3. The gravitational attraction between the Himalaya Mountains and the plumb bob causes the plumb bob to be deflected from the true vertical. The deflection is less than expected, suggesting that mountains are underlain by light rocks.

1 cubic foot of water weighs 62.4 pounds

1 cubic inch of water weighs $\frac{1}{1728} \times 62.4 = 0.036$ pound

∴1 cubic inch of granite weighs $2.7 \times .036 = 0.097$ pound

1 mile = 5280 feet = 63,360 inches

∴A mile-high column of granite 1 inch in cross-section weighs $63,360 \times .097 = 6,145.92$ pounds

Therefore, the weight of a column of granite a mile high produces a pressure of 6,146 pounds per square inch at its base. Thus, if the granite approaches four miles in thickness, its compressive strength is exceeded and the base will crumble under its own weight. The continental crust exceeds this thickness, and some mountain ranges are higher. How is this possible? A factor we have so far neglected, but which will not basically alter the problem, is the change in strength of granite due to the increase in temperature and pressure at depth. The specific gravity of rock will increase due to pressure, the pressure will tend to strengthen the rock, and the higher temperature will weaken it. The total of these effects is to increase greatly the height of the column of rock that is self-

supporting, but clearly a limit will be reached. Because the conditions deep in the crust and below it cannot be duplicated in the laboratory, and because the exact composition of the rocks is not known, an accurate height cannot be calculated. It is clear, however, that the continents cannot be supported by their rigidity or strength.

Then what does support the continents? Mountain ranges stand high because they have light roots, much as a big iceberg stands higher above the water than a small iceberg because its root is deeper. In a similar way, the crust of the earth may float on the mantle, and the lighter rock of the continents floats higher than the thinner, denser rocks underlying the oceans.

Apparently, then, the higher temperature and pressure at some depth make the rocks plastic, and the crust floats on these plastic rocks. Because the rocks in this zone are weak, probably because they are near the melting point for the pressure there, this could be the zone where magmas are generated. The theory that the crust floats on this viscous substratum is called *isostasy*. The depth at which the rocks become plastic is not easy to determine. It is not at the Mohorovičić discontinuity, which marks the base of the crust, but is

between 60 and 125 miles below the surface, and so is in the mantle. It is deeper under the continents than under the oceans. Apparently below this depth the increased pressure increases the strength of the rocks and they behave as solids, at least to rapid deformations such as seismic waves generated by deep earthquakes. A low-velocity layer in the upper mantle was, on the basis of sparse seismic data, thought to exist and was proved in several ways. Study of the seismic waves generated by underground nuclear blasts revealed shadow zones due to refraction of waves by the low-velocity zone (Fig. 14-5); the intense Chilian earthquake of 1960 set the zone above this layer vibrating, much like the ringing of a bell; and reflections from this zone have been detected on new sensitive seismographs. Proof of the long-suspected weak layer adds much credence to the isostasy hypothesis. This low-velocity zone is apparently important in the other processes that form the earth's surface structures that are described in later chapters.

Further evidence for isotasy can be found in areas recently covered by continental glaciers, around some recent volcanoes, and by measurements of the gravitational attraction near most mountain ranges. Parts of the earth that were covered by thick accumulations of ice during the recent glacial period were apparently

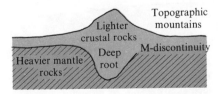

Figure 14-4. The root of a mountain range.

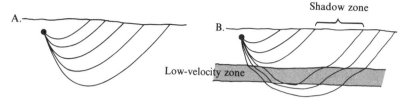

Figure 14-5. The recognition of the low-velocity zone in the upper mantle. Part A shows the paths of seismic waves with no low-velocity zone. Part B shows diagrammatically how refraction in the low-velocity zone produces a shadow zone at the surface. In both diagrams, for simplicity, the refraction that occurs at the M-discontinuity is omitted.

depressed by the weight, and now that the ice has melted, they are rising. This rising may be due in part to elastic or plastic compression of the crust, but calculation suggests that these effects are not enough to account for the post-glacial uplifts and that isostasy was involved. The Hawaiian Islands are a group of young volcanoes, and their weight has apparently depressed the crust, creating a moat of deeper water surrounding them, as shown in Fig. 14-6. On the continent, erosional debris would mask a similar occurrence where a rapidly emplaced volcano depresses the crust. Measurements of the gravity force — which is, of course, affected by the mass distribution under the point — when corrected for such effects as elevation and nearby mountain mass, show that most of the topographic features of the earth are compensated at depth, and thus tend to

prove the hypothesis of isostasy. This suggests that movements in the plastic zone that tend to restore isostatic balance are fairly rapid in geologic terms.

Deformation of Rocks

The continents display structures on many scales. Bent, broken, and otherwise deformed rocks are seen in small and large exposures. Study of these deformed rocks reveals much information on the kinds of forces that caused the deformation. The occurrence of these deformed rocks in distinct belts shows larger, more fundamental structures that occupy many thousands of square miles. These larger structures are best developed in mountain belts. Further study reveals many other unusual features of these large belts of deformed rocks and suggests that

they may show how continents form. Thus, this chapter progresses from the small, easily studied features of deformed rocks to much larger, more conjectural structures.

When rocks are subjected to forces, they may either break or be deformed. Abundant evidence of both kinds of response is common as faults and folds, respectively. Whether a rock breaks or is folded depends on its strength, which depends on the type of rock and the prevailing temperature and pressure. Thus weak rocks tend to be folded and brittle rocks tend to be faulted. Folding is favored at depth, and faulting is more common near the surface.

Folds

Some general facts about the nature of the deforming forces can be learned from the study of folded and faulted rocks. Folded rocks generally record compressive forces, and folded rocks occupy less length than they did before they were folded (Figs. 14-7 and 14-8).

Folds may be upright or overturned, open or tight. The basin-like folds are called *synclines* and the hill-like folds are called *anticlines* (Fig. 14-9). A better definition, especially if the folds are overturned

Figure 14-6. Depression of the crust into the viscous mantle by addition of a weight such as a volcano. As erosion removed the weight, the crust would return to its original position. The Hawaiian Islands and some other volcanic piles in the Pacific Ocean have apparently depressed the crust in this manner, because they are surrounded by a ring of deeper water.

Figure 14-7. Folded rocks occupy less length than unfolded rocks.

(as shown in Fig. 14-9), is that in anticlines the oldest rocks are in the center and in synclines the youngest rocks are in the center. Remember that the oldest rocks are at the bottom of horizontal sedimentary rocks. A bed is overturned (upside-down) if it has been bent more than 90 degrees.

Folds are not all horizontal but may plunge, as shown in Fig. 14-10. Notice the difference in the pattern of the ridges between the horizontal folds in Fig. 14-9 and the plunging folds shown in Fig. 14-10. A short anticline that plunges in both directions can be called a dome. (See Fig. 14-11.)

All rocks do not respond in the same way to the same forces. For example, if a series of sandstone and shale beds is folded, the weaker shales will flow into the crests of the anticlines and the bottoms of the synclines.

A special type of fold, caused in most cases by a vertical movement, is called a monocline. (See Fig. 14-12.) A fault is generally found below most monoclines. In the usual case, brittle basement rocks respond to the forces by faulting, and the more pliable overlying sediments respond by bending.

Faults

A fault is a break in rocks along which movement has taken place. Four types of faults can be distin-

Figure 14-8. Folds in Rocky Mountains in southern British Columbia. Photo from Geological Survey of Canada.

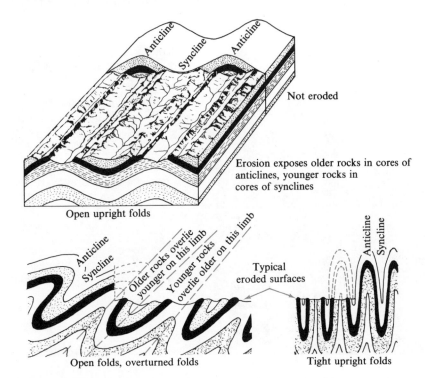

Not eroded

Erosion exposes older rocks in cores of anticlines, younger rocks in cores of synclines

Open upright folds

Open folds, overturned folds

Tight upright folds

Typical eroded surfaces

Figure 14-9. Types of folds.

Figure 14-10. Plunging folds. In synclines erosion exposes younger rocks in the core and in anticlines, older rocks.

guished, but there is much gradation among these types. Two of these fault types have predominantly vertical movement, and two have mainly horizontal movement.

The two types of faults with mainly vertical movement are shown in Figs. 14-13 and 14-14. Note that *reverse* faults (see Fig. 14-15) are caused by compression and that the crust is

Figure 14-12. Monocline. This type of fold is commonly caused by faulting at depth.

Figure 14-11. Oblique aerial photograph of a dome at Sinclair, Wyoming. Photo by J. R. Balsley, U. S. Geological Survey.

Original block

Compressive or reverse fault shortens block

Tensional or normal fault lengthens block

Figure 14-13. Normal and reverse faults.

A.   B.

Figure 14-14. A. A small normal fault. B. A thrust fault at Torquay, Devon, England. The flat upper beds have been thrust faulted above the folded limestones. A. by D. E. Winchester, U. S. Geological Survey; B. Crown Copyright, Geological Survey, reproduced by permission.

Figure 14-15. Development of a reverse fault. Step B is never found in nature because erosion would remove the overhang. In C the bending of beds in the fault zone is shown.

Figure 14-16. Development of a normal fault. In C the bending of beds in the fault zone is shown.

shortened as a result of this type of faulting. *Normal* faults (see Fig. 14-16) lengthen the crust and so are tensional features. The term normal is poor because normal faults are not normal in the sense of being more common than others. At the time the term was first applied they were thought to be more common, so the term is an historical mistake. Compressive features such as folds and reverse faults are much more common, suggesting that most orogenic forces are compressional. Normal faults are generally associated with the stretching that accompanies vertical forces, such as uplifts, rather than horizontal tension in the crust.

Very few measurements of stress in rocks have been made; but where the vertical stress has been measured, it is generally more than the theoretical value, which is the weight of the overlying rocks. This suggests that vertical forces do exist in the crust.

Thrust faults are reverse faults that are nearly flat and so have mainly horizontal movement. Although this definition may lead one to think that they are merely a special case of reverse fault, in practice they are very different. This is not to imply that there are no borderline cases where either the term *reverse* or *thrust* can be used. Many thrust

faults have relatively thin upper plates that have moved long distances. Such instances pose real problems because it can be calculated that the friction on the fault plane is so great that the force necessary to overcome this friction and move the upper plate would crumple the upper plate. The problem is discussed below. At other places thrust faults involve basement rocks and clearly reveal crustal compression and shortening. Some thrust faults are formed by extreme folding where the lower limb of the fold is pushed out as a thrust fault. In some cases this has occurred at such depth that the rocks were plastic, and metamorphic rocks are in the cores of the folds. The most extreme example of thrust structure is in the Alps of Europe. Here, in the various ranges of the Alps, thrust sheets and great overfolds have moved many miles and have piled up, one on the other, as shown in Figs. 14-17 and 14-18.

The recognition of the great thrusts in the Alps began in 1841 when Arnold Escher von der Linth, who was a professor at the University of Zürich, saw that at Segnes Pass in Glarus, older rocks overlay younger rocks with fault gouge or mylonite on the contact. He interpreted the relationship as thrust faulting, and he was able to con-

Figure 14-17. Cross-section through part of the Alps. Nappe is the term applied in the Alps to the overthrust rocks. Note the large distances that the overthrust rocks have moved. From J. C. Crowell, *Bulletin,* Geological Society of America, Vol. 66, 1955.

Figure 14-18. Mount Pilatus in the Alps south of Lucerne, Switzerland. A syncline apparently modified by faulting forms the summit of the mountain. The discordant beds on the left are probably thrust faulted against the syncline. Photo from Swissair Photo Ltd.

Figure 14-20. San Andreas and related faults in California show locations of historic surface movement. From G. B. Oakeshott, California Division of Mines and Geology, *Bulletin* 190, 1966.

Figure 14-19. Lateral fault.

vince other geologists; but such a structure was unheard of and he did not publish his conclusions. It was not until 1884 that Marcel Bertrand recognized and published the true thrust relation at Glarus, and at the same time he hinted that such thrusts might explain other problems in Alpine geology. In the years from 1865 to 1887, geologic maps of the Alps had been made. In the 1890's, the many, far-travelled thrusts piled one on the other were recognized. In the early 1900's, thrusts were recognized in the Austrian Alps, then in southern France, the Pyrenees, Scotland, and in the northern Rocky Mountains.

The second type of horizontal-moving faults will be called *lateral faults* (see Fig. 14-19), although

many other terms, such as strike-slip and wrench have been applied to these faults. Lateral faults are generally nearly vertical and may have considerable displacement. The San Andreas fault in California is of this type although it may also be a transform fault, a type of fault considered in Chapter 16. (See Figs. 14-20 and 14-21.) This fault is over 600 miles long, and the total lateral movement has been estimated at 350 miles, although many geologists who have studied it believe the lateral movement to be much less, and some be-

Figure 14-21. Aerial photo of the San Andreas fault, California. The crushed fault zone forms the diagonal line on the photo. The fault is crossed by a highway in the foreground. Photo by J. R. Balsley, U.S. Geological Survey.

lieve the main movement has been vertical. (See Fig. 14-22.) Interestingly, the earthquake foci on the San Andreas are at depths of five or six miles, perhaps indicating that this long fault is rather shallow. Most lateral faults are much smaller, and it is very difficult to prove long movements on any fault. In the Sangre de Cristo Mountains of New Mexico, a vertical fault of this type is believed to have a 23-mile horizontal displacement, based on the separation of a distinctive body of metamorphic rock. Lateral faults that apparently offset oceanic features will be discussed in Chapter 16.

Most faults have neither all horizontal nor all vertical movement, but have diagonal movement. (See Figs. 14-23 and 14-24.) This is known from study of the grooves and scratches on fault planes that reveal the direction of the last movement. It should be clear from Chapter 13 that most faults have recurring movements; these scratches show only the last movement. It is not generally possible to determine the exact amount of movement on a fault. To determine the exact movement, a point that was once joined must be found on both sides of the fault. (See Fig. 14-25.) Generally, only planar features, such as dikes or sedimentary beds, are available to determine the movement, and such planar features do not always permit unique solutions to the amount of actual movement. Also many fault planes are curved. The movement problem is further complicated on most faults by the deformation of the rocks near the fault. On the fault plane the rocks are ground and milled by the movement, creating what is called *fault gouge*. This is a type of metamorphism described in

Figure 14-22. Evidence for movement on the San Andreas fault. Map A shows an area east of Pilarcitos fault in Figure 14-20, and Map B is located at the junction with the Garlock fault. The geologic time terms are defined in Figure 1-1. Middle Pleistocene was about one-half million years ago and middle Miocene about 20 million years ago. Do you feel that this evidence proves conclusively that large movements have occurred on the San Andreas fault? From T. W. Dibblee, Jr., California Division of Mines and Geology, *Bulletin* 190, 1966.

A. Normal fault with diagonal movement   B. Reverse fault with diagonal movement

Figure 14-23. Faults with oblique movements.

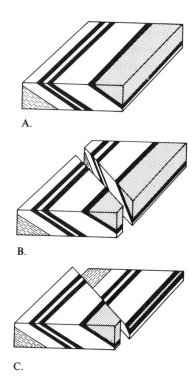

A.

B.

C.

Figure 14-24. Map patterns in faulted rocks. Part A shows the original block. Part B shows the oblique-vertical movement on the fault. Part C shows the same area after erosion has leveled the uplifted block. Can the actual movement on the fault be determined from this map pattern alone?

A and A′ locate the points of intersection of the near margin of the dike with the top of bed at the fault plane

Offset of dike gives horizontal component of movement

Actual movement

Displacement of horizontal bed gives vertical component of movement

Figure 14-25. Determination of the actual movement on a fault. In most cases only one planar feature such as the displacement of the dike or of one of the sedimentary beds is available, and such features only reveal one component of the movement. In this figure the actual movement can be determined by finding the displacement of the intersection of the dike and one of the beds. Practically, in this case, the fault movement can be determined by finding the horizontal component by measuring the offset of the dike, and the vertical movement from the displacement of the horizontal sedimentary beds. This figure shows a highly idealized case. In most instances the rocks are more deformed, especially near the fault plane.

Chapter 5; on deep-seated faults some recrystallization may occur. The rocks near the fault are generally bent or dragged by the movement, making reconstruction of the prefaulting geometry very difficult.

Joints are cracks or breaks in rocks along which no movement has occurred. It is possible in some cases to show that the jointing was caused by the same forces that caused faulting or folding. Most rocks are jointed, and the origin of the joints is not always clear. Deep-seated rocks such as batholiths commonly have joints more or less parallel to the surface, suggesting that such joints form when

the rock expands when the overlying rocks are eroded away. (See Fig. 4-4.)

Inferences from Deformed Rocks

Some inferences about crustal forces can be made solely from study of deformed rocks. One of these is the importance of compression in production of folded and thrust-faulted rocks. This conclusion seems clear, especially in areas where rigid metamorphic basement rocks have been folded or thrust along with the more

plastic sedimentary rocks. At other areas, only the near-surface rocks have been compressed; the deeper rocks are not deformed. This is the case of some thrust faults where the upper plate is not strong enough to withstand the pressure necessary to push it. In such cases, perhaps uplift produces a slope down which the block slides under the influence of its own weight. Such a mechanism is mechanically possible on rather gentle slopes. Folding can also be

Figure 14-26. Gravity thrust fault. The block broke away and slid down the slope.

Figure 14-27. Gravity folding. Plastic sediments tend to flow downslope, producing folds.

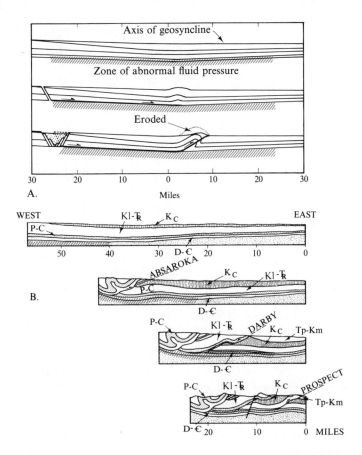

Figure 14-28. Possible role of fluid pressure in overthrust faulting. A. Idealized diagram showing zone of abnormal fluid pressure and incipient fault. B. Suggested example of a thrust belt in western Wyoming that formed as a result of abnormal fluid pressure. The letters designate various sedimentary rock units. From W. W. Rubey and M. K. Hubbert, *Bulletin*, Geological Society of America, Vol. 70, 1959.

caused by plastic beds slumping or sliding downslope. This can be seen, on a small scale, in many sedimentary beds. (See Figs. 14-26 and 14-27.) Another possible way to explain thrust faults is that the rocks may in some way be buoyed up or floated, at least partly, by water in the rocks. This process seems possible, and the high water pressures encountered in some wells are also suggestive. If the upper plate were unweighted in this way, the friction force would be reduced and movement could occur with a very small force. This hypothesis is currently

being tested. (See Fig. 14-28.)

Vertical movements on the continents are easy to prove. Marine sedimentary rocks high on mountain ranges certainly prove such movements. Igneous and metamorphic rocks that formed deep in the earth and are now exposed in mountain ranges also show uplift. Other evidences of vertical movements are the uplifted, wave-cut beaches on many seacoasts, as discussed in Chapter 11. Areas below sea level, such as Death Valley, imply downward movement. The only agent that could have eroded Death Valley below sea

level is the wind, and the wind is much too feeble an eroder to have done this.

Although most of our evidence of forces in the crust comes from study of ancient rocks, some movements can be seen directly. Recent earthquake faults show rapid movements, both horizontal and vertical. Slow creep measured by surveying also is currently occurring on some faults, as described in Chapter 13.

Dating folding and faulting is discussed in Chapter 17, but the principles can be seen in question 28 at the end of this chapter.

Other types of deformation are the igneous and metamorphic processes considered earlier. These processes are commonly part of the mountain-building processes that create folds and faults. Some of the relationships among these processes will be seen in the next section.

Larger Features of the Continents

The most obvious features of the continents are relatively flat continental platforms and linear belts of mountains. (See endpapers.) The earliest idea on the origin of the continents was that they formed during the initial solidification of the earth. Study of the older parts of the continents disclosed that the oldest rocks are metamorphosed sedimentary rocks that are themselves the products of erosion of a still earlier terrane. Radioactive dates have revealed that, in a general way, the oldest rocks are in long belts. Because each of these belts has structures similar to parts of some of our present mountain ranges, they may be the eroded stumps of older moun-

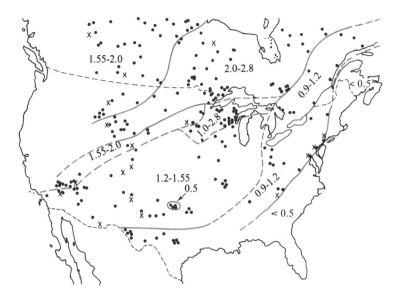

Figure 14-29. Distribution of ages in crystalline rocks from the central part of North America. Circles: ages that are within the limits specified for a given zone on the map; crosses: ages that are outside the limits. Age limits are given in billions (1,000,000,000) of years. After G. R. Tilton and S. R. Hart, *Science* (April 26, 1963), Vol. 140, p. 364.

tain ranges. (See Fig. 14-29.) Thus, mountains may be the key to understanding continental structure.

This discovery suggests that the present processes going on at the surface have been going on since the formation of the crust, and that the continents may have formed by accretion. The oldest rocks so far discovered are about 4000 million years old, which suggests a total age of the earth between 5000 and 6000 million years. Thus, the process of continent formation may still be going on. We can look for evidence of this in the mountains where much of the present volcanic and earthquake activity occurs.

Mountains

A mountain is a topographic feature that rises above the level of the surrounding area, more or less abruptly. Various types of mountains are shown in Fig. 14-30. The definition and the illustration do not emphasize the most important geologic aspect of mountains, which is the deformation of the rocks that form the mountains. Most mountains occur in ranges of the complex type, and although single isolated mountains of the other types shown in Fig. 14-30 do occur, they are relatively rare.

Thus, typical mountains have complex internal structures formed by folding, faulting, volcanic activity, igneous intrusions, and metamorphism. In a sense, complex mountains are residual, too. The origin of the internal mountain structure is different from the origin of the topographic expression of mountains. Topographic mountains are formed by uplift and erosion, and so are residual. Thus most mountain ranges have complex internal structure formed at an earlier time, and the topographic expression of mountains is the result of later uplift and erosion — a two-step history. When the work of erosion is completed, only deformed rocks remain as evidence of an earlier mountain range.

Dating mountain building is difficult because three dates are involved:

The age of formation of the rocks,

The age of the internal structure, and

The date of uplift when erosion formed the topographic mountains.

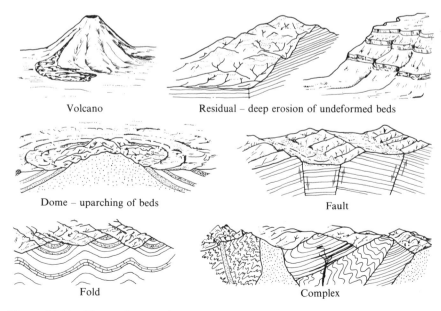

Volcano

Residual – deep erosion of undeformed beds

Dome – uparching of beds

Fault

Fold

Complex

Figure 14-30. Types of mountains.

Commonly the internal structure is formed in several stages, all of which should be dated. Determining each of these three dates in actual cases involves many difficulties.

Geosyncline Theory—Origin of Mountains

In 1859, James Hall, an American geologist who made important studies of fossils, made an important observation. He discovered that rocks in the northern part of the Appalachian Mountains were much thicker and more clastic than were rocks of the same age located further in the interior of the continent. In other words, the present mountains are on the site of a much deeper, earlier basin than is the area to the west. (See Fig. 14-31.)

This observation was extended by James Dwight Dana, who recognized that it was true of the whole Appalachian Mountains. His studies culminated in 1873 when he applied the name *geosyncline* to the great, elongate basin of deposition that was later folded, faulted, and uplifted to form a mountain range. Similar observations have been made all over the world, namely, that many of the present mountain ranges were once elongate basins, or geosynclines.

It has also been noted that parts of some of the geosynclinal mountain ranges are composed of metamorphic rocks and large intrusive bodies of granitic rock termed batholiths. Such rocks form the oldest parts of the continents, and are sometimes called basement rocks because they generally underlie the thin veneer of sediments on the stable parts of the continent.

These observations have been combined to form the geosyncline theory. The following description is of a very idealized geosyncline. The

Figure 14-31. Sedimentary rocks of the same age are thicker and more clastic in mountain ranges than in the interior of continents.

geosyncline begins as a broad, elongate downwarp, generally at the edge of a continent, that is flooded by sea water. Because the geosyncline acts as a trap for sediments, the sedimentation stage begins with this downwarp. In general, the part of the geosyncline near the ocean is more active than the rest, is depressed deeper, and is the site of volcanic activity. These geosynclines are large, the order of many hundred miles long and up to several hundred miles across.

In spite of the great thicknesses of rocks in geosynclines, there is abundant evidence that the water was at no time very deep. The cross-bedding, ripple-marks, mudcracks,

and types of fossil life displayed in these rocks are all features that form today in relatively shallow water. Thus, the reason for the great thicknesses of sedimentary rocks must be that the geosynclines continued to sink as they were filled with sediment. In some places the features in the volcanic, or at least more active, part of the geosyncline suggest that the water may have been deep there at times. In addition to the evidence of deep-water fossils, such features as graded bedding and the channeling due to mud and muddy water sliding down slopes suggest some deep water. Such slides are known to occur today on the slope between the continental shelf and the deep

1. Initial sedimentation stage (above)

2. Orogeny accompanied by plutonic activity in the volcanic or more active part of the geosyncline (not shown).
Sedimentary rocks are folded, thrust faulted, and uplifted.

3. A new depositional basin is produced by downwarp of folded rocks in front of the orogenic area. Clastic fan in this foredeep is deposited in part during deformation (above).

4. Main orogenic phase accompanied by more plutonic activity.

5. Uplift and erosion, see Fig. 14 – 33.

Figure 14-32. Stages in development of a geosyncline.

ocean. It is interesting that almost no deep-water deposits of the type now forming in the deep oceans far away from the continents, oozes and red clay, have been recognized in old rocks, perhaps because such deposits are thin and difficult to recognize.

Differences in environment are shown by differences in sedimentation. As shown in Fig. 14-32, the geosyncline gradually tapers off to form the shelf that borders the continent. In most cases the continent has little relief; this is shown by the sediments (shale, limestone, and fairly pure quartz sandstone) that form on the shelf. Such sediments reveal rather complete weathering. This same suite of sedimentary rocks also develops nearby in the geosyncline, but here the rocks are much thicker because of the greater downsinking. The volcanic part of the geosyncline, on the other hand, is the scene of more rapid sedimentation as a result of deeper and more rapid downwarping. In addition to volcanic rocks, the sediments here are rapidly deposited, poorly sorted, quickly buried clastic rocks such as impure shales, dark sandstones with clay matrix (graywacke), and coarser clastic rocks.

The source of these "dumped-in" clastics is a problem. Some of the material may come from the continent, but the volume generally requires a source on the other side of the geosyncline. The problem is further complicated, as we shall see, by the later igneous and metamorphic activity, which obscures the early situation in this part of the geosyncline. Because a geosyncline commonly forms at the edge of a continent with the volcanic part on the ocean side, some geologists have suggested that the source of these

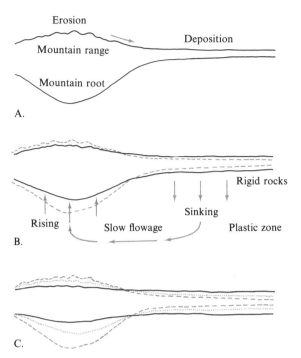

Figure 14-33. Removal of a mountain root by erosion and isostasy. Erosion removes material from the mountains, thereby lightening them. This can result in isostatic uplift that reduces the depth of the root. The place where the eroded material is deposited may also be depressed by the additional weight. Several stages in the process are shown in the diagram.

clastics was a continent now below sea level; but recent studies show that this is not indicated by the structures of the continental margins. Possible present-day geosynclines will be studied in Chapters 15 and 16.

The sedimentation phase of a geosyncline ends when it stops sinking and stops receiving sediments, generally after 30,000 to 50,000 feet of sedimentary rocks have been deposited. The next phase is the orogenic (mountain-making) phase during which (1) the sediments are deformed to make mountain structure and (2) the volcanic part of the geosyncline may be changed by igneous and metamorphic activity into a rigid mass of granitic composition, thus becoming continent-like. The history of actual geosynclines is

generally more complex, with overlap and repetition of these three phases. Some geologists believe that this is the way that the continents are formed — but not all geosynclines are at the edges of continents, and other theories have been suggested. The basement on which geosynclines form is exposed only in the less active parts of geosynclines, and here the basement is continental. No pre-geosyncline basement is exposed in the more active volcanic part of geosynclines, and here the geosyncline could have formed over oceanic crust.

The orogenic phase of the geosyncline is probably the most interesting, most important, most studied, and least understood phenomenon in geology. In this phase the geosyncline is folded, apparently by compressive forces. In the volcanic part of the geosyncline, the folding may be accompanied by metamorphism and emplacement of granitic rocks. In this way part of the geosyncline may be transformed into a rigid mass. It should be noted that granitic rocks and metamorphism are not always confined to only the volcanic part of the geosyncline, and, at places, parts of the volcanic geosyncline may not be affected at all. At a few places, batholiths, generally small ones, occur in non-geosynclinal areas. At this stage a new depositional basin generally develops over the border between the old geosyncline and the shelf or continent, and a new sedimentation cycle may ensue. This may be followed by other episodes of folding, faulting, and thrust faulting in this or other parts of the geosyncline. Metamorphism and emplacement of batholiths may or may not accompany these orogenic events. These events may be spread over a few hundred million years. Geosynclinal folding and faulting occur, probably episodically, over long lengths of time, but emplacement of batholiths appears to be much more rapid. There is an unfortunate tendency to date mountain building from the easily obtained radioactive dates of batholith crystallization, even though many batholiths are clearly earlier or later than the deformation.

The final pulse may be the emplacement of later, small granitic batholiths. These late batholiths are emplaced either before or during the post-orogenic volcanism, if any, and are similar in chemical composition to those volcanic rocks.

After the deformation the geosynclinal areas generally are uplifted to form mountain ranges. The geosynclinal deformation produces mountain structures, but topographic mountains do not exist until this uplift occurs. The uplift is probably caused by the development of a light granitic root by igneous and metamorphic processes. Then the light rocks tend to float up under the influence of isostasy. The uplift of some mountain ranges occurs much later than the deformation. The development of the root and the uplift are apparently late features and do not occur until after the forces that caused the initial downwarp and those that caused the deformation cease to operate.

Once a mountain range exists, isostasy may control its future to a large extent. As soon as it rises, a mountain range is subjected to erosion. As erosion removes weight from the mountains, the buoyancy of the root probably causes the range to rise, as shown in Fig. 14-33. This process should go on until the root has risen to the level of the rest of the continent and the topographic mountains have been reduced by erosion to the level of the rest of the continent. In this manner a new section of the continental platform may be formed. The rocks of this section would display mountain structures, and this process may produce an age distribution similar to that shown in Fig. 14-29.

Why geosynclines form is not known, but some possible hypotheses will be considered in Chapter 16. The formation of continents is an important problem. Are they original features of the crust merely modified by geosynclines, or have they grown during geologic time? The most common theory of origin of continents is that they have been distilled some-how from the upper mantle. It is clear from earlier chapters that unless the continents had been rejuvenated, they would have been eroded to base level soon after their formation. This rejuvenation could have occurred either by building up the continents from below or by lateral spreading. The radioactive dates of basement rocks and the geosyncline theory suggest that the continents have grown through geologic time.

Examples of Continental Structure

The main structural provinces of the United States will show that the distribution of deformed rocks is not random. The structural provinces considered are shown in Fig. 14-34, which should be compared with the landforms map. (See endpapers.)

The *Continental Interior* is the stable part of the continent between the geosynclinal mountains. For the most part this province is low lying and is covered by a thin veneer of sedimentary rocks. To the northeast the sedimentary rocks are thin and the ancient metamorphic rocks of the Canadian Shield are exposed. To the west the elevation increases on the Great Plains. The Continental Interior is, for the most part, only slightly deformed and contains a number of gentle basins and domes. (See Fig. 14-35.) Faulting is minor, folds are very gentle, and igneous rocks are few and far between. As the mountains in either direction are approached, the sedimentary rocks become thicker, folds and faults can be easily seen in the outcrops, and intrusive and volcanic rocks become more common. In addition, the Con-

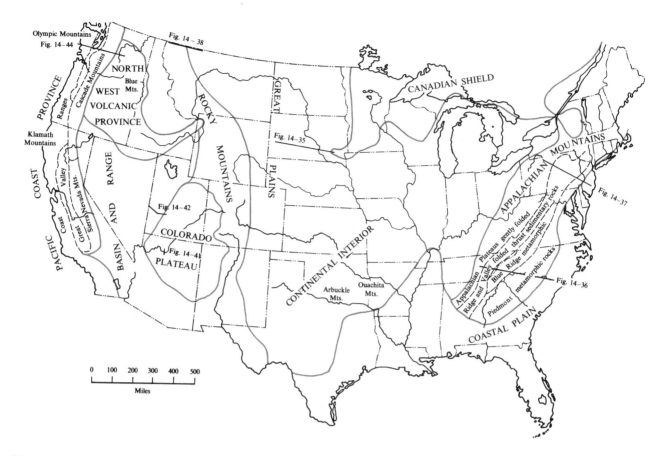

Figure 14-34. Geologic provinces of the United States. Based on geologic structure and surface topography. The boundaries of such provinces are necessarily arbitrary. Cross-sections showing geologic structures are located on the map.

tinental Interior is interrupted by the strongly deformed rocks of the Arbuckle and Ouachita Mountains in southern Oklahoma and adjacent Arkansas.

The *Appalachian Mountains* are typical geosynclinal mountains. The eastern side is dominated by igneous and metamorphic rocks, and the west is mainly sedimentary rocks. In the southern Appalachian Mountains, metamorphic rocks at low elevations form the Piedmont, and similar rocks at higher elevations form the Blue Ridge. In this area the sedimentary rocks are mainly thrust faulted in the Ridge and Valley area. (See Fig. 14-36.) Further north, in Pennsylvania, the sedimentary rocks are mainly folded. (See Fig. 14-37.) West of the Ridge and Valley area is the Appalachian Plateau where the sedimentary rocks are only mildly deformed. New England has mainly igneous and metamorphic rocks, and the folded and thrust belt narrows.

The *Atlantic Coastal Plain* is a thin cover of undeformed sedimentary rocks that presumably overlies eroded metamorphic and batholithic rocks similar to those of the Piedmont.

Figure 14-35. Cross-section across the Great Plains from the Black Hills of South Dakota to the Canadian Shield in eastern South Dakota. After David White and others, U. S. Geological Survey, *Bulletin* 691, 1919.

Figure 14-36. Cross-section through the southern Appalachian Mountains. The Piedmont and Blue Ridge areas contain mainly metamorphic rocks, and the Valley and Ridge Belt is largely thrust-faulted sedimentary rocks. From P. B. King, Geological Society of America, *Guides to Southeastern Geology*, 1955.

The *Rocky Mountains* are also geosynclinal mountains. The northern Rocky Mountains are folded and thrust faulted like the western part of the Appalachians (see Figs. 14-38 and 14-39), and the many thrust faults indicate considerable shortening. The western part of the northern Rocky Mountains is the Idaho batholith and the batholiths of northeastern Washington. Again this is similar to the Appalachians. The rest of the Rocky Mountains is different. In Wyoming the mountains are mainly uplifts with some faulting on the margins, and there are large sedimentary basins between the mountain ranges. These structures are something like those of the Basin and Range Province but differ in that the ranges and the basins are much larger in Wyoming. The age of faulting, and the rock types and the internal structures exposed in the mountains, are also very different. In the central Rocky Mountains ancient rocks like those of the Canadian Shield are exposed. In Colorado, unlike Wyoming, there are more ranges than basins. (See Fig. 14-40.) In New Mexico these structures merge with those of the Basin and Range Province.

The *Colorado Plateau* may be the cause of the narrowing of the Rocky Mountains. This is an unusual structural feature in that it is an area at high elevation with a fairly thick accumulation of sedimentary rocks that is relatively undeformed. Why this province is undeformed is unknown because the rocks surrounding it are deformed. The sedimentary rocks are flat-lying and are broken by a few nearly vertical faults and monoclines. (See Fig. 14-41.)

The *Basin and Range Province* is an area of many relatively small block-faulted mountain ranges. (See Figs. 14-42 and 14-43.) The rocks that make up these ranges are greatly

Appalachian Plateau | Ridge and Valley   Piedmont | Coastal Plain
A.

B.

Figure 14-37. A. Cross-section showing the structures in the Appalachian Mountains in Pennsylvania. Here the sedimentary rocks are folded rather than thrust as in Figure 14-36. After Douglas Johnson. B. The Appalachian Mountains in Pennsylvania from 570 miles high. The plunging folds of the ridge and valley area are in the upper left. The Susquehanna River flows from upper left to lower right. ERTS photo.

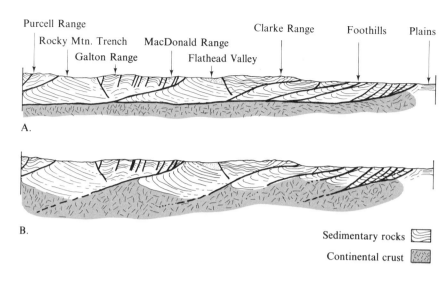

Purcell Range

Rocky Mtn. Trench

Galton Range

MacDonald Range

Flathead Valley

Clarke Range

Foothills

Plains

A.

B.

Sedimentary rocks

Continental crust

Figure 14-38. Cross-section through the northern Rocky Mountains near the International Boundary. The sedimentary rocks here are thrust faulted, and the total movement may be many miles. Two interpretations of the structure are shown. A assumes that the thrusts join, and the deformation is mainly in the sedimentary rocks. Recent seismic data support this interpretation. B assumes that the basement rocks are involved in the deformation.

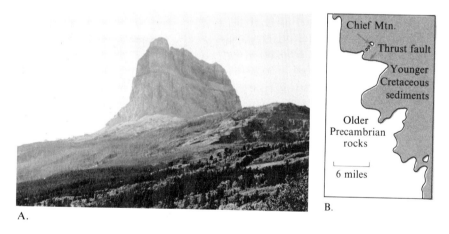

Chief Mtn.

Thrust fault

Younger Cretaceous sediments

Older Precambrian rocks

6 miles

A.

B.

Figure 14-39. A. Chief Mountain, Glacier National Park, Montana. Chief Mountain is an erosional remnant of a thrust plate of older rocks thrust over much younger rocks. B. Map showing thrust fault near Chief Mountain. The minimum displacement on the thrust is the distance between the farthest east and the farthest west that the thrust fault is exposed, a distance of about 12 miles (not shown here). The movement was slightly north of east as shown by gouges on the fault plane. Part A from National Park Service. Part B from C. P. Ross and others, *Geologic Map of Montana*, U. S. Geological Survey.

deformed by folding, faulting, and igneous and metamorphic processes. Thrust faults with many miles of displacement can be traced from range to range. The farthest west of these ranges is the Sierra Nevada in California that is the batholithic part of the geosyncline in this area.

The *Pacific Coast Province* is a very complex, geologically young area. It is geologically active with much seismic activity and with the only active volcano in conterminous United States. The San Andreas fault, described earlier, is in this area. The Coast Ranges, consisting of complexly folded and faulted rocks, form the western part of this province. The central part of the province is a lowland except in northern and southern California. The eastern part of the province is, in California, the Sierra Nevada Mountains that are related to the Basin and Range Province. North of the Sierra Nevada is the Cascade Range. This range is composed of volcanic and continental sedimentary rocks capped by very young, high volcanoes. The volcanic rocks of the Cascade Mountains have more in common with the Northwest Volcanic Province to the east than with the Pacific Coast Province, but other structural and sedimentary features are related to the rest of the province.

The *Northwest Volcanic Province* consists of volcanic rocks of several ages. (See Fig. 14-44.) The Columbia River Basalt covers southeastern Washington and adjacent Oregon with several thousand feet of basalt flows. Much of Oregon and the Snake River Valley in Idaho are

Figure 14-40. Looking north at the Rocky Mountain front south of Denver, Colorado. The ranges on the left are composed of metamorphic and igneous rocks. In the foreground and middleground the eroded remnants of sedimentary rocks are prominent. Thus the basic structure shown is one flank of a large anticline. In the distance on the right is Table Mountain, formed by a resistant lava flow. Photo by T. S. Lovering, U. S. Geological Survey.

Figure 14-41. Cross-section of Grand Canyon from Hopi Point, on the south rim, to Tiyo Point, on the north rim. From U. S. Geological Survey.

covered by younger volcanic rocks. In this area are exposed many recent volcanic features not yet old enough to be much affected by weathering. The volcanic rocks extend into the northwest corner of the Basin and Range Province and are here considered part of that province because they have been deformed by basin

Figure 14-42. Diagrammatic cross-section of part of the Basin and Range Province. From J. H. Mackin, *American Journal of Science*, Vol. 258, 1960.

Figure 14-43. Oblique aerial view of a range in the westernmost part of the Basin and Range Province in southern California. Erosion has removed much of the range front, and recent movement on the fault has revealed its location far in front of the range. Note the difference in the dissection on both sides of the fault scarp. Compare with Figure 9-4. Photo from U. S. Geological Survey.

and range faulting. The Blue Mountains in Oregon are composed of much older rocks that give our only indication of the structures of the geosynclinal rocks in this region.

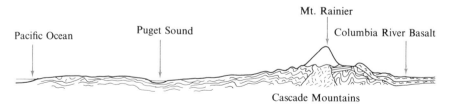

Figure 14-44. Cross-section through central Washington. The Cascade Mountains are composed of volcanic and continental sedimentary rocks that grade into marine rocks to the west. Small batholiths invade the Cascade Mountain rocks, and the large volcanoes such as Mount Rainier were built on top of the existing range. The Columbia River Basalt covers the area to the east.

Summary

The main topographic features of the earth are:

Continental features — platforms and mountain ranges

Oceanic features — platforms, submarine mountains, and mid-ocean ridges

Continent-ocean borders — continental slopes, volcanic island arcs, and ocean deeps

Isostasy is the name given to the theory that the crust of the earth floats in a viscous layer.

The difference in elevation between ocean floors and continents results from each floating at a different height because of differences in composition and thickness.

The continents are made of granite, and the oceans are floored by a thinner and denser layer of basalt.

Mountains are higher than their surroundings because they have deep roots.

The low-seismic-velocity zone, between 60 and 125 miles down, is believed to be the level at which floating occurs.

Folded rocks generally record compressive forces.

Synclines are basin-like folds and are best characterized by having younger rocks in their cores. See Fig. 14-9.

Anticlines are hill-like folds and are best characterized by having older rocks in their cores. See Fig. 14-9.

Reverse faults are caused by compression. See Fig. 14-13.

Normal faults are caused by tension or uplift. See Fig. 14-13.

Thrust faults are nearly flat reverse faults.

Lateral faults are nearly vertical and have horizontal movements.

Some folding and thrust faulting may be the result of downslope movement.

The oldest rocks so far found are 3,980 million years old.

Radioactive dates reveal that, in a general way, rocks of similar age occur in bands, and the oldest rocks are near the centers of continents.

Mountains are of several types, and, in general, the age of the deformation of the rocks is earlier than the time of the uplift that creates the mountains.

Geosynclines begin as elongate basins of deposition, and they are subsequently deformed by folding, faulting, intrusion of batholiths, and metamorphism. Later uplift creates mountains.

Continents may have formed by the intrusion of granitic batholiths and metamorphism in the geosynclinal process.

In the Continental Interior, a thin veneer of nearly undeformed sedimentary rocks overlies ancient igneous and metamorphic rocks that are exposed on the Canadian Shield to the north.

The Appalachian Mountains are composed of igneous and metamorphic rocks to the east and folded and thrust faulted sedimentary rocks to the west.

The Atlantic Coastal Plain is a thin cover of sedimentary rocks that overlie igneous and metamorphic rocks of the Appalachian Mountains.

The northern Rocky Mountains are similar to the Appalachians, with folded and thrust sedimentary rocks, and igneous and metamorphic rocks.

In Wyoming, the Rocky Mountains are ranges with large sedimentary basins separating the ranges.

The Colorado Plateau is an area of slightly deformed sedimentary rocks surrounded by deformed rocks.

The Basin and Range Province is an area of relatively small, block-faulted mountain ranges.

The Pacific Coast Province is a complex area with the Coast Ranges on the west, large valleys further east, and the Sierra Nevada and Cascade Mountains forming the eastern portions.

The Northwest Volcanic Province is composed of young volcanic rocks. The Columbia River Basalt covers southeastern Washington and adjacent Oregon with several thousand feet of basalt flows.

Questions

14-1. What are the main topographic features of the earth's surface, and what is the geologic significance of the two most important features?

14-2. The continents are composed of _____.

14-3. The oceans are underlain by_____.

14-4. What is the evidence for mountain roots?

14-5. How is the crust of the earth supported?

14-6. Why do the continents stand higher than the ocean floors?

14-7. Is it the strength of rocks that enables mountains to stand above the general level of the continents? Prove it.

14-8. At what depth does isostatic compensation occur? Why?

14-9. Can isostasy account for the formation of sedimentary basins? To answer this question, a simple calculation is necessary. Assume an initial depth of 100 feet of water that is completely filled with sediment. Calculate how much the mantle is depressed by equating the net gain in weight in the basin to the weight of the depressed mantle. (Ans. $h = 240$ feet)

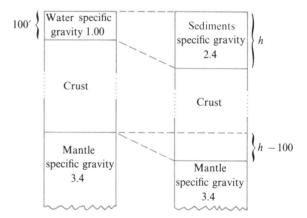

14-10. What are the ages of the basement rocks forming North America?

14-11. How are these rocks distributed?

14-12. Why have geologists concentrated much of their attention on mountain ranges?

14-13. What steps are involved in the formation of mountains?

14-14. How are mountains dated?

14-15. Outline briefly the history of a geosyncline.

14-16. Discuss the origin of continents.

14-17. What type of force produces folding?

14-18. What type of force produces normal faults?

14-19. What type of force produces reverse faults?

14-20. What is an anticline? Discuss several ways one can be distinguished from a syncline.

14-21. What is a thrust fault?

14-22. What evidence would you look for in the field to tell whether a crack in the rocks is a fault or a joint?

14-23. What determines whether a rock folds or faults if it is subjected to a compression? List all factors.

14-24. How could you tell in the field whether a fault is still active?

14-25. How does one determine the amount of movement on a fault?

14-26. Prove to your girlfriend (boyfriend), who is an art major and has taken no science courses, that crustal movements have occurred.

14-27. Compare the continental structure in the Mississippi Valley with the Rocky Mountains.

14-28. The following diagram is a cross-section through an area. Starting with the earliest event, list the sequence of events that have occurred in this area. You may wish to review your answers after reading Chapter 17.

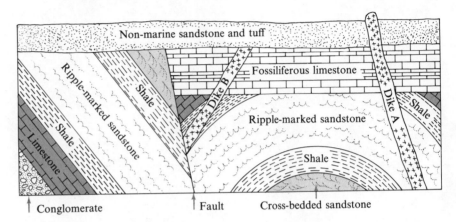

Supplementary Reading

Earth's Size, Shape, and Weight

Daly, R. A., *Strength and Structure of the Earth*. New York: Prentice-Hall, Inc., 1940, 434 pp.

Ewing, Maurice, and Leonard Engel, "Seismic Shooting at Sea," *Scientific American* (May, 1962), Vol. 206, No. 5, pp. 116-126.

Heiskanen, W. A., "The Earth's Gravity," *Scientific American* (September, 1955), Vol. 193, No. 3, pp. 164-174. Reprint 812, W. H. Freeman and Co., San Francisco.

Oliver, Jack, "Long Earthquake Waves," *Scientific American* (March, 1959), Vol. 200, No. 3, pp. 131-143. Reprint 827, W. H. Freeman and Co., San Francisco.

Continents

Bailey, E. B., *Tectonic Essays Mainly Alpine*. Oxford: Clarendon Press, 1935, 200 pp. Reissued by University Microfilms, Ann Arbor, Michigan, 1967.

Engel, A. E. J., "Geologic Evolution of North America," *Science* (April 12, 1963), Vol. 140, No. 3563, pp. 143-152.

Kay, Marshall, "The Origin of Continents," *Scientific American* (September, 1955), Vol. 193, No. 3, pp. 62-66. Reprint 816, W. H. Freeman and Co., San Francisco.

King, P. B., *The Evolution of North America*, Chapter 4, pp. 41-75. Princeton, N.J.: Princeton University Press, 1959, 190 pp.

Milne, L. J., and others, *The Mountains*. New York: Life Nature Library, Time, Inc., 1962, 192 pp.

The Oceans and the Continent-Ocean Borders

15

H. H. Hess, "History of Ocean Basins," p. 599, in *Petrologic Studies, A Volume to Honor A. F. Buddington*, Geological Society of America, November, 1962, pp. 599-620.

Oceans

Until quite recently very little was known about the more than seventy per cent of the earth's crust that is covered by ocean water. The continents, except for the continental shelves that are parts of the continents covered by ocean, can be studied directly. The oceanic crust must be studied by such indirect means as depth soundings and seismic studies and by bottom samples and cores. This difference in the kinds of data available results in knowing different things about these two major parts of the earth's crust. One very obvious aspect of this is that, with a few exceptions, only large structures are known in the oceans. Hopefully, future, more detailed, studies and new methods will provide as much information about the oceanic crust as is known of the continental crust.

This chapter describes the features of the ocean and the next chapter will discuss their relationship to continental structures.

Oceanography began in the 1870's with the voyage of the British ship *Challenger,* the first oceanographic research vessel; however, very little was learned about the shape of the ocean bottoms until the invention of the echo sounder in the 1920's (Fig. 15-1.) Prior to that time, soundings had to be taken with a weight on a line. It was not until World War II that enough soundings were made to discover that some submarine mountains have flat tops. During the late 1950's, in connection with the International Geophysical Year, enough soundings were made to discover the unity of the mid-ocean ridges. These discoveries were made possible by the development of the continuous echo sounder, which can operate while a ship is under way. Using sound waves, it is possible to obtain a topographic profile of the bottom. By using other energy sources, such as dynamite or electric sparks, it is possible to penetrate to various depths into the bottom rocks and so to learn more about the structures. These techniques have been used mainly on the continental shelves in exploration for potential oil-bearing structures. (See Fig. 15-2.)

Beginning in 1970, deep-sea drill-

Atolls in Tuamotu Archipelago, South Pacific. Photo from NASA.

291

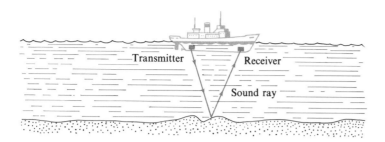

Figure 15-1. An echo sounder. Depth = ½ speed of sound × echo time.

ing became a reality. The drilling ship *Glomar Challenger* can drill deep holes in deep water and recover cores from the holes. This enabled our first samples of rocks below the surface of the ocean floors and helped to confirm many of our ideas about continental drift discussed in Chapter 16.

Oceanic Platforms

The deep ocean platforms or deep ocean floors cover about forty per cent of the earth's surface. (See Fig. 15-3.) The reason for this vast platform was explored in an earlier section; it was found to result from the level at which the denser basaltic crust under the ocean floats isostatically.

Seismic studies have revealed a remarkably uniform structure of the oceanic platforms. (See Fig. 15-4.) Under a depth of two and three-quarter miles of water lies a thin layer of unconsolidated sediments averaging about 1,000 feet thick. This first layer is composed of red clay and oozes. Below this is the second layer of irregular thickness, averaging about a mile and a quarter thick. This layer is believed to be composed of basalt and consolidated sediments. The third layer extends to the Moho and is everywhere very close to three miles thick. The composition of this layer is not known but is traditionally thought to be basalt.

The oceanic platforms are generally fairly smooth surfaces. They are interrupted at places by low hills, submarine mountains, mid-ocean ridges, and fracture zones, all of which will be described below. The submarine mountains indicate some volcanic activity. The low hills are probably also volcanic, but may in part be caused by faulting or other processes. Erosion is very slow on the oceanic platforms, so minor features tend to be preserved. Sedimentation, too, is slow for the most part and in areas far from land is limited to dust settling on the ocean and biologic material, especially shells, sinking from the near surface. Perhaps the smoothest surfaces on the earth are parts of the oceanic platforms near the continental slopes. The smoothness is probably the result of burial of the surface irregularities by sediments moving down the continental slopes, perhaps as density currents. (See Fig. 15-5.)

The age of the ocean floors is remarkably young in geologic terms —

Figure 15-2. Sparker profile near Santa Cruz Island, California. An electric arc created waves that penetrated the bottom sediments and were reflected back to the receiver. The sea floor is shown by the strong reflection at a depth of about 300 feet on the right side. The sedimentary rocks that form most of the cross-section are in fault contact with the volcanic rocks on the left (major fault). Three other faults in the sedimentary rocks are also labeled. Note that the sea floor is displaced about 8 feet by the fault labeled on the top. Courtesy of R. F. Herron who made the profile using EG&G International, Inc., equipment.

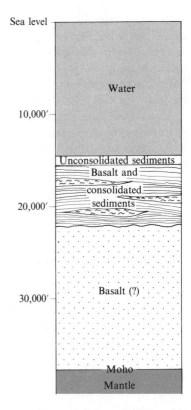

Figure 15-3. A. Continental margin showing typical continental shelf, continental slope and oceanic platform. B. Actual profiles of continental margins with no vertical exaggeration. Part B from K. O. Emery and others, U. S. Geological Survey, *Professional Paper*, 260-A, 1954.

Figure 15-4. Typical section of rocks on the oceanic platforms.

less than 160 million years. The youngest parts are near the mid-ocean ridges, and the oldest near continents. The distribution of ages will be discussed further in the next chapter.

Submarine Mountains

The submarine mountains are mainly in the Pacific Ocean. These mountains, rising thousands of feet above their bases, are apparently basaltic volcanoes that tend to form in lines. Some of these, such as the Hawaiian Islands, extend above sea level.

In some areas these submarine mountains are flat topped. (See Fig. 15-6.) These flat tops are now at depths up to a mile below the surface of the ocean. They are especially interesting because their flat tops were probably produced by wave erosion at sea level, although some types of submarine volcanism may

produce initially flat-topped mountains. Wave-cut benches are being formed now on the present islands, and if this process continues, such islands will become flat topped. Lowering sea level or uplifting an island can preserve such benches, and this can be seen on present-day islands and coastlines. (See Fig. 11-18). Similar benches are revealed in profiles of some of the flat-topped seamounts (Fig. 15-6). Samples dredged from flat-topped seamounts contain basalt fragments and Cretaceous and Early Cenozoic shallow-water fossils, suggesting the age of the wave erosion.

The coral atolls of the Pacific may also reveal subsidence. The corals live in the near-surface zone and form an offshore reef. (See Fig. 15-7.) If the island sinks slowly enough, the corals will build up the reef so that they remain near the surface where their food supply is. This

process may continue until the island is completely submerged, perhaps quite deeply, and only a coral reef remains. At Eniwetok the coral is 4,500 feet deep.

The submarine mountains reveal much of the history of the oceans. Because they are basalt volcanoes, they record the importance of volcanic activity in the oceans. Only about 40,000 submarine mountains have been discovered. This is a small number compared to the number of continental volcanoes, which suggests that either volcanoes form at a slow rate on the oceans, or that the ocean floors are young features. The flat-topped seamounts and the atolls reveal large changes in elevation of the deep ocean floors. Figure 15-8 shows that the distribution of flat-topped seamounts and atolls is not random, but rather that a large elongate area in the southwest Pacific Ocean has subsided, probably indicating a former position of the mid-ocean ridge or rise. The Cretaceous and Early Cenozoic shallow-water fossils dredged from the flat tops suggest the time when this subsidence occurred. Note that the atolls are mainly on the south flank of this area, suggesting that climate may have controlled the growth of coral. Present areas of coral growth are restricted to about 30 degrees on each side of the equator. The present depths below sea level of the flat-topped seamounts are not regular; therefore, as suggested above, differing rates of subsidence may also have influenced coral growth.

Mid-ocean Ridges

The mid-ocean ridges form the longest continuous topographic mountain range on the earth. Their extent can be seen on the map in Fig. 15-9. Their length is about 40,000 miles; their width is 300 to 3,000 miles; 'they rise up to about 10,000 feet above their base; and at places, the highest points form islands. Two parts of this ridge are fairly well known: the Mid-Atlantic Ridge and the East Pacific Rise.

The Mid-Atlantic Ridge has been known for many years because islands such as Iceland, the Azores, St. Paul Rocks, Ascension, St. Helena, and Tristan da Cunha are parts of it. That it is part of a much larger feature was not known until the 1950's; at that time, many soundings greatly increased our knowledge of its nature. It consists of numerous ridges and lines of peaks paralleling the main ridge that extend outward for hundreds of miles on the flanks (Fig. 15-10). The ridges and peaks probably are caused mainly by faulting.

Figure 15-5. The continental margin and ocean floor off eastern United States. The continental shelf and slope are cut by submarine canyons. The smooth surfaces in the deep ocean below the continental slope are probably depositional. The rough topography to the southeast and the scattered hills are from volcanic activity. This is a portion of the Physiographic Diagram of the North Atlantic Ocean published by the Geological Society of America © Copyright by Bruce C. Heezen and Marie Tharp. Reproduced by permission.

Figure 15-6. Flat-topped seamount. Note the small wave-cut bench.

The central crestal zone has one or more discontinuous rift valleys that are up to thirty miles wide. The seismic activity of the ridge is shown by the numerous earthquake epicenters located on it. The bedrock, revealed in dredged samples, is mainly basalt, some of which is pyroclastic with some metamorphic equivalents of basalt such as greenschist and amphibolite, and some serpentine. Some of the peaks that form in lines are apparently basalt volcanoes that

formed along the faults. (See Fig. 15-11.) At many places these features are offset by what appear to be cross faults that, as can be seen in Fig. 15-9, seem to have large horizontal displacements. The unusual nature of these faults is described in the next chapter.

The Mid-Atlantic Ridge has been studied directly on the few islands that are part of the ridge. The islands are mainly basaltic, and recent volcanic activity on Tristan da Cunha

and Iceland has also been basaltic. St. Peter and St. Paul Rocks are tiny islands on the ridge, and sheared peridotite (mylonite) is found there. Radioactive dates on these rocks suggest that they may be as old as 4500 million years. As will be seen in Chapter 18, this is about the time the earth is believed to have crystallized. Thus the composition and the age of these rocks suggest that they may be mantle material somehow brought to the surface by the process that formed the ridge. Peridotite has also been dredged from the Indian Ocean. Elsewhere the rocks are mainly basaltic (including gabbro) with some serpentine. The central rift valley passes through Iceland; and though the evidence is not clear, dilation of up to 240 miles may have occurred.

The East Pacific Rise is similar to the Mid-Atlantic Ridge. It has similar topography except that it lacks the rift valley at its crest. It is also an active seismic area. It rises up to three miles above its base and forms the Galápagos and Easter Islands. It meets North America just south of Baja California, leaves in northern California, and extends to British Columbia. Measurements on the East Pacific Rise show that the heat flow near the crest is up to five times greater than average, and on the flanks is lower than average (Fig. 15-12). The average heat flow on the rise is about twice the normal oceanic heat flow. This high heat flow extends into the Gulf of California. Seismic studies using underwater explosions have revealed that the crust under the East Pacific Rise is only about three-quarters of its thickness elsewhere in the Pacific.

The mid-ocean ridge appears to

Figure 15-7. Coral atolls. A. Development of a coral atoll. B. Perspective diagram of Bikini Atoll and Sylvania flat-topped seamount adjoining it on the west. Bikini Atoll formed on a flat top, unlike the atoll shown in A. Most of Bikini Atoll is submerged and only the small black areas in the diagram are islands. Part B. from K. O. Emery and others, U. S. Geological Survey, *Professional Paper*, 260-A, 1954.

go inland in Africa where it may form the African rift valleys that, up until the discovery of the mid-ocean ridge, were thought to be unique. These African areas are now being studied closely in the hope of learning more about these ridges. The African rift valleys are unsymmetrical but similar to the rifts on the mid-ocean ridges, and are associated with high plateaus, indicating uplift like the mid-ocean ridges. The African rift valleys are not all the same age, and most show evidence of recent geologic activity, such as volcanoes and changes in elevation shown by drainage features. The volcanoes are not ordinary basalt, but are alkali-rich rocks, some of which are extremely rare types. (See Fig. 15-13.) Nearby in the Red Sea, unusual hot brines containing metals are found on the mid-ocean ridge.

The mid-ocean ridges are young, active parts of the crust, and they may play an important role in the development of the crust. Mid-ocean ridges will be considered further in the next chapter.

Fracture Zones

Fracture zones and faults, both of which are associated with mid-ocean ridges, are prominent oceanic features. Some of the faults are indicated by the apparent offsets of the mid-ocean ridges shown in Fig. 15-9. The total displacement of the southern part of the Mid-Atlantic Ridge may be quite large, and recently a probable 300-mile offset northeast of Greenland was discovered. In the northeastern Pacific Ocean, similar features associated

Figure 15-8. Map of the Pacific Ocean showing atolls, seamounts, and flat-topped seamounts. The coral atolls are almost entirely confined to the warm waters within 30 degrees of the equator. The flat-topped seamounts also form distinct clusters. The data on atolls, seamounts, and flat-topped seamounts is from H. W. Menard, *Marine Geology of the Pacific*, McGraw-Hill, 1964.

with the East Pacific Rise are found. In this area great fracture zones and the San Andreas fault are also found. These faults are different from those described in Chapter 14, and they are the *transform faults* discussed in the next chapter.

The longer fracture zones of the Pacific are from 2,500 to 6,000 miles long and are shown in Fig. 15-14. The Mendocino fracture zone is the most northerly of the long fracture zones and departs most from a great circle route. It is over 3,500 miles long and has a south-facing scarp up to 8,600 feet high. The Murray fracture zone is about 2,500 miles in length and is a north-facing scarp up to 7,000 feet high. The Molokai fracture zone is less regular than the others. The Clarion fracture zone is

a trough up to a mile deep for much of its length. The Clipperton fracture zone is over 6,000 miles long, about one-quarter of the earth's circumference. It is a zone about 50 miles wide of irregular relief of 1,000 to 2,000 feet, and is less pronounced than the other fracture zones. Seismic studies show that the crustal thickness changes abruptly at some of these fracture zones. Possible displacements of features are indicated in Fig. 15-14. Magnetic data, which will be discussed further in Chapter 16, also suggest faulting. The fracture zones do not extend inland. Some of the volcanic islands, such as the Hawaiian Islands, cross the fracture zones without displacement, suggesting that the islands are younger than the fractures. Debris

CONTINENTS   OCEAN

☐ Edge of continental block   ☐ Mean position of rift valley   ☐ Major strike-slip faults

☐ Deep-sea trenches   ☐ Mid-oceanic ridge

Figure 15-9. The mid-ocean ridge and associated features. After B. C. Heezen, in *Continental Drift*, Academic Press, 1962.

from such islands apparently masks the fracture zones in some areas. All of these features are believed to be transform faults, a type of fault that will be considered in the following chapter.

Continent-Ocean Borders

The change in elevation between the continents and the oceans is one of the main features on the elevation diagram, Fig. 14-1. Two types of continent-ocean borders can be easily distinguished. They are the continental slopes that are geologically relatively stable and the active volcanic island arcs. Although these

two border types seem distinct, studies suggest that they are related.

It also seems likely that both types of border are related to geosynclines. This relationship will be explored in the following chapter. Sedimentation on the shelf and slope was described in Chapter 4.

Continental Slope

The continental slope appears to be the simpler of the border types, but this may be only apparent. The continental slope begins abruptly at the edge of the continental shelf. The shelves of the world average about 45 miles wide, but the range is from near zero to 800 miles. Their average depth is near 200 feet, and their edge is generally at 400 to 500 feet, although the range is from near sea level to 2,000 feet. The slope angle changes from about 10 feet per mile on the shelf to between two to six degrees on the slope in a distance of under 20 miles. Some continental slopes are as steep as 45 degrees. Compared to the average slope of most mountain ranges, five degrees is very steep. The average drop on

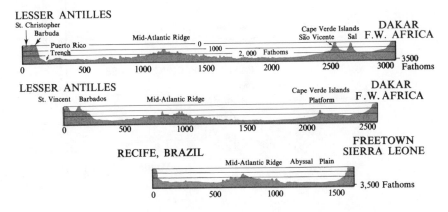

Figure 15-10. Transatlantic topographic profiles showing the Mid-Atlantic Ridge. Vertical exaggeration 40:1. From B. C. Heezen, Geological Society of America, *Special Paper* 65, 1959.

Figure 15-11. The ocean floor on the Mid-Atlantic Ridge. Basalt rocks and some bottom-dwelling organisms can be seen. Courtesy R. M. Pratt.

the continental slope is about 12,000 feet, and the slope angle lessens with depth, making the distinction between the ocean platforms and the slope difficult. Thus, the slopes are a few tens of miles on the average.

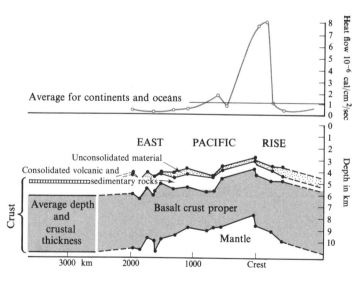

Figure 15-12. Cross-section and heat flow across the East Pacific Rise. The heat flow is very high on the crest of the rise but lower than normal on the flanks. The crust is thin on the crest of the rise. The section on the left is average for the ocean. From H. W. Menard, *Science* (December 9, 1960), Vol. 132, p. 1741.

The region between the slope and the ocean platform is called the *continental rise*. The rise is apparently composed of sediments.

The east coast of the United States is typical of continental slopes (Fig. 15-3). Most areas of the west coast are similar, but the East Pacific Rise and the fracture zones are also present.

Continental slopes are composed of sediments. One question is whether the slope is an area of erosion or deposition. Submarine canyons are cut into the continental slope at many places, as described in Chapter 4. This implies that, at least at these places, erosion is at work to lower the angle of the continental slopes. However, some deposition is taking place also and the sediments were described in Chapter 4. The deltas built by large rivers modify the continental slopes.

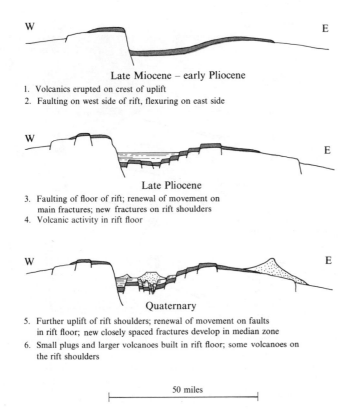

W   E

Late Miocene – early Pliocene

1. Volcanics erupted on crest of uplift
2. Faulting on west side of rift, flexuring on east side

W   E

Late Pliocene

3. Faulting of floor of rift; renewal of movement on main fractures; new fractures on rift shoulders
4. Volcanic activity in rift floor

W   E

Quaternary

5. Further uplift of rift shoulders; renewal of movement on faults in rift floor; new closely spaced fractures develop in median zone
6. Small plugs and larger volcanoes built in rift floor; some volcanoes on the rift shoulders

|——————— 50 miles ———————|

Figure 15-13. Development of the Central Rift in Kenya. From B. H. Baker in *East African Rift System*, University College, Nairobi, 1965.

The average angle of deltas is less than one and one-half degrees, much less than the continental slopes, which implies that the continental slopes are not entirely constructional. Some of the erosion may be caused by deep currents parallel to the slopes as well as by turbidity currents moving down the slopes. At places, the slope is bare of unconsolidated sediments. Thus, both erosion and deposition occur on the continental slopes. (See Fig. 15-15.)

Some continental slopes are in part formed by reefs, as shown in Fig. 15-15. The barrier reef of Australia is an example.

Seismic studies show that the continental crust ends at the continental slope, but the details of the structure in this area are not clear. Faulting has been suggested by some as the origin of continental slopes because many are seismically active. In the next chapter the origin of geosynclines will be considered, and one possibility is that the continental shelf may be one part of a geosyncline and the continental slope may be the more active part.

Island Arcs

At other places, particularly on the rim of the Pacific Ocean, the border is more active with chains of volcanic islands forming arcs. (See Fig. 15-16.) The island arcs were noted early in the history of geologic oceanography, but recognition of many of their features came much later. Deep trenches exist on the ocean side of the volcanic island chains. During the 1920's and early 1930's, the Dutch geologist F. A. Vening-Meinesz devised a method of measuring the value of gravitational attraction at sea. Using a method that measured the period of a pendulum, he had to work in Dutch and U.S. Navy submarines to avoid the effects of ocean waves. He made the extremely important discovery that the acceleration of gravity is low (negative gravity anomaly) over these deep trenches. This discovery shows that the formation of these deep trenches was an anti-isostatic process that upset isostasy locally. (See Fig. 15-17.) The island arcs are also active seismically, and about the same time it was discovered that some of the earthquakes come from very deep in the mantle. (See Figs. 15-16 and 15-18.) Such a concentration of features on the narrow border between the continent and the ocean must have important geologic implications.

The volcanic island arcs pose many problems, but also give much insight into earth processes. The volcanoes of the island arcs are andesitic in composition. Small batholiths of similar chemical composition are found on some, such as the Aleutian Islands. This is also the composition of the volcanoes in the Andes Mountains of South America and the Cascade Range in North America that, together with the island arcs, form what has been somewhat poetically called the "rim of fire" of the Pacific Ocean. The deep oceanic trough off

South America and the deep-focus earthquakes inland suggest that the structure there may be similar to that of the island arcs of the western Pacific. On the oceanward side of this rim of andesite volcanoes, all of the volcanic islands are basaltic (see andesite line in Fig. 15-16). Shallow (35 miles or less) and intermediate (35 to 180 miles) depth earthquake foci parallel the island arcs and lie on planes that dip about 35 degrees under the continents.

Deep-focus earthquakes (180 to 400 miles) occur only on the Pacific rim. (See Figs. 15-16 and 15-18.) Their foci lie on a plane continuous with the shallow and intermediate earthquakes, but dipping about 60 degrees. The change in slope occurs at about 180 miles.

In the next chapter the probable relationship among the features described in this chapter will be considered.

Figure 15-14. Principal fracture zones of the Pacific plotted on a great-circle projection. A straight line on this projection is part of a true great circle. Cross hatching indicates areas of submarine volcanoes. Apparent displacements on the fracture zones are indicated by the displacements of the edge of the East Pacific Rise. From H. W. Menard, *Science*, January 6, 1967, Vol. 155, p. 73.

Figure 15-15. Continental slopes. A. Depositional. B. Erosional. C. Faulted. D. Reef.

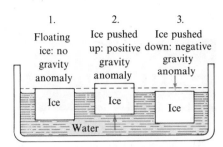

Figure 15-17. Gravity anomalies: Station 1 gravity is normal. At Station 2 the attraction of gravity is higher because more mass is present below the station. At Station 3 there is less mass below the station because the lighter ice displaces the heavier water, so the gravity force is less.

Figure 15-16. Map of the Pacific Ocean showing the relationships among island arcs, deep-focus earthquakes, and deep trenches. The volcanic rocks on the continent side of the andesite line are andesite, and basalt occurs on the oean side. The epicenters of shallower earthquakes lie between the deep earthquakes and the island arcs.

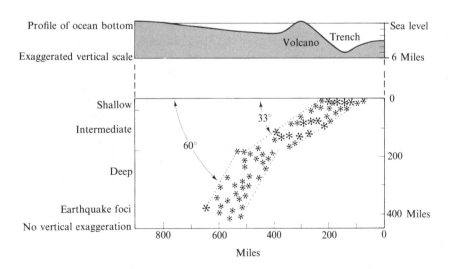

Figure 15-18. Locations of earthquake foci associated with a volcanic island arc (Kamchatka-Kurile area). The foci may locate a fault plane. After Hugo Benioff.

Summary

The *oceanic platforms* cover about forty per cent of the earth's surface. They are covered by about two and three-quarters miles of water, about 1,000 feet of unconsolidated sediments, about a mile and a quarter of basalt and consolidated sediments, and finally three miles of basalt that ends at the Moho.

The ocean floors are remarkably young geologically and are youngest near mid-ocean ridges.

Submarine mountains are basalt volcanoes.

Flat-topped seamounts record wave erosion followed by subsidence.

Coral atolls may form in tropical areas if the corals are able to grow as rapidly as the island subsides.

Mid-ocean ridges form the longest continuous mountain range on the earth. They are about 40,000 miles long, 300 to 3,000 miles wide and rise up to 10,000 feet above their base. They are composed of basalt and have a rugged topography, generally with a rift valley at the crest. Heat flow is high, and many earthquakes occur on mid-ocean ridges.

Fracture zones offset mid-ocean ridges, and some are up to 6,000 miles long. They have high scarps and irregular topography. They are *transform faults* that will be described in the next chapter.

Continental slopes connect the continental shelves and the ocean platforms.

Continental shelves have continental structure and average about 200 feet deep.

The *continental rise* is the region between the continental slope and the ocean platform.

Continental slopes are composed of sediments and are dissected by *submarine canyons*.

Volcanic island arcs are composed of andesite volcanoes and have submarine trenches on their oceanic sides.

Negative gravity anomalies occur over the trenches, implying that the crust is thinner.

Deep-focus earthquakes, up to 400 miles deep, occur on a plane sloping about 60 degrees beneath the continent.

Questions

15-1. What kinds of sediments are found on the oceanic platforms?

15-2. Describe the submarine mountains of the Pacific (composition, arrangement, etc.).

15-3. How do flat-topped seamounts form? What does this suggest about the history of the oceans?

15-4. How long are the mid-ocean ridges?

15-5. Describe a cross-section of the mid-ocean ridges. (A sketch might help.)

15-6. What is the evidence of faulting on the mid-ocean ridges?

15-7. Describe the fracture zones of the Pacific Ocean.

15-8. How is the distinction between continent and ocean made?

15-9. Sketch the ocean-continent boundary at a stable place, and label the parts.

15-10. What does a negative gravity anomaly mean?

15-11. Describe a typical island arc and sketch a cross-section.

15-12. It has been suggested that the East Pacific Rise runs inland near the west coast of North America. How do the structures in that area compare with the structures found on the mid-ocean ridges?

15-13. Describe the similarities between island arcs and geosynclines.

15-14. What are the similarities and differences between geosynclines and continental slopes?

Supplementary Reading

Dietz, R. S., "Origin of Continental Slopes," *American Scientist* (March, 1964), Vol. 52, No. 1, pp. 50-69.

Emery, K. O., "The Continental Shelves," *Scientific American* (September, 1969), Vol. 221, No. 3, pp. 107-122.

Ewing, Maurice, and Leonard Engel, "Seismic Shooting at Sea," *Scientific American* (May, 1962), Vol. 206, No. 5, pp. 116-126.

Fisher, R. L., and Roger Revelle, "The Trenches of the Pacific," *Scientific American* (November, 1955), Vol. 193, No. 5, pp. 36-41. Reprint 814, W. H. Freeman and Co., San Francisco.

Heezen, B. C., "The Rift in the Ocean Floor," *Scientific American* (October, 1960), Vol. 203, No. 4, pp. 98-110.

Heezen, B. C., and I. D. MacGregor, "The Evolution of the Pacific," *Scientific American* (November, 1973), Vol. 229, No. 5, pp. 102-112.

Menard, H. W., "The East Pacific Rise," *Scientific American* (December, 1961), Vol. 205, No. 6, pp. 52-61.

Menard, H. W., *Marine Geology of the Pacific.* New York: McGraw-Hill Book Co., 1964, 271 pp.

Raff, A. D., "The Magnetism of the Ocean Floor," *Scientific American* (October, 1961), Vol. 205, No. 4, pp. 146-156.

Stetson, H. C., "The Continental Shelf," *Scientific American* (March, 1955), Vol. 192, No. 3, pp. 82-86. Reprint 808, W. H. Freeman and Co., San Francisco.

Tazieff, Haroun, "The Afar Triangle," *Scientific American* (February, 1970), Vol. 222, No. 2, pp. 32-40.

A World System of Crustal Structure— Sea-Floor Spreading and Continental Drift

16

Continental drift was first seriously proposed in 1910 by Alfred Wegener, a German meteorologist who marshalled evidence from geology and meteorology. His theory was not accepted, however, because his mechanism was not adequate. Much of the geologic evidence for continental drift comes from the southern hemisphere, so geologists there tended to accept drift while European and American geologists did not. In 1937, A. L. Du Toit, a South African, published a book, *Our Wandering Continents,* with a better but still unconvincing mechanism. Thus it was, until after World War II when new evidence reopened the question. In the 1950's two discoveries were made. P. M. S. Blackett and S. K. Runcorn in England showed, as will be described below, that the magnetism in rocks strongly suggests that the continents have moved relative to each other. At the same time, the American oceanographers Bruce Heezen and Maurice Ewing showed the extent of the mid-ocean ridge.

In 1960 Harry Hess of Princeton University proposed sea-floor spreading, a theory somewhat similar to one proposed in 1928 by the Englishman Arthur Holmes. Almost at once, many geologists began to take continental drift seriously. The early and middle 1960's was a time of active study of the magnetism of the ocean floors. As results of these studies were reported, more and more geologists accepted continental drift and sea-floor spreading. In 1965, J. Tuzo Wilson, a Canadian, proposed transform faults, and in 1968, Bryan Isacks, Jack Oliver, and Lynn R. Sykes of Columbia University supported the theory with evidence from seismology. In 1970, the deep-sea drilling project of *Glomar Challenger* started returning more evidence for the theory.

Thus, in less than ten years, geology has undergone a revolution in thinking. Perhaps the reason that sea-floor spreading was so quickly adopted is that it is a unifying theory in the sense that it brings together observations from many fields in earth science into a single theory. Some of those who are not convinced say that Neptunism was also a theory that explained many observations and so simplified and unified geology.

Continental Drift

The idea of continental drift has been with us since the first accurate maps showed that Africa and South America fit together. During the last

The world's highest mountains along the border of India and Tibet. The Himalaya Mountains may have been formed by the collision of India and Asia. Photo from **NASA.**

A.

B.

C.

D.

Figure 16-1. Evidence for continental drift. A. *Glossopteris*. This leaf has been found in beds overlying ancient glacial deposits in India, Australia, South Africa, and Antarctica. B. Glacial polish in Australia giving clear evidence of ancient glaciers there. Ice moved toward upper right. C. Striated boulder from ancient glacial moraine in Australia. D. Glacial polish from glaciers of the same age in Africa. Etching by Bushmen on left. Parts B., C., and D., photos by Warren Hamilton.

part of the nineteenth century, European geologists were deciphering the thrust structures in the Alps and American geologists were mapping the West, which was just opening up. As a result, they took little notice of some startling discoveries made in India, Australia, and South Africa. At each of these places a peculiar flora, called the *Glossopteris* flora after its most abundant genera, was found in beds overlying glacial deposits. Since then the same sequence has been found in South America and Antarctica. The glacial deposits are from continental, not mountain, glaciers and rest on polished and striated basement rocks. (See Fig. 16-1.) The striations suggest that in many cases the ice moved from areas that are now occupied by oceans. The moraines also suggest continental drift. They contain many boulders of rock types not found on that continent, and in some cases the foreign boulders match well with outcrops on the adjoining continent. The questions immediately arise: How could extensive glaciers form in these tropical areas, and how did the *Glossopteris* flora spread across the oceans now separating these areas? (See Fig. 16-2.) These areas are so widespread that shifting the rotational axis of the earth does not help because even in the most favorable location, some of the glacial areas are still within ten degrees of the equator. If all of the continents of the southern hemisphere had been in contact at one time, these occurrences would be explained; but the continent would have to break up and drift to explain the present geography. (See Fig. 16-3.) Thus began the idea of continental drift. Today the evidence is even more compelling

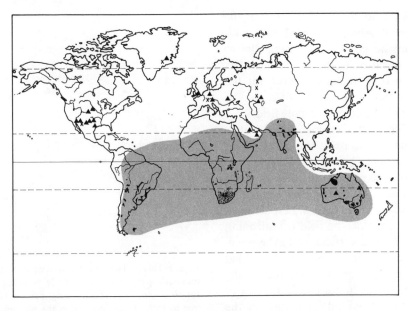

Figure 16-2. Map showing the present distribution of the ancient glaciation of the southern hemisphere. The area involved in the hypothetical southern continent existing before drifting is stippled. The areas of glacial deposits are indicated in black and extend from north of the equator to near the south pole. Reefs (crosses) and evaporites (triangles) far to the north indicate much warmer climates. Reefs and evaporites from Opdyke, 1962.

because of the discovery of *Glossopteris* and other fossils, indicating a humid, temperate climate at one time in Antarctica. There is much more evidence, but before evaluating it, we will discuss the earth's magnetic field because magnetic data are used extensively in reconstructing continental drift.

The Earth's Magnetic Field

It is the earth's magnetic field that aligns a compass needle and makes it point toward magnetic north. The earth behaves much as if it had a giant bar magnet in its core. (See Fig. 16-4.) However, even though the core is made of iron and nickel, no such magnet can exist at the temperatures there. Iron loses its magnetism when heated above a certain

temperature — well below its melting point. Also the earth's magnetic field is not that of a simple dipole, which is the field of a bar magnet. (See Fig. 16-5A.) The earth's field consists of a dipole field and a non-dipole field. One effect of the non-dipole field is

Figure 16-3. Reconstruction of the ancient continent of the southern hemisphere. The glaciers and their directions of movement are indicated.

that the south magnetic pole, at the edge of Antarctica, is 1,300 miles from the point that is antipodal to the north magnetic pole, which lies north of Canada. However, for the most part, the earth's field can be considered nearly a dipole field.

The earth's magnetic field causes a compass to point toward the magnetic pole, but if the compass is pivoted to move in the vertical direction also, it will indicate inclination as well. (See Fig. 16-4.) At the magnetic equator, the field is horizontal; at the poles, vertical; and in between, it varies. The intensity of the field at the poles is about twice that at the equator. At some places records of the direction and inclination of the earth's field have been kept for hundreds of years, and they reveal changes in direction and inclination. (See Fig. 16-5.) More recent studies

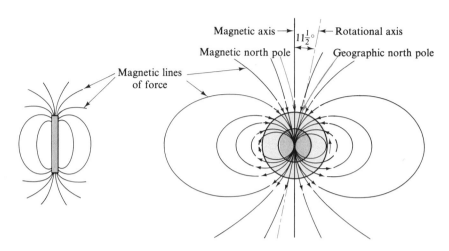

Figure 16-4. The earth's magnetic field is much like that of a bar magnet. The magnetic poles do not coincide with the geographic poles that are determined by the rotation axis. The short, heavy arrows show the vertical direction in which a compass free to move in the vertical plane will point.

also show variation in intensity. Study of the remanent magnetism of rocks, to be discussed below, indi-

cates that several times in the geologic past the direction of the earth's field has reversed, i.e., the north pole of a compass needle would point toward Antarctica. The timing of these reversals in the last 150 million years is well enough known that they can be used to date rocks. (See Fig. 16-6.) Any theory for the origin of the earth's magnetic field must explain these observations.

The earth's magnetic field is believed to originate from electric currents in the core. It is a well-known law in physics that a moving current creates a magnetic field, and conversely, a current is induced into a conductor moving in a magnetic field. Electrical currents in the core could start in a number of ways, such as through a weak battery action caused by compositional differences. Once started, these currents could be amplified by a dynamo action in the earth's core. A dynamo produces electric current by moving a conductor in a magnetic field. In the

Figure 16-5. Variations in the earth's magnetic field. A. Variations in position of magnetic north pole computed in 1945 from direction and inclination of the earth's magnetic field at various places. These calculations were made assuming a dipole field and show the effect of the non-dipole part of the earth's field. B. Changes in the position of the magnetic north pole with time. Time between points is 40-50 years. In both figures the cross shows the position of the north geographic pole, and the X, the average 1945 position of the north geomagnetic pole. From Allan Cox and R. R. Doell, *Bulletin*, Geological Society of America, Vol. 71, 1960.

Normal polarity   Reversed polarity

0   1   2   3   4

Age in millions of years from radioactive dates

Figure 16-6. A few of the many known reversals of the earth's magnetic field. From data by Cox.

earth the core is the conductor, and because it is fluid, it can move. The mechanism envisioned is similar to a dynamo which produces an electric current, and this current is fed to an electric motor that drives the dynamo. Friction and electrical resistance would prevent such a perpetual motion machine from continuing to operate unless more energy is added. In the earth this additional

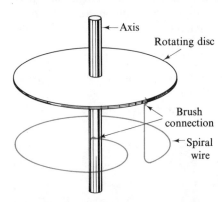

Figure 16-7. The disc dynamo illustrates in a general way how the earth's magnetic field is believed to originate. The disc is attached to the axis and rotates with the axis. The spiral wire is in electrical contact with the axis and the disc by means of brushes. To start this self-sustaining dynamo, suppose that a current is flowing in the spiral wire. This current produces a magnetic field; the disc is a conductor moving through the field, and so a current is induced in the disc. This current flows to the spiral wire by means of the brush and so continues the processes.

energy can come from movement of the fluid core. Calculation shows that very little additional energy needs to be added in the case of the earth; and the necessary energy for core motion, probably in the form of convection, could come from many sources.

Thus, a current is somehow started in the earth's core, and this current produces a magnetic field. The conducting core moves through the magnetic field, and a current is induced in this part of the core, starting the whole process over. (See Fig. 16-7.) In a stationary earth, the movements of the core would probably be more or less random and the magnetic fields produced would probably cancel each other. The rotation of the earth tends to orient the motions of the core so that the motions and, therefore, the currents are in planes perpendicular to the rotational axis. Viewed from outside the earth, the overall current produced would be parallel to latitudes. Such currents produce a dipole magnetic field centered on the rotational axis.

A field produced in this way would have all of the characteristics of the earth's magnetic field. The main field would be a dipole field; superimposed on this would be other fields caused by variations in the core's motion. This can account for

the variations in intensity and location of the earth's field. The earth's dipole field does not coincide with its rotational axis, however, as this theory predicts, although if averaged over a long enough period, the dipole field would perhaps appear centered on the rotational axis. The positions of the magnetic pole deduced from remanent magnetism of Pleistocene and younger rocks cluster around the rotational axis, suggesting that this is true. The present dipole field is 11.5 degrees from the rotational axis and lies northwest of Greenland, some distance from the magnetic north pole north of Canada.

Remanent Magnetism

Much of the recent data on continental drift have come from measuring the remanent magnetism of rocks. Remanent magnetism is the magnetism frozen into a rock at the time the rock forms. For instance, in the cooling of basalt, when the temperature reaches a certain point, which is well below the crystallization temperature, the magnetic minerals, mainly magnetite, are magnetized by the earth's field. On cooling below this temperature, their magnetism is not greatly affected except by very strong fields, many times stronger than any in the earth. Thus it is possible to measure the direction and inclination of the earth's field at the time the basalt cooled. The inclination can be used to calculate the distance to the magnetic pole.

Remanent magnetic data for rocks of different ages have been collected on each of the continents. The study of magnetic data from old rocks is called *paleomagnetism* (old magnetism). The magnetic poles for each age (geologic period) on each continent group in a fairly small area. However, in general, for any given continent, each geologic period has a different location for the magnetic pole, and the poles for any period fall in different locations for each continent. (See Fig. 16-8.) Here, then, is evidence of continental drift. If the magnetic pole moved, it should be at the same place during each period for each continent; but instead, each continent gives a different location for each period. Thus, if the interpretation of the paleomagnetic data is correct, the continents must have moved relative to each other. (See Fig. 16-9.) The magnetic pole may also have moved. Note in Fig. 16-8 that the magnetic data indicate no drift of North America relative to Europe since Eocene.

Paleomagnetic data have certain limitations. They tell the direction and distance to the magnetic pole. Thus they show the rotation of a continent and its latitude. (See Fig. 16-10.) They do not indicate the longitude, and the continent may be anywhere on the indicated latitude. This gives the imagination rather free rein in proposing reconstructions, particularly since many of the rocks are incompletely mapped and dated. (See Figs. 16-11 and 16-12.) In spite of the close grouping of poles for a given period, the possibility remains that the magnetic pole may not have been coincident with the geographic pole. At present, they differ by 11.5 degrees.

We will now examine the evidence for and against continental drift.

Pro

The evidence for continental drift includes the following.

(1) Europe	
(2) North America	
(3) Australia	
(4) India	
(5) Japan	

Pl Pliocene	*About 6 million years ago*
M Miocene	*About 18 million years ago*
E Eocene	*About 50 million years ago*
K Cretaceous	*About 100 million years ago*
J Jurassic	*About 160 million years ago*
Ŧ Triassic	*About 200 million years ago*
P Permian	*About 250 million years ago*
C Carboniferous	*About 310 million years ago*
D Devonian	*About 375 million years ago*
S Silurian	*About 420 million years ago*
Є Cambrian	*About 550 million years ago*

Figure 16-8. On this diagram, for each continent the position of the magnetic pole relative to the present position of each continent for each geologic period is indicated by the letters. The lines drawn through these locations show the apparent drift of the magnetic pole for each continent. The diagram indicates that movement of the continents relative to each other must have occurred but does not show the route of drifting because of the limitations of paleomagnetic data shown in Figure 16-10. From Allan Cox and R. R. Doell, *Bulletin,* Geological Society of America, Vol. 71, 1960.

Paleomagnetism is discussed above.

Fit of the Continents. It has been said that it is too good to be accidental, and recent matches made by computers on submarine contours to include the continental shelves seem to bear this out. Folding, faulting, and intrusion since drifting apart (about 250 million years) would have modified the shapes of the continents.

Climate. Glaciation was discussed above.

In Antarctica, before that glacia-

A.

P = Permian
C = Carboniferous
D = Devonian
S = Silurian

B.

Figure 16-9. A. The curves show the position of the magnetic pole for each geologic period relative to the present positions of the continents. B. Shows that if South America and its polar curve are moved so that the two polar curves match, the two continents join. The position at which they join is geographically and geologically the most likely.

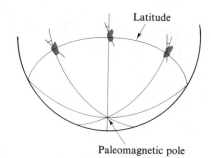

Figure 16-10. Paleomagnetic data can only show the latitude and orientation of a continent. The horizontal component of the paleomagnetism points the direction toward the pole, and the vertical inclination gives the latitude. The continent could have been in that orientation anywhere on that latitude.

tion, the climate was much different from the present. Rocks bearing shallow marine fossils have been found, and at another place a thousand miles away dune sands suggesting deserts are widespread.

It is hard to imagine that the *Glossopteris* flora recently found 210 miles from the South Pole in Antarctica would grow there in that region of cold, months-long nights as well as in tropical India. Other temperate- and warm-climate-indicating fossils, including coal beds, have been found in Antarctica. Some plants do grow inside the Arctic circle at the present time.

The distribution of evaporites through geologic time also suggests drift and appears to be in harmony with the magnetic data. Evaporites suggest hot arid conditions, and deserts occur today between 30 degrees north and south of the equator. Extensive salt deposits are not forming at present, however. Ancient salt deposits of approximately the same age as the glaciation are found in northeast Greenland, arctic Canada, Siberia, Germany, Kansas, and New Mexico, as well as other places. (See Fig. 16-2.)

Coral reefs are also currently restricted to about 30 degrees north and south of the equator, but ancient

Figure 16-12. Movement of continents. A. Late Paleozoic time, 250 million years ago. B. End of Jurassic Period, 136 million years ago. C. End of Cretaceous Period, 65 million years ago, is shown dotted and present positions are shown by the solid lines. Based on R. S. Deitz and J. C. Holden, 1970.

Figure 16-11. Reconstruction of the positions of continents in the geologic past from paleomagnetic data. A single geologic period (Jurassic) is shown. On the left the continents are shown in their present locations with the Jurassic latitudes computed from paleomagnetic data indicated. Obviously many locations of each continent are possible on the basis of these data. On the right the probable locations are shown. They are based on paleomagnetic and other data such as reasonable movements from the position during the last and to the next geologic period. Figure 16-12 shows the positions of the continents during other geologic periods. From D. Van Hilten, *Tectonophysics,* Vol. 1, 1964.

reef deposits are found far to the north.

One thing to keep in mind in evaluating the climatic evidence is that the last glaciation, the Pleistocene, occurred in what is now the temperate zone. Thus in the last million years the climate has changed; and so the climate could have changed in the past although it is difficult to imagine continental glaciation at the equator, even though mountain glaciers occur there now.

Apparent Rifting of the Continents

is the probable origin of the Red Sea. (See Figs. 16-13 and 16-14.) The Gulf of California and the African Rift valleys may also have originated in this way. If a small sea can form in this way, it seems possible that a larger ocean, such as the Atlantic, could also.

Rock Layers and their Fossils.

Piecing together the continents is like putting together a jig-saw puzzle whose pieces do not quite match. Fitting together the various layers of sedimentary rocks is like putting together a multi-layer jig-saw puzzle with many of the pieces missing. The generalized stratigraphy of the southern continents is shown in Fig. 16-15. The similarity among these continents in the Carboniferous Period is striking. This similarity persists through the Jurassic in all but Australia. This stratigraphy and fossils in Australia suggest that it became isolated from the other continents after the glaciation and that it has remained isolated to the present. The Australian fauna and flora of the present appear to have migrated from time to time very selectively from mainland Asia. The present flora and the marsupial fauna are

Figure 16-13. View northeast across the Sinai Peninsula taken from Gemini XI. The southeast end of the Mediterranean Sea is in the upper left, and the north end of the Red Sea is in the lower right. The Sinai Peninsula is bordered by the Gulf of Suez and the Gulf of Aqaba. The Gulf of Aqaba is part of a linear depression on which the Dead Sea can be seen. The Red Sea may be a rift that originated when Africa and Arabia separated, and the two Gulfs shown here may be related to that fracture. Photo from NASA.

very different from those of the other continents. The placental mammals that populate the other continents did not reach Australia.

The *Glossopteris* flora found just above the glacial deposits was cited earlier as evidence of previous connection among the southern continents. It is possible that seeds (spores) could have floated across the oceans to distribute these plants, but it seems rather unlikely. Today, a newly formed volcanic island soon develops plants, but the seeds are carried by birds. The first flying animals did not appear until tens of millions of years after the supposed separation and drift. Land bridges and island chains have been sug-

gested in the past, but the known oceanic structure precludes this possibility.

Glossopteris leaves a foot long, tree trunks two feet in diameter, and coal seams 25 feet thick in Antarctica, within three degrees of the south pole, indicate, at the least, a marked change in climate. The similarity of the faunas in the underlying (Devonian) rocks strongly suggest that Antarctica, Africa, and South America were once a single continent. These faunas are so similar that the species cannot be distinguished.

The swimming reptile *Mesosaurus*, found in both South America and south Africa, is also strong evidence favoring connection of these two con-

Figure 16-14. Looking northeast at the south end of the Red Sea and the Gulf of Aden, taken from Gemini XI at an elevation of 470 nautical miles above Africa that forms the foreground. The Indian Ocean is to the right, and the southwest end of the Arabian Peninsula forms the top of the picture. The Red Sea and the Gulf of Aden may be fractures that formed when Africa and Arabia separated. The part of Africa where the Red Sea narrows (Afar Triangle) is recently uplifted ocean floor. Photo from NASA.

tinents. Although a swimmer, this aquatic reptile lived in fresh water, and it is very unlikely that it could swim across an ocean. Other reptiles that lived during the time of the alleged connection have been found on the southern continents. The distribution of some of these reptiles requires a single connected continent.

Radioactive dates, although few in number, show a good match between the basement rocks of west Africa and northern Brazil. The dates obtained on the basalts shown in Fig. 16-15, however, do not suggest a close correspondence.

Structures. Another test is to see if the folds, faults, and other structures of appropriate age match when the continents are put together. Much diversity of opinion exists here, in part because results depend to a great extent on how the continents are joined, and there are many opinions on this latter point. In all of the reconstructions, some of the structures match and some do not, so here again it becomes somewhat subjective as to which structures are considered most important. Dating the formation of structures is difficult in many cases, and allowance for modification by the proposed drift and by later folding and faulting is a problem. (See Fig. 16-16.)

Oceanic Sediments. The continental shelf and slope off the east coast of North America has about six times more volume of sediment than does the west coast. This is in agreement with both continental drift and ocean-floor spreading if the North American continent is moving westward, covering the sediments shed from the continent into the Pacific.

Con

The evidence against continental drift will now be considered. Some points against drift were included with the arguments for drift to keep the discussion of each type of evidence in one place.

Heat Flow. The heat flow on continent and ocean is about the same. The continental heat flow is the amount expected from the known amounts of radioactive elements in the continental crust. The oceanic crust is thinner and is known to contain smaller amounts of radioactive elements; thus, the oceanic heat flow must come from the mantle. This suggests that the continents were formed from the mantle beneath them because if the mantle contributes to the heat flow under continents, the flow would be higher than it is. Secondly, if drift has occurred, one would expect to find areas where the uniform pattern of heat flow is upset; that is, the heat flow might double where continent covers former ocean. It should be pointed out that heat flow data are meager, and this argument, therefore, should be held in abeyance until more data are collected.

Undeformed Sediments in Trenches. The sediments in some of the volcanic island arc trenches that have been studied appear to be flat-lying. If the sea floor is moving

AGE	ANTARCTICA Transantarctic Mountains	SOUTH AMERICA Southern Brazil	PENINSULA INDIA	AUSTRALIA	SOUTH AFRICA
Cretaceous	*(no rocks)*	Basalt (Radioactive date)	Volcanic rocks; Marine sediments	*(no rocks)*	Some marine rocks
Jurassic	Basalt (Radioactive date)	Sandstone	Sandstone and shale; Volcanic rocks; Sandstone and shale	Mainly emergent with some coal basins and marine embayments	Basalt
Triassic	Sandstone, shale, and coal with *Glossopteris*	Sandstone, shale with reptiles	Sandstone and shale with reptiles	*(no rocks)*	Sandstone and shale; Sandstone and shale with reptiles
Permian	Tillite (may in part be much younger)	Shales and sandstones with *Glossopteris*; Marine shale; Shale with *Mesosaurus*; Sandstone, shale, and coal with *Glossopteris*	Sandstone, shale, and coal with *Glossopteris*; Shale	Sandstone, shale, and coal, and marine sediments; Sandstone, shale and coal with *Glossopteris*; Tillite	Shale; Sandstone, shale, and coal with *Glossopteris*; Shale with *Mesosaurus*; Sandstone, shale, and coal with *Glossopteris*
Carboniferous (Mississippian and Pennsylvanian in U.S.)		Tillite	Tillite		Tillite

Figure 16-15. Simplified stratigraphic columns in the southern hemisphere. The age terms are described in Chapter 17. Vertical lines indicate no rocks of that age.

downward into the trench, it seems likely that the sedimentary rocks would be deformed. The structure of the near-bottom sediments is determined by seismic methods and sparker profiles, and it is possible that the deeper sediments are deformed.

Undeformed Sediments in Fracture Zones. The sediments in the fracture zones or transform faults also appear to be undeformed. This suggests that the movement in those areas is slow. Seismic data do reveal movements on these faults.

Geographic Problems. The fitting together of the main continents was described above and is fairly good.

Problems arise for some smaller but significant areas. In most reconstructions of the pre-drift continents, there is no room for Central America because North and South America are too close together. Spain also overlaps, and so in reconstructions it must be bent into the area now occupied by the Bay of Biscay.

Structures of Continental Interiors. One of the important failings of the theory of sea-floor spreading is that it does not explain structures not on the margins of continents. Sea-floor spreading can explain geosynclines on continental margins, but similar structures also occur in the interior of continents. An example is the basin that occupied the area of the

Rocky Mountains in part of the Mesozoic Era. This feature had many similarities to geosynclines, and parts of it were later deformed. Thus not even all geosynclines or geosyncline-like areas can be explained by sea-floor spreading, but most true geosynclines do fit into the theory.

Sea-floor Spreading

Age

The evidence that new sea floor is created at the mid-ocean ridges is abundant. Radioactive dating of basalt samples shows a progressively older age away from the ridges. The sediments on the sea floor also

Figure 16-16. West Africa and South America fitted together, showing the coincidence in both structural trends and radioactive ages of the rocks (m. y. is million years). From P. M. Hurley and others, *Science* (August 4, 1967), Vol. 157, p. 496.

thicken away from the ridges, and examination of drill cores reveals that the bottom sediments are older further away from the ridges. The sea floors are young geologically, and only a few samples older than 130 million years have been found. The oldest sediments so far found are sandstones about 160 million years old off South America and in the western Pacific.

Magnetic Studies

The earth's magnetic field is impressed on a rock at the time the rock is formed. Basalt, because it contains small amounts of magnetite, an especially magnetic mineral, faithfully records the earth's magnetic field at the time the basalt cools through a certain temperature (Curie point) that is well below its melting point.

Magnetic surveys near the mid-ocean ridge show bands of alternating high and low magnetic intensity that parallel the ridge as shown in Fig. 16-17. The meaning of these magnetic bands was not clear until magnetic studies were made on the continents. It was soon discovered that many rocks recorded a magnetic field opposite or reversed to the present magnetic field. Magnetic studies of each flow in areas of thick accumulations of basalt flows showed that these reversed magnetic field rocks did not occur in a random way but, rather, that if radioactive dates of the rocks were also obtained, the magnetic reversals occurred simultaneously all over the earth. Starting with the youngest rocks, it has been possible to determine the times of reversals accurately enough to use the magnetic direction to date some young rocks. (See Fig. 16-6.) These

reversals suggest that the bands of alternating high and low magnetic intensities paralleling the mid-ocean ridge could be caused by bands of ocean floor basalt that have normal and reversed magnetic fields. The normally magnetized basalt would reinforce the present magnetic field, giving a high magnetic intensity; and the reverse magnetized rocks would partially cancel the present field, giving a lower than normal magnetic intensity. Careful comparison of the magnetic intensity on each side of the mid-ocean ridge shows that the magnetic intensity on one side is nearly the mirror image of that on the other side. (See Fig. 16-18.) The final argument is to construct a model of the ocean floor by calculating the positions of bands of normal and reversed magnetic basalt and then to calculate the theoretical magnetic field produced by such a model. As shown in Fig. 16-19, this model produces a field almost identical to that found in nature.

Movement of the Sea Floors

Crustal Plates

Much of the evidence for sea-floor spreading just described indicates that, in a general way, the newly created sea floor moves away from the mid-ocean ridge at rates between ½ and 7 inches per year. Seismic activity shows where this motion occurs. In the earlier discussion of earthquakes, it was noted that earthquakes are believed to result from the sudden movement of one segment of crust past another. Figure 16-20 shows where earthquakes occur, and their concentration is at the deep-sea trench—island arc areas, to a lesser extent on the mid-ocean ridge, and at many places on the faults or fracture zones that offset the ridge. Near the island arcs the earthquake foci tend to be deep and elsewhere they are generally less than 40 miles deep. These earthquakes suggest that the earth can be thought of as consisting of seven main plates and a few smaller plates, each one moving from the mid-ocean ridge to a trench or island arc. These plates are indicated in Fig. 16-20. Basaltic ocean crust is created at the ridge and absorbed at the trenches. (See Fig. 16-21.)

Transform Faults

More detailed seismic studies reveal the nature of the movements on the faults that offset the mid-ocean ridge. It is possible to determine the relative direction of the movement of the plates on each side of the fault from the direction of the first movement on seismograms. Such studies show that the movement on each block is away from the mid-ocean ridge, as expected if new crust is created there. Note, however, as shown in Fig. 16-22, that the relative movement is opposite to that expected by ordinary fault movement. This special kind of fault is termed a *transform fault*, and, so far, it is only known to occur in association with mid-ocean ridges. The San Andreas fault and the oceanic fracture zones are believed to be transform faults.

Origin of Crustal Structures

In this section an attempt will be made to find a mechanism that will account for the structures of the crust. Folded rocks suggest shorten-

Figure 16-17. Magnetic anomalies in part of the ocean along the west coast of North America. Positive magnetic anomalies are indicated. They are believed to be formed where the remanent magnetism of the underlying rocks reinforces the earth's field. The intervening white areas have lower than normal magnetism and may be places where the underlying rocks have reversed magnetism. The straight lines are faults that displace the pattern. The arrows indicate short lengths of oceanic ridge, and the pattern is parallel and symmetrical to the ridges. After F. J. Vine, *Science,* Vol. 154, p. 1406, December 16, 1966.

ing, and one of the first theories suggested that this was caused by shrinking of the earth due to cooling. This contraction hypothesis had to be abandoned when calculation showed that thermal contraction could not be great enough, and when very old glacial deposits were discovered,

Figure 16-18. Formation of magnetic anomalies at mid-ocean ridges. In this hypothesis the lavas that form the ocean floor cool at the ridge and are magnetized by the earth's field at the time they cool. Convection currents carry the lavas away from the ridges and so create magnetic anomaly patterns parallel and symmetrical to the ridges.

showing that the climate has not changed radically since then. The features that need to be explained, in addition to the compressive structures of the continents, include geosynclines and the features of ocean-floor spreading.

Relationship of Structures to Sea-floor Spreading

Geosynclines are the main features of continental geology and appear to be involved in continent formation. Sea-floor spreading is the main process in the oceans and accounts for the formation of oceanic crust. Geosynclines form on ocean-continent boundaries and appear to be related to sea-floor spreading as shown in Fig. 16-23. The sedimentation phase of the geosyncline is the development of the familiar continental shelf, continental slope, and continental rise. The sedimentary rocks on the shelf and the slope are similar in type and thickness to those found on the landward sides of geosynclines—that is, beds of shale, sandstone, and limestone. The rocks of the continental rises are very different and are largely emplaced by turbidity currents that flow down the continental slope. These rocks are poorly sorted, have graded bedding and other features mentioned in Chapter 4, as well as some features of deep-ocean sediments. The rocks of continental rises resemble the rocks on the oceanward side of geosynclines in all features except their lack of volcanic rocks.

The development of a trench and a volcanic island arc can be the answer to that problem. To form a trench, the oceanic crust must move downward. It seems likely that the oceanic crust could not bend down sharply near the trench but would move downward under the continent in the zone of the deep-focus earthquakes (Fig. 16-23B.) The higher temperature in the mantle would melt the crustal material and so produce the magma that rises to form the volcanic island arc. It seems likely that some deformation of the sedimentary rocks would occur in the formation of a volcanic island arc and by subsequent movement of the descending oceanic plate.

The collision of a second continental plate with a geosyncline is another way in which a geosyncline can be deformed. In this case, the compression will result in thickening of the crust, and isostasy will cause a high mountain range to form. The Himalaya Mountains are believed to have formed in this way. The heat and friction of such a collision could also cause metamorphism and generation of magma. In this way new continental crust could be created. The oceanic side of geosynclines is generally most deformed and the site of most metamorphism and igneous activity. The sedimentary rocks on the continental side tend to be folded, and in some cases thrust faulted, but they lack the extreme deformation of the oceanic side. This may be because they were deposited on continental shelves and so were above continental crust that protected them

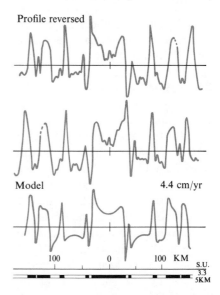

Figure 16-19. Magnetic profiles across the East Pacific Rise. The two upper curves are an actual magnetic profile, shown both as recorded and reversed. The lower curve was computed from a model, assuming a spreading rate of 4.4 cm per year. After F. J. Vine, *Science* Vol. 154, p. 1409, December 16, 1966

from extreme deformation.

Thus crustal deformation may occur where two rigid plates, moving in opposite directions, meet. If the collision is between two plates of oceanic crust, one will move downward, forming a trench. If one of the plates is continental crust, it will not move down because of its buoyancy; the oceanic plate will move downward, forming a trench. If two continental plates meet, a large mountain range, such as the Himalaya Mountains, will form because of the compressive thickening of the crust. Where an oceanic plate sinks, it and its associated oceanic sediments will be partially melted by the higher temperatures in the mantle and by the frictional heat generated. The result will be the andesitic volcanic rocks that form the volcanic islands associated with trenches. (See Fig. 16-24.) Subduction is the name given to the process by which one crustal plate descends below another plate.

Convection Currents

Convection currents have been proposed as the cause of most geological features. Convection cells develop when a fluid is heated at the bottom. As the warmed fluid expands, it becomes less dense and tends to rise; denser, cooler fluid at the top moves down to take the place of the rising, warm fluid, as shown in Fig. 16-25. In this process heat is transported upward by the movement of the warmed fluid.

Although there is disagreement on the viscosity of the mantle, convection has been explored theoretically by a number of investigators who have established that it could exist in the mantle, even though the mantle

Figure 16-20. The crustal plates. In general, each plate begins at a ridge (heavy line) and moves toward a trench (hatching). Earthquake epicenters (dots) tend to outline the plates. Transform faults are shown by thin lines. The dashed lines show the ages (in millions of years), based on magnetic data. The information and interpretation are from many sources, mainly workers at Lamont-Doherty Geological Observatory.

is viscous enough to behave as a solid in transmission of earthquake waves. A substance may respond elastically to a rapid stress but may be deformed viscously by a stress applied over a long period of time. The latter case could occur in the mantle, especially in view of the temperatures and pressures there. Thus it is possible for a solid to flow at temperatures well

The basalt would become new ocean floor. (See Fig. 16-26.) Horizontal currents could cause transform faults, continental drift, and sea-floor spreading. Descending currents could cause trenches and volcanic island arcs.

The low seismic velocity zone in the upper mantle that was described earlier may be involved in sea-floor spreading. If the movement is due to convection, the convection could involve the whole mantle or only a part of it. Convection could be confined to the low seismic velocity zone, but such thin cells would not be stable and would tend to break up into much smaller cells.

It has been suggested that the plates, extending down to the weak zone in the upper mantle, move as units, and the return flow is in the weak zone of the upper mantle. Such movement resembles convection (Fig. 16-27). The movement of the plates may be caused by the addition of material to the crust at the mid-ocean ridges and/or by the weight of the down-sinking leading edge of the plate at the island arc pulling the plate (Fig. 16-26).

If convection is involved in sea-floor spreading, it is not simple convection as suggested in the diagrams. Study of the world map showing the locations of mid-ocean ridges and volcanic island arc-trench systems (Fig. 16-20) shows that there is not a trench opposite every ridge. There is no trench anywhere on the margin of the plate that contains Africa, suggesting that that plate may be growing. Thus much more study is needed before any complete theory can be presented.

One interesting speculation about processes in the upper mantle is that

below the melting point. A possible example of this behavior is the slow movement of glacial ice under the influence of gravity and its very different behavior when subjected to a rapid stress. Convection currents are believed to move at the rate of one to a few centimeters per year.

Many features of the mid-ocean ridges can be explained by convec-

tion currents. The high heat flow suggests upwelling convection currents in the mantle. The convection currents could cause the rise and the consequent crustal thinning, as well as the faulting and earthquake epicenters. The upwelling mantle material could melt partially as a result of the reduced pressure near the crust-mantle boundary, forming basalt.

there may be a few places other than at mid-ocean ridges where basalt magmas form. Many of the islands of the Pacific are in lines. The Midway Islands and the Hawaiian Islands are such a line, and they are progressively younger toward the southeast. The possibility is that a hot area may occur in the mantle and this hot area is now below the active volcanoes in the Hawaiian Islands. As the Pacific plate moved over this melting spot, the line of volcanoes formed.

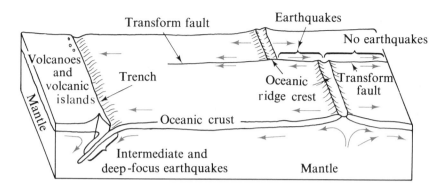

Figure 16-21. Schematic representation of the oceanic crust. Note the formation of crust at ridges and destruction in trenches. From Gross, *Oceanography*, 2nd edition.

Energy

Large amounts of energy are necessary to deform rocks, generate magma, and cause metamorphism. The heat flow from the earth indicates the presence of sources of heat energy large enough to account for geologic processes. The total heat flow from the earth is about 1,000 times more than the energy of earthquakes and 100 times more than volcanism, and is enough to cause geologic processes. Heat flow is calculated from temperature measurements in drill holes and on the bottom of the ocean. Reliable heat flow measurements were first made on the continents in the mid 1930's and at sea in 1950. An obvious source of the heat flow is the heat generated by the decay of radioactive elements. The known radioactive content of crustal rocks of the continents is enough to account for the continental heat flow. The average heat flow from the oceans is about the same as the average from the continents. However, the thin layer of oceanic basalt does not contain enough radioactive elements to account for the oceanic heat flow. Therefore, some of the heat flow

from the oceans must come from the mantle, and this is especially true of the mid-ocean ridges such as the East Pacific Rise, where the heat flow is very high.

Radioactive elements are fairly evenly distributed in the continental crust; and geologically active areas, such as volcanoes and mountain belts, are no more radioactive than other areas. Radioactive energy is enough to cause all geologic processes, but the distribution of radioactivity suggests that it is not the

energy source, so other sources must be considered.

Some of the heat reaching the surface seems to come from deep in the earth. The deep source of heat is believed to be the liquid core. The heat could be released by crystallization of the core, and the inner core, believed to be solid, may be slowly crystallizing. The internal heat could also come from recrystallization in the mantle, but we know very little about either the composition or the physical conditions there.

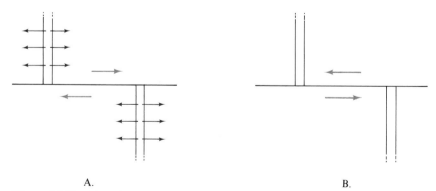

Figure 16-22. Transform fault compared with simple fault. A. Transform fault. The relative movement on the transform fault is caused by the new sea floor created at the two segments of the ridge. The transform fault is shown by the single line, and the ridge by the double line. B. Two segments of once-continuous linear feature separated by a simple fault. Note that the relative movement, shown by the arrows, is opposite to that of the transform fault shown in A.

A. Ocean-ocean plate collision

B. Continent-ocean plate collision

C. Continent-continent plate collision

Figure 16-24. Types of collisions of plates. A. Ocean-ocean. B. Continent-ocean. C. Continent-continent.

Figure 16-23. Geosynclines and sea-floor spreading. A. Continental shelf, slope, and rise. B. Development of volcanic island arc. C. Collision of continents. D. Separation of continents by the development of a new oceanic rise.

Figure 16-25. A. Two simple convection cells produced by heating at the bottom. Warm fluid at ① rises and cool fluid from ② falls to take the place of the rising warm fluid. B. Convection in a fluid whose viscosity depends on temperature and pressure. Heating is on the bottom.

Figure 16-26. Formation of oceanic ridge, trench, and volcanic island arc by convection currents or sea-floor spreading.

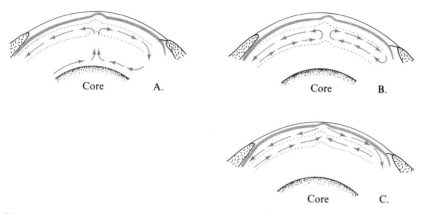

Figure 16-27. Convection currents. A. Deep convection involving much or all of the mantle. B. Thin convection confined to the low-velocity zone of the mantle. C. A convection-like flow in which the rocks above the low-velocity zone move and the return flow is in the low-velocity zone.

Table 16-1. Summary of heat flow measurements. The units are microcalories per square centimeter per second (10^{-6} cal/cm²/sec). From W. H. K. Lee and Seiya Uyeda, "Review of Heat Flow Data," in *Terrestrial Heat Flow,* American Geophysical Union, Geophysical Monograph 8, 1965.

AVERAGE: 1.5 ± 0.15 (Same on continents and oceans)	
Continents	
Undeformed areas such as interior lowlands	1.54 ± 0.38
Areas of metamorphic rocks deformed more than 600 million years ago	0.92 ± 0.17
Areas deformed between 600 and 225 million years ago	1.23 ± 0.40
Areas deformed less than 225 million years ago	1.92 ± 0.49
Oceans	
Trenches	0.99 ± 0.61
Floors	1.28 ± 0.53
Ridges	1.82 ± 1.56

Summary

The earth's magnetic field is nearly a dipole field, but the departures from a simple dipole are enough so that the north magnetic pole is $11\frac{1}{2}°$ from the geographic pole, and the south magnetic pole is not antipodal to the north magnetic pole.

The dynamo theory explains the earth's magnetic field. In this theory, an electric current moves through the earth's conducting core, creating a magnetic field. The conducting core moves through this magnetic field, and a current is induced. This current starts the whole cycle over again.

Remanent magnetism is the magnetism frozen in a rock at the time the rock forms.

Paleomagnetic (old magnetic) data reveal the north magnetic direction and latitude of ancient rocks, but not where on that latitude the rock formed.

The evidence for continental drift includes paleomagnetic data, apparent fit of the continents, climatic data such as glaciation, apparent rifting, fossils such as *Glossopteris* and *Mesosaurus*, and age of the oceanic sediments.

The evidence against continental drift includes similar heat flow of oceans and continents, undeformed sediments in some trenches and fracture zones, poor fit of some continents, and structures of continental interiors.

Sea-floor spreading is the theory that new oceanic crust is being created at the mid-ocean ridges, is then moving away from the ridges, and is being consumed at the submarine trenches.

The bottom sediments on the sea floor are younger and thinner near the mid-ocean ridges and become thicker and older away from the ridges. The oldest sediments so far found are latest Jurassic (about 160 million years old).

Basalt and other rocks are magnetized by the earth's magnetic field when they cool through a certain temperature (Curie point) that is well below their melting points.

The earth's magnetic field has reversed its polarity numerous times. These changes in magnetic direction are impressed on the new ocean floor created at the mid-ocean ridges and appear as symmetrical bands of alternating magnetic intensity parallel to the mid-ocean ridges.

Crustal plates are bounded by mid-ocean ridges, submarine trenches, and transform faults. They are the units of the earth's crust that move in the theory of sea-floor spreading. Plates can be composed of either continent or ocean or both.

Transform faults are the surfaces where one plate moves past another plate. The direction of movement revealed by seismic study is in the opposite sense to the apparent offset of the mid-ocean ridges associated with the two plates involved.

Geosynclines may be related to sea-floor spreading and may form at volcanic island arcs and/or continental shelf-slopes.

Collisions between plates are of three types. In an ocean-ocean plate collision, one of the oceanic plates moves downward, creating a trench; in an ocean-continent collision, the oceanic plate moves down, creating a trench; and in a continent-continent collision, neither moves down.

Convection currents may occur in the mantle and could account for sea-floor spreading.

The energy that causes deformation of the crust may come from radioactivity or may come from crystallization of the core.

Questions

16-1. What are the components of the earth's magnetic field?

16-2. What changes have occurred in the earth's magnetic field?

16-3. How is the earth's magnetic field believed to be caused?

16-4. How is remanent magnetism impressed on rocks?

16-5. What data on the previous position of a continent is obtained from remanent magnetism?

16-6. How does remanent magnetism favor sea-floor spreading?

16-7. What evidence suggests that the southern continents were once joined?

16-8. What evidence suggests that drift alone may have occurred (as distinct from showing that the continents were once joined?

16-9. What evidence do you feel is strongest in suggesting continental drift? Which against?

16-10. Discuss whether all of the lines of evidence that point toward continental drift should be considered as proof, or whether each piece of evidence must be evaluated individually and must stand alone or be dropped.

16-11. What is the evidence for sea-floor spreading?

16-12. What are crustal plates? How many are there?

16-13. Describe transform faults.

16-14. Where do magmas originate? What is the evidence?

16-15. Where does the energy come from that causes the deformation of rocks?

16-16. What are convection currents? What is the evidence for them in the mantle?

16-17. How might geosynclines be formed? What kind of forces cause the deformation of geosynclines?

Supplementary Reading

Bullard, Sir Edward, "The Origin of the Oceans," *Scientific American* (September, 1969), Vol. 221, No. 3, pp. 66-75.

Cox, Allan, G. B. Dalrymple, and R. R. Doell, "Reversals of the Earth's Magnetic Field," *Scientific American* (February, 1967), Vol. 216, No. 2, pp. 44-54.

Cox, Allan, and R. R. Doell, "Review of Paleomagnetism," *Bulletin,* Geological Society of America (June, 1960), Vol. 71, No. 6, pp. 645-768.

Dalrymple, G. B., and others, "Origin of the Hawaiian Islands," *American Scientist* (May-June, 1973), Vol. 61, No. 3, pp. 294-308.

Dewey, J. F., "Plate Tectonics," *Scientific American* (May, 1972), Vol. 226, No. 5, pp. 56-68.

Dickinson, W. R., "Plate Tectonics in Geologic History," *Science* (October 8, 1971), Vol. 174, No. 4005, pp. 107-113.

Dietz, R. S., "Geosynclines, Mountains and Continent-Building," *Scientific American* (March, 1972), Vol. 226, No. 3, pp. 30-38.

Doumani, G. A., and W. E. Long, "The Ancient Life of the Antarctic," *Scientific American* (September, 1962), Vol. 207, No. 3, pp. 169-184. Reprint 863, W. H. Freeman and Co., San Francisco.

Elders, W. A., and others, "Crustal Spreading in Southern California," *Science* (October 6, 1972), Vol. 178, No. 4056, pp. 15-24.

Elsasser, W. M., "The Earth as a Dynamo," *Scientific American* (May, 1958), Vol. 198, No. 5, pp. 44-48. Reprint 825, W. H. Freeman and Co., San Francisco.

Hallam, A., "Continental Drift and the Fossil Record," *Scientific American* (November, 1972), Vol. 227, No. 5, pp. 56-66.

Heirtzler, J. R., "Sea-Floor Spreading," *Scientific American* (December, 1968), Vol. 219, No. 6, pp. 60-70.

Hurley, P. M., and J. R. Rand, "Pre-drift Continental Nuclei," *Science* (June 13, 1969), Vol. 164, No. 3885, pp. 1229-1242.

Jacobs, J. A., *Earth's Core and Geomagnetism.* New York: Pergamon Press, Inc., 1963, 201 pp. (paperback).

James, David E., "The Evolution of the Andes," *Scientific American* (August, 1973), Vol. 229, No. 2, pp. 60-69.

Kurtén, Bjorn, "Continental Drift and Evolution," *Scientific American* (March, 1969), Vol. 220, No. 3, pp. 54-64.

McKenzie, D. P., and J. G. Sclater, "The Evolution of the Indian Ocean," *Scientific American* (May, 1973), Vol. 228, No. 5, pp. 62-72.

Maxwell, J. C., "Continental Drift and a Dynamic Earth," *American Scientist* (Spring, 1968), Vol. 56, No. 1, pp. 35-51.

Menard, H. W., "The Deep Ocean Floor," *Scientific American* (September, 1969), Vol. 221, No. 3, pp. 126-142.

Meyerhoff, A. A., and H. A. Meyerhoff, " 'The New Global Tectonics': Major Inconsistencies," *Bulletin,* The American Association of Petroleum Geologists (February, 1972), Vol. 56, No. 2, pp. 269-336. The other papers in this issue are concerned with the pros and cons of crustal plates and sea-floor spreading.

Orowan, Egon, "The Origin of the Oceanic Ridges," *Scientific American* (November, 1969), Vol. 221, No. 5, pp. 102-119.

Rona, P. A., "Plate Tectonics and Mineral Resources," *Scientific American* (July, 1973), Vol. 229, No. 1, pp. 86-95.

Takeuchi, Hitoshi, Seiya Uyeda, and Hiroo Hanamori, *Debate about the Earth.* San Francisco: Freeman Cooper & Co., revised edition, 1970, 281 pp. (paperback).

Tazieff, Haroun, "The Afar Triangle," *Scientific American* (February, 1970), Vol. 222, No. 2, pp. 32-40.

Valentine, J. W., and E. M. Moores, "Plate Tectonics and the History of Life in the Oceans," *Scientific American* (April, 1974), Vol. 230, No. 4, pp. 80-89.

Vine, F. J., "Sea-floor Spreading — New Evidence," *Journal of Geological Education* (February, 1969), Vol. 17, No. 1, pp. 6-16.

Wilson, J. T., "Some Aspects of the Current Revolution in the Earth Sciences," *Journal of Geological Education* (October, 1969), Vol. 17, No. 4, pp. 145-150.

Wyllie, P. J., *The Dynamic Earth: Textbook in Geosciences.* New York: John Wiley & Sons, Inc., 1971, 416 pp.

Principles of
Historical Geology
17

The result, therefore, of our present enquiry is, that we find no vestige of a beginning,—no prospect of an end.

James Hutton, 1788

Geologic Dating

Rocks can be dated both relatively and absolutely. Relative dates are of two types. The simpler of these is the determination of the local sequence based on relationships such as the fact that the oldest beds are at the bottom of undisturbed sedimentary rocks. Question 28 at the end of Chapter 14 illustrates the method. Determining these relationships by mapping the rocks—that is, plotting on an accurate map the positions of rock bodies — is the main method of geologic study. The second way of dating rocks is by referring them to a geologic time scale based on the study of the fossils in the rocks. Absolute dating is a more recent development and is based on radioactive decay of elements in

Historical geology, the history of the earth, is perhaps the most important goal of geology. In this chapter the methods used to determine geologic history will be described. Some of these methods have already been discussed, such as interpretation of the formation of a rock based on study of the rock itself and its relationship to other rocks. Simple examples are the interpretation of a sandstone composed of well-rounded quartz grains as probably the product of deep weathering and long transport, and recognition that an intrusive rock is younger than the rock it intrudes. Geologic history is determined from interpretation of the way rocks were formed and the dating of these events. This chapter is concerned mainly with methods of dating so that rocks and events can be placed on a time scale.

rocks; this method has been used to calibrate the relative geologic time scale developed from the study of fossils.

Geologic Time

The recognition of the immensity of geologic time and the development of methods for subdividing geologic time are among the great intellectual accomplishments of man. It is difficult to discuss the development of ideas in generalities, but a brief outline is presented here. Discoveries made near the end of the eighteenth century led to rapid progress, although there had been earlier studies. The culmination of this early work was the theory now known as *uniformitarianism,* published by the Englishman James Hutton in 1785

Unconformity between conglomeratic sandstone and overlying flat limestone. Antigonish-Mabou Basin, Nova Scotia. Photo from Geological Survey of Canada.

and 1795. Uniformitarianism simply means that the present is the key to the past; that is, the history of old rocks can be interpreted by noting how similar rocks are being formed today. Hutton's book was poorly written, and the theory did not gain wide acceptance until it was popularized by his colleague Playfair in 1802. Uniformitarianism is a simple idea, but it is the keystone of modern geology. A simple example will show the power of this theory. Many varieties of shell-bearing animals can be seen in the mud along the seashore at low tide. A similar mudstone with similar shells encountered in a canyon wall is interpreted, using uniformitarianism, as a former sea bottom that has been lithified and uplifted. If this seems too elementary, we need only remember that just a few years before Hutton's time, only a few people recognized that fossil shells were evidence of once-living organisms.

Recently uniformitarianism, as presented above, has been criticized as being oversimplified. The objection is that geological processes have modified the earth; therefore, at least in some cases, these processes do not operate now in the same ways that they did in the past. Thus, the present may not be the key to the past. For example, it will be shown that after the origin of life, organisms changed the composition of the atmosphere, making further creation of life impossible. Thus, though uniformitarianism in its simplified version can explain many geologic phenomena, the doctrine has limitations that must be taken into account.

Geologic history is based largely on the study of sedimentary rocks and their contained fossils. Such studies have led to the development of a geologic time scale showing the relative ages of rocks. In recent years

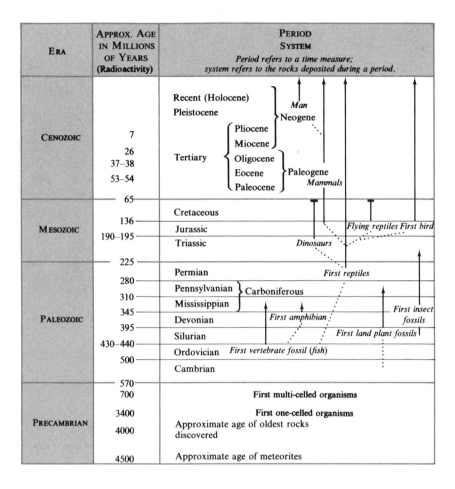

Figure 17-1. The geologic time scale. Shown to the right is a very simplified diagram showing the development of life. Not included on the diagram are many types of invertebrate fossils such as clams, brachiopods, corals, sponges, snails, etc., which first appeared in the Cambrian or Ordovician and have continued to the present.

it has been possible, using measurements of the radioactive decay of certain elements in rocks, to determine the absolute ages of some rocks. This ability to date igneous and metamorphic rocks has expanded the range of historical geology.

After several false starts, historical geology developed in the last part of the eighteenth century and the first half of the nineteenth. In 1782 the great French chemist Lavoisier demonstrated that near Paris the quarries dug for pottery and porcelain clay all exposed the same sequence of sedimentary rocks. Georges Cuvier

and Alexandre Brongniart published maps in 1810 and 1822 showing the distribution of the various rock types around Paris, and in 1815 William Smith published a geologic map of England. Cuvier and Brongniart studied the fossils in each of the sedimentary layers that they mapped and discovered, as did Smith, that each layer contained a different group of fossils. They discovered that a sedimentary bed could be identified by its fossils. That work set the scene for the next 20 years, for during this period the geologic time scale was developed.

The geologic time scale, or column, was the result of individual work by a number of people in western Europe. These men studied the rocks and their fossils at well-exposed places, generally near their homes, and described these rocks and their fossils in books and papers. They called these sequences of rocks *systems* and named each of the systems. As the studies proceeded in this haphazard manner, the general sequence of older to younger was recognized and the gaps were filled in. It was also shown that rocks in similar stratigraphic position, although far removed from the type area, contained the same or very similar fossils. Thus it took work by many people spread over a wide area to demonstrate that fossils can be used to date rocks, not only in local areas, but all over Europe. In this way a geologic column, or time scale, was developed as a standard of reference; fossils from other areas could be correlated or dated by comparison with fossils on which the time scale was based. The geologic time scale in current use is shown in Fig. 17-1. Almost all of the systems or periods were originally defined in the first half of the nineteenth century.

The units used in historical geology are of several types. The *systems* just described are the rocks deposited during a time interval called a *period*. Thus the Cambrian System was deposited during the Cambrian Period. The systems and periods are arbitrarily defined, and their recognition away from the type area where they were defined is based on interpretation of fossils. A geologist mapping an area locates on his map the occurrence of easily recognized, objective units called *formations*. Formations are rock units named for a location, a type area, where they are

well exposed, such as the Austin Chalk and the Tensleep Sandstone. All parts of a formation are not necessarily the same age. The sandstone bed in part B of Fig. 17-8, for example, might be defined as a formation. Thus there are three kinds of units, time units such as periods, time-rock units such as systems, and rock units such as formations. Table 17-1 summarizes the terms used.

The changes in marine invertebrates during geologic history are obvious to even a casual observer; for this reason such fossils can be used to date the enclosing rocks. (See Fig. 17-2). The changes between periods are not as great but are just as obvious to a trained observer. The most marked changes between adjacent periods are between Cambrian and Ordovician, Permian and Triassic, and Cretaceous and Cenozoic. Marked changes also occur at the end of the Ordovician, Devonian, and Triassic.

The geologic column or time scale provided only a relative time scale to which rocks could be referred. It also showed the general development of life and so set the stage for Darwin, who published his theory of evolution in 1859.

Although the idea of evolution was not new to Darwin, his book started a controversy that is still continuing in non-scientific circles. Until his time no controversy existed between geology and religion because even the geologists interpreted geologic history in biblical terms. Biblical scholars thought that the earth was about 6000 years old, based on their interpretation of the Old Testament. However, in the hundred years before Darwin, some geologists began to recognize that a much longer time was necessary for the deposition of the sedimentary rocks and other geologic events than students of the Bible would allow. Before evolution was accepted, the changes in fossil life throughout the geologic column were explained by a catastrophic dying out of all life and a new creation of the animals found as fossils in the succeeding bed. This idea had its roots in the biblical Noachian Flood and the widely held theory of Werner, a late eighteenth century German, who believed that all surface rocks, including igneous rocks, were precipitated in a universal sea. Darwin had a pronounced influence on the development of geologic thought, but his theory also needed the long duration of geologic time envisioned by the uniformitarianists.

Thus the central point became the

Table 17-1. Stratigraphic units.

A.	TIME AND TIME-ROCK UNITS	
	Time Units	*Time-Rock Units*
	Eon	—
	Era	Erathem
	Period	System
	Epoch	Series
	Age	Stage
	(Phase)	Chronozone
B.	ROCK UNITS	
	Group	
	Formation	
	Member	

A.

B.

Figure 17-2. Brief comparison of these two fossiliferous slabs of rock show the great changes in life that have occurred during geologic time. A. Pennsylvanian fossils from Kansas. Bryozoa and brachiopods are most abundant. Photo from Ward's Natural Science Establishment, Inc., Rochester, N. Y. B. Miocene fossils from Virginia. Snails and clams are most abundant. Photo by W. T. Lee, U.S. Geological Survey.

age of the earth. The geologists and the evolutionists needed a very old earth. Joly in 1899 calculated the age at 90 million years by dividing the total amount of salt in the ocean by the amount of salt annually brought into the ocean by the rivers. Estimates of the age of the earth based on the time necessary to deposit the rocks of the geologic time scale were made by many geologists in the latter part of the 19th century; and although they varied between 3 and 1584 million years, about 100 million years was an average figure. The lack of precision and the subjective nature of the assumptions made these estimates suspect. Lord Kelvin, one of the most influential physicists of the day, entered the fray and attempted to calculate the age of the earth from thermodynamics. He assumed that the earth began as a melted body and has been cooling ever since. He calculated that 20 to 40 million years had passed from the time when the earth had cooled enough for life to exist to the present. About the turn of the century, when Kelvin was an old man but still refining his calculations, radioactivity was discovered. As noted earlier, radioactivity is a source of heat within the earth, and so Kelvin's heat flow calculations were based on an incorrect assumption.

Radioactivity was discovered by Becquerel in 1896. In 1902 Rutherford and Soddy showed that by radioactive decay an element is changed in part to another element. This sounded much like the quest of the alchemists, who tried to change lead into gold, and was not immediately accepted by many scientists. However, in 1905 Boltwood noted the association of lead and uranium in uranium minerals and that the ratio of lead to uranium was higher in the older minerals. Boltwood used this information to calculate the age of these minerals, and one of them had a computed age of over 2000 million years. This greatly extended the possible age of the earth and was quickly accepted by a few geologists. Present methods of radioactive dating show the earth to be about 5000 million years old. These methods will be described later in this chapter. Thus, in a single stroke, radioactivity provided the energy source not taken into account by Kelvin and a means to find the age of the earth.

Unconformities

Once the time scale was set up, geologists traced the systems into previously unmapped areas. One problem that soon appeared was that the systems were originally defined, in many cases, as the rocks between two important breaks in the geologic record. In other areas there were no breaks, but continuous deposition. Thus, between the original systems were intervals not included in either system; and arguments about the assignment of the neglected intervals are, in some cases, still in progress. The breaks are of several types, such as change from marine to continental deposition, influx of volcanic rocks, and unconformities. (See Fig. 17-3.) The latter is the most important and requires discussion.

All breaks in the geologic record that indicate a time interval for which there is no local record are called *unconformities* (see Figs. 17-4 and 17-5); they may take several

Figure 17-3. Natural breaks in the stratigraphic record. A-A and D-D are unconformities because the time necessary for erosion is not represented by sedimentary rocks. The other breaks may or may not be unconformities, depending on whether there is a time interval between the different rock types or continuous deposition. D-D is an angular unconformity.

different forms. If sedimentary rocks overlie metamorphic rocks, clearly the sediments were deposited after the metamorphism; and because no sedimentary rocks representing the time of metamorphism are present, this is a type of unconformity. In this case, if the metamorphic rocks were formed at considerable depth in the crust, as is generally the case, then the period of erosion which uncovered the metamorphic rocks prior to deposition of the sediments is also missing from the sedimentary record. Another type of unconformity may consist of a change in fossils, representing a short or a long period of time, that occurs between two beds in the sedimentary section. Such an unconformity may be due to erosion of previously deposited beds or to nondeposition, and may not be at all conspicuous until fossils are studied. The third type of unconformity is more obvious and consists of folded and eroded sedimentary rocks that are overlain by more sedimentary rocks. Such an unconformity is called an *angular unconformity* because the bedding of the two sequences of sedimentary rocks is not parallel. In this case, the

time of folding and erosion is not represented by sedimentary rocks.

Correlation and Dating of Rocks

So far we have considered how the geologic systems were defined, and it is obvious that they are recognized in previously unstudied areas by comparing the contained fossils with the fossils from originally defined type areas. In this manner the systems defined in Europe were recognized in North America and the other continents. This type of correlation by fossils is also most important in establishing the time equivalence of nearby beds.

Another type of correlation is to establish that two or more outcrops are parts of a once continuous rock body. The simplest case of such rock correlation is actually to trace the beds from one area to the other by walking out the beds. In other cases, distinctive materials in a bed, such as fragments of an unusual rock type or a distinctive or uncommon mineral or assemblage of minerals, may be used to identify a bed. Another method uses the sequence of beds. (See Fig. 17-6.)

By correlating from one area to another, the total thickness of sedi-

mentary rocks in a region can be determined as shown in Fig. 17-7.

A.

B.

Figure 17-4. Unconformities. A. An angular unconformity in the Grand Canyon. The rocks beneath the unconformity are Precambrian in age and those just above it are Cambrian. Photo by L. F. Noble, U. S. Geological Survey. B. An erosional unconformity in Silurian rocks in Niagara Gorge, N. Y. Photo by G. K. Gilbert, U. S. Geological Survey.

Note that the methods of correlation, with the exception of the use of fossils, establish only the continuity of a sedimentary bed. Fossils establish the fact that both beds are the same age. The difference between these two types of correlation can be seen by considering an expanding sea, a not uncommon occurrence in geologic history. As the sea advances, the beach sands, for example, form a continuous bed of progressively younger age. This is shown diagrammatically in Fig. 17-8.

This diagram also illustrates another difficulty in the use of fossils. Note that limestone, shale, and sandstone are all being deposited at the same time in different parts of the basin. Each of these different environments will attract and be the home of different types of organisms. Thus, the fossils found in each of these environments will be different, even though they are of the same age. This means that environment must be taken into account in establishing the value of a fossil or group of fossils in age determination. For this reason, free-swimming animals make the best fossils because their remains are found in all environments. Fossils of this type are very useful and have been called *index fossils* if they establish clearly the age of the enclosing rocks. Ideally, an index fossil should be widespread in all environments, be abundant, and have a short time span. Very few fossils meet all of these requirements, and generally, groups of fossils are used to establish the age of a bed. (See Figs. 17-9 and 17-10.)

Time-lines, or surfaces, can be established in sedimentary rocks by physical means alone in some rare instances. For example, a thin ash bed from a volcano may be spread over a very large area by wind. The deposition of such an ash bed may occur over a number of days or even weeks, but in terms of geologic time it is instantaneous. These rare occurrences help in local correlation problems and show the validity of the methods of dating and correlation described here.

The only rocks that have reasonably abundant fossils are marine sedimentary rocks, and not even all of these have enough fossils to establish their age clearly. On the previous pages, we have discussed dating of fossiliferous marine rocks. Similar techniques can be used to date continental sediments, but here the number of fossils is generally much smaller. This fact is easy to understand if one compares the number of easily fossilized animals exposed at low tide at the seashore with the much smaller number of easily fossilized animals living in a similar-sized area of forest or grass land, and if one considers how easy it is for a clam shell to be buried and so preserved, especially when compared to the chance that a plant leaf or a rabbit skeleton will be buried and so preserved as a fossil. A land organism, even if buried, may not be preserved, as oxidizing conditions that destroy organic material exist even at some depth. Also, a dead rabbit is apt to be eaten and dismembered by scavengers. Another problem in the use of continental fossils is that they are controlled even more by climate than are marine organisms. The progression

C.
Figure 17-4. Continued. C. Cambrian sedimentary rocks overlying Precambrian metamorphic and igneous rocks. Note that the bedded sediments were deposited on a hilly terrain. Photo was taken during construction of Taum Sauk project, about 100 miles south of St. Louis, Missouri. Photo from Union Electric.

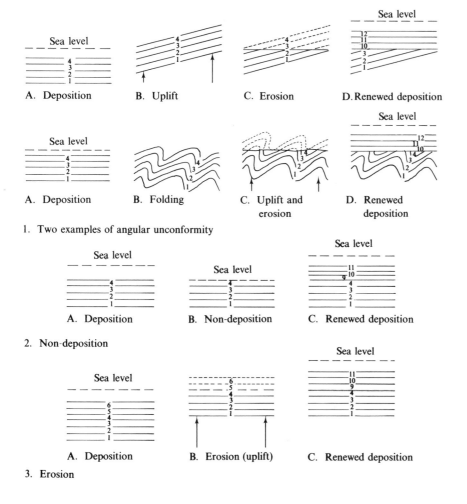

A. Deposition   B. Uplift   C. Erosion   D. Renewed deposition

A. Deposition   B. Folding   C. Uplift and erosion   D. Renewed deposition

1. Two examples of angular unconformity

A. Deposition   B. Non-deposition   C. Renewed deposition

2. Non-deposition

A. Deposition   B. Erosion (uplift)   C. Renewed deposition

3. Erosion

Figure 17-5. Development of unconformities. In each case a time interval is not recorded in the bedded rocks. It is commonly difficult to distinguish between cases 2 and 3. In each case, for simplicity, sea level is assumed to be the base level that controls whether erosion or deposition takes place.

= ammonite
= coral
= snail

Correlation by fossils

xxxx = distinctive volcanic ash

Correlation by sequence of beds. Note the changes in thickness in the two areas.

Correlation by lithologic similarity. Also shows use of a keybed (an easily recognized bed)

Figure 17-6. Types of correlation.

of types of plant and animal life seen during a mountain climb will illustrate this point.

Dating igneous and metamorphic rocks presents even more problems because such rocks almost never contain fossils. They must be dated by their relationships to fossil-bearing rocks. Volcanic rocks are dated by the fossiliferous rocks above and below and, in some cases, by interbedded sedimentary rocks. Intrusives must be younger than the rocks that they intrude; they cannot be dated closer unless the intrusive has been uncovered by erosion and, then, unconformably covered by younger rocks. (See Fig. 17-11.)

Metamorphic rocks are even more difficult to date because both the age of the original rock and the date of the metamorphism should be obtained. The age of the parent rock may be found by tracing the metamorphic rock to a place where fossils are preserved. Finding the date of metamorphism may be more difficult because, in general, an unconformable cover, if present, gives only the upper limit.

Radioactive Dating

Some of these problems in dating igneous and metamorphic rocks can be overcome by modern radioactive methods. Such methods produce an absolute age in years and so have been used, in addition, to date the geologic time scale in years. (See Fig. 17-12.) In spite of the seeming precision of radioactive dating, it is not yet possible to date a rock as accurately this way as with fossils. This is because of inherent inaccuracies in the measurements of the amounts of the elements produced by radioactive decay. These inaccuracies produce an uncertainty in

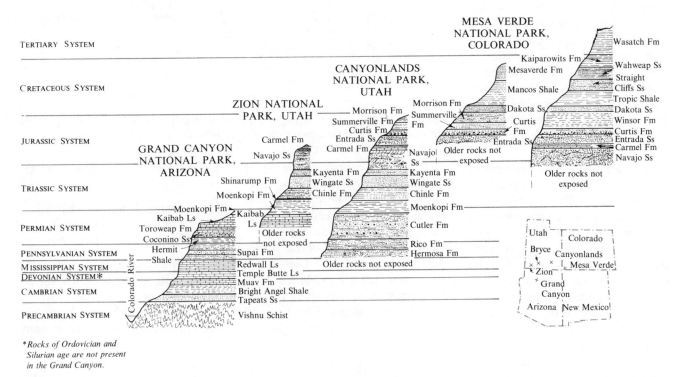

TERTIARY SYSTEM

CRETACEOUS SYSTEM

JURASSIC SYSTEM

TRIASSIC SYSTEM

PERMIAN SYSTEM

PENNSYLVANIAN SYSTEM

MISSISSIPPIAN SYSTEM

DEVONIAN SYSTEM*

CAMBRIAN SYSTEM

PRECAMBRIAN SYSTEM

*Rocks of Ordovician and
Silurian age are not present
in the Grand Canyon.*

Figure 17-7. Correlation of the strata at several places on the Colorado plateau reveals the total extent of the sedimentary rocks. After U.S. Geological Survey.

the radioactive date that is more than the span of zones based on fossils, especially in those parts of the geologic column that contain abundant short-lived fossils. As methods of analysis are improved, this limitation may be overcome and the present high cost of dating may also be reduced.

Radioactive dates are obtained by studying the daughter elements produced by a radioactive element in a mineral. The principles of radioactivity were covered in Chapter 2. Sometimes more than one element is studied in an effort to provide a check on accuracy. A radioactive date tells the time of formation of that mineral and so tells the time of crystallization of an igneous rock, or

the time of metamorphic recrystallization in the case of a metamorphic

rock. In some favorable cases, in which microscopic studies show that

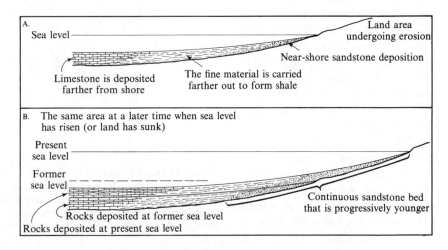

Figure 17-8. Deposition in a rising or expanding sea. A. The deposition at the start. B. Deposition at a later time after sea level has risen.

different minerals formed at different times in a rock that has undergone more than one period of metamorphism, it is possible to date the periods of metamorphism by dating elements in minerals formed during each period of metamorphism. In many radioactive decays a gas is one of the products, and heating during metamorphism may drive off the gas, thus resetting the radioactive clock to give the time of metamorphism.

A radioactive element changes to another element by spontaneously emitting energy. The rate of this decay is unaffected by temperature or pressure; hence, if we know the rate of formation of the daughter products, all we need to find is the ratio of original element to daughter product to be able to calculate the time of crystallization of the mineral containing the original radioactive element. The main assumptions, then, in using radiodates are that the decay rate is known and constant, and that no daughter or parent elements are lost. The latter assumption is the least sound because of weathering and metamorphism.

Several radioactive elements are used to date rocks, but only the most important are reviewed here. The methods in current use are tabulated in Appendix C (see also Chapter 2). A few minerals contain uranium, and these generally also contain thorium, which is about the same atomic size. These minerals are rare and with a few exceptions are almost completely confined to pegmatite veins in batholiths. Because such veins are probably the last to crystallize, these occurrences give only the upper limit of the age of the batholith. Each of the three radioactive isotopes that

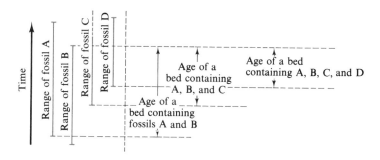

Figure 17-9. The use of overlapping ranges of fossils to date rocks more precisely than can be accomplished by the use of a single fossil.

these minerals contain — the two of uranium as well as thorium — produces a series of radioactive daughter products; each member of each series decays to the next daughter until, finally, a stable isotope of lead is produced. (See Fig. 2-10.) These many steps constitute a possible source of inaccuracy, as it is possible for any one of the daughters to be removed by leaching or some other process.

The elements and the final products are:

Uranium 238 → Lead 206
Uranium 235 → Lead 207
Thorium 232 → Lead 208

Thus, in this case, it is possible to measure three ratios which should all agree in age.

A problem with age determinations involving lead is that many

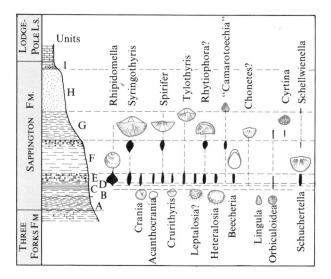

Figure 17-10. An example of the distribution of brachiopods in the Sappington Formation in western Montana. The relative number of each fossil found in the beds is indicated by the thickness of the black lines. The occurrence of these animals was at least in part controlled by the sedimentary environment. From R. C. Gutschick and Joaquin Rodriquez, *Bulletin*, American Association of Petroleum Geologists, Vol. 51, 1967.

Figure 17-11. Dating of intrusive rocks. Intrusive body ① is younger than beds A; no upper limit can be determined. Intrusive ② is younger than beds A and older than beds B. Closer dating may be possible if the conglomerate composed of pebbles of ② in the basin at the right contains fossils. The contact between beds A and B is an angular unconformity.

rocks contain some lead of nonradioactive origin. The amount of nonradiogenic lead is quite small but may require an appreciable correction in calculating the age of very old rocks. This ordinary lead contains the isotope lead 204, so its presence can be easily recognized. The problem is that along with the minor amounts of lead 204 are the three isotopes, lead 206, lead 207, and lead 208, which also occur radiogenically, and the ratio of all four isotopes varies. Thus to correct for the pre-existing ordinary lead, one must analyze nonradioactive rocks in the area to find the ratio between lead 204 and each of the other isotopes. In this way the nonradiogenic

lead 206, lead 207, and lead 208 can be excluded from that due to radioactive decay. Meteorites contain lead 204, and the correction for the original lead is a problem. As discussed later, meteorites are used to date the formation of the earth, so this problem results in some uncertainty in this important date.

Rubidium is a radioactive element that is generally present in small amounts in any mineral that contains potassium, because both elements are about the same size. Rubidium decays in a single step to strontium (rubidium 87 → strontium 87). Because potassium minerals that contain rubidium are so common (feldspar and mica) and only

a single radioactive decay is involved, this process may become the most useful method of radioactive dating. However, at present the half-life of this decay cannot be measured accurately because the low energy of the beta particles emitted makes them difficult to detect.

Potassium also has a radioactive isotope that is useful in dating, especially since it is such a common element. The small amount of radioactive potassium decays to form two daughters:

Calcium 40 is ordinary calcium and so is present in most minerals. Therefore, the method used is to measure the amounts of potassium 40 and argon 40. The main problem is finding minerals that retain the argon gas. It appears that micas, in spite of their cleavage, retain most of the argon but feldspar loses about one-fourth of the argon.

The natural radioactive isotope of carbon is effective in dating carbonaceous material, such as wood or coal, less than 40–50,000 years old. This method depends upon the fact that all living matter contains a fixed ratio of ordinary carbon 12 and radioactive carbon 14. Cosmic rays react with nitrogen in the atmosphere to form the carbon 14, which ultimately gets into all living matter via the food cycle. The carbon 14 decays back to nitrogen. Therefore, the ratio of carbon 12 to carbon 14 can be used to calculate how long an organism has been dead, because a dead organism takes in no more carbon 14. Extensive use of this method is made in archaeology and

Figure 17-12. Determination of absolute age of one of the geologic periods. The sandstone contains Cambrian fossils. Dikes A and B have been dated by radioactive methods. The Cambrian Period is younger than dike A and older than dike B. Can the absolute age of the Cambrian be determined from a single occurrence such as this?

in glacial studies, although carbon 14 dates do not always agree with dates obtained by other means.

Another type of radioactive dating is to measure the damage done to a crystal lattice by the particles released by radioactivity. These methods, mentioned in Chapter 2, are relatively inaccurate.

Ages of rocks can also be determined by chemical means. At present these methods yield mainly relative ages, but new advances may result in absolute ages. An obvious example is the relative dating of glacial moraines on the basis of weathering. Rates of weathering are dependent on so many factors that absolute ages are not possible in this way and great care must be exercised in application. Another example is the crystallization of volcanic glass. Obsidian, or any other glass, is not stable but tends to crystallize; however, the rate of crystallization depends on the composition and amount of water in the glass, as well as on other factors. A promising chemical method of dating is the breakdown of amino acids in fossils. The breakdown is dependent on many factors, such as temperature and composition of formation water, but gives fairly accurate results in young rocks.

Age of the Earth

The age of the earth itself is a geological problem to which radioactive dating has been applied. So far, the oldest rocks found are about 4,000 million years old. This, then, is the youngest that the earth can possibly be, and is probably about the time the crust formed; the earth itself is probably older. Meteorites were discussed earlier and are thought to be

the material from which the solar system was formed. Radioactive dates from meteorites give ages of around 4,550 million years. This suggests that the earth may have formed 4,500 to 5,000 million years ago.

The first abundant fossils occur in rocks of the Cambrian system although a few fossils have been found in much older rocks. The base of the Cambrian has been dated radioactively as about 600 million years. Why abundant, relatively advanced animals appeared suddenly just prior to this time is a problem, but these fossils may record only the development of easily preserved hard parts such as shells. (See Fig. 17-1.) Because fossils are so much easier to use than radioactive methods in dating and correlating rocks, we know many more details of geologic history from Cambrian time onward than Precambrian time. We know only the broadest outlines of the history of the Precambrian, even though it includes almost nine-tenths of the earth's past.

Interpreting Geologic History

Deciphering the history of the earth or any part of it is a difficult reiterative process. It has many steps, and each affects every other step. Multiple working hypotheses must be used to test each step. The process begins with study of the rocks themselves and their relationships to other rocks. This leads to some ideas about their mode of origin. The rocks must be dated and correlated, using the methods described in this chapter. If the rocks have been deformed by folding and faulting, their

original relationships and positions must be determined. Each layer or rock unit can now be reconstructed to show the geography at its time of formation. Remember that the rock record is incomplete because of erosion and because many rock layers are exposed only at a few places. The final step is to reconstruct the events that caused the changes between each of the steps.

This brief outline shows how the incomplete rock record is interpreted. Note the various levels of abstraction that are involved. It is no wonder that the history of the earth is not completely known. Interpreting the history of the earth or a small area is difficult, and revision is constantly necessary as more data are obtained. Thus geology is an ongoing process that offers great challenge; many fundamental problems remain to be solved.

Interpreting Ancient Environments

The environment of the geologic past can be determined at least partially from study of rocks. Such features as coral reefs, evaporites, desert deposits, and glacial deposits are a few obvious examples. Some of the features that indicate environment were described in earlier chapters. Fig. 17-8A shows in a generalized way how the rock type changes from sandstone near shore to shale and limestone farther from shore. Fig. 4-38 is an example of a reconstruction of an ancient sedimentary basin from the study of scattered outcrops. Many of the features used in such studies are also described in Chapter 4. The grains that compose a sedimentary rock, such as a sandstone, also reveal much environmental information. The composition of the grains in a sandstone is influenced by both the nature of the source rocks and the amount of weathering. The topography and the climate greatly influence the weathering. The size and shape of the grains are influenced by the topography, and the distance and method of transportation. Obviously, only some of the factors involved in interpretation of ancient environments have been

mentioned here, and a little thought will reveal many more aspects.

Isotopes can be used to determine ancient temperatures as well as age. Oxygen isotopes are used in temperature determination. Oxygen has three isotopes: oxygen 16, oxygen 17, and oxygen 18. When water evaporates, the lightest isotope — common oxygen 16 — evaporates at a slightly higher rate than the heavier isotopes. This difference in evaporation rates is proportional to temperature. The ratio of oxygen 16 to oxygen 18 can

be measured and used to determine the temperature. In practice, this ratio is measured in fossil shells made of calcium carbonate ($CaCO_3$). There are many pitfalls in this method, but so far the results have agreed fairly well with other geological evidence.

Other means of estimating temperatures are coral reefs, evaporites, desert deposits, and glacial deposits. Fossils, both plant and animal, have also been used very successfully to reconstruct climates and ecology.

Summary

Rocks can be dated both relatively and absolutely.

Relative dates can be determined by the relationships of the rock units to each other *and* by referring the rocks to a geologic time scale based on the fossils in the rocks.

Absolute dating is based on radioactive decay of elements in the rocks and has been used to calibrate the relative geologic time scale based on fossils.

Uniformitarianism is the basis of geological interpretation and means that ancient rocks are interpreted as having formed in the same ways that similar rocks are presently forming, or "the present is the key to the past."

Systems are the rocks deposited during a time interval called a *period*.

Formations are rock units that can be mapped easily. The other stratigraphic units are summarized in Table 17-1.

The *geologic time scale* is shown in Fig. 17-1.

The most marked changes in life

occurred at the end of the Cambrian, Permian and Cretaceous Systems.

The development of life through geologic time was important evidence for Darwin's evolution theory and the recognition of the immensity of geologic time.

Unconformities are breaks in the geologic record that indicate a time interval for which there is no local depositional record. If the older rocks were deformed so that their bedding planes are not parallel to the younger rocks, an *angular unconformity* results.

Correlation and *dating* of rocks can be accomplished in many ways. Comparison of fossils can establish the time equivalence of beds. Rock correlation can establish that two outcrops are parts of a single rock body. This can be accomplished if the bed has distinctive properties that can be recognized, or from a similar sequence of beds, or by actually tracing the beds on the ground.

A *lithic body*, such as a sandstone unit, is not necessarily the same age throughout.

Index fossils should be found in all environments, be abundant, and have a short time span. Very few fossils meet these requirements, so most dating uses groups of fossils. Fossils are

most abundant in marine sedimentary rocks.

Igneous rocks are dated by their relationships to sedimentary rocks or by radioactive methods. An intrusive rock is younger than the rocks that it intrudes.

In dating metamorphic rocks, the age of the parent rock and the age of metamorphism should be determined. They are dated by tracing them to places where they are not metamorphosed, by their relationships to sedimentary rocks, and by radioactive methods.

A radioactive element spontaneously changes into another element at a rate that is unaffected by temperature and pressure.

The more important methods of radioactive dating are:

Uranium 238 → Lead 206
Uranium 235 → Lead 207
Thorium 232 → Lead 208
Rubidium 87 → Strontium 87
Potassium 40 ⟨ Calcium 40 / Argon 40
Carbon 14

Meteorites are about 4,550 million years old, suggesting that the earth formed about 4,500–5,000 million years ago.

The oldest rocks so far found on the earth are about 4,000 million years old, suggesting that the crust formed about this time.

The first abundant fossils are found in Cambrian rocks which are about 600 million years old.

Geologic history is interpreted by determining how the rocks were formed and when, how they have been deformed and when, and how and when they have been eroded. Multiple hypotheses must be used to test each step in the reconstruction of events.

Ancient environments are interpreted from the rocks themselves, the fossils, and the ratio of oxygen 16 to oxygen 18.

Questions

17-1. How does one interpret the depositional environment of a layer of ancient rocks?

17-2. Describe in general terms how the geologic column was established and discuss whether it consists of natural subdivisions in North America.

17-3. What are unconformities? Describe the various types.

17-4. How was the value of fossils in determining geologic age established?

17-5. What is meant by correlation?

17-6. List and briefly describe the methods of correlating rocks.

17-7. Why is it generally easier to date marine rocks than continental sediments?

17-8. How are metamorphic rocks dated? What dates are required?

17-9. Which method of radioactive dating is best and why?

17-10. What are index fossils?

17-11. How can metamorphic events be dated radioactively?

17-12. What effect would a small amount of modern root material mixed with a sample to be dated by the carbon 14 method have?

17-13. How can an angular unconformity be distinguished from a low angle thrust fault?

17-14. Fossils aid in determining the age of a rock unit. How else do they aid in geologic interpretation?

17-15. Prove to your girlfriend (boyfriend), who is an art major and has not taken any science, that the earth is very old.

Supplementary Reading

Deevey, E. S., Jr., "Radiocarbon Dating," *Scientific American* (February, 1952), Vol. 186, No. 2, pp. 24-33. Reprint 811, W. H. Freeman and Co., San Francisco.

Eicher, D. L., *Geologic Time*. Englewood Cliffs, New Jersey: Prentice-Hall, Inc., 1968, 150 pp. (paperback).

Eiseley, L. C., "Charles Lyell," *Scientific American* (August, 1959), Vol. 201, No. 2, pp. 98-106. Reprint 846, W. H. Freeman and Co., San Francisco.

Emiliani, Cesare, "Ancient Temperatures," *Scientific American* (February, 1958), Vol. 198, No. 2, pp. 54-63. Reprint 815, W. H. Freeman and Co., San Francisco.

Engel, A. E. J., "Geologic Evolution of North America," *Science* (April 12, 1963), Vol. 140, No. 3563, pp. 143-152.

Faul Henry, *Ages of Rocks, Planets, and Stars*. New York: McGraw-Hill Book Co., 1966, 109 pp. (paperback).

Fenton, C. L., and M. A. Fenton, *Giants of Geology*. Garden City, New York: Doubleday & Co., 1952, 333 pp. (Also published in paperback, New York: Dolphin Books and Dolphin Masters, 1952, 318 pp.)

Hay, E. A., "Uniformitarianism Reconsidered," *Journal of Geological Education* (February, 1967), Vol. 15, No. 1, pp. 11-12.

Hurley, P. M., "Radioactivity and Time," *Scientific American* (August, 1949), Vol. 181, No. 2, pp. 49-51. Reprint 220, W. H. Freeman and Co., San Francisco.

McAlester, A. L., *The History of Life*. Englewood Cliffs, New Jersey: Prentice-Hall, Inc., 1968, 152 pp. (paperback).

Mintz, L. W., *Historical Geology*. Columbus, Ohio: Charles E. Merrill Pub. Co., 1972, 786 pp.

Nichols, R. L., "The Comprehension of Geologic Time," *Journal of Geological Education* (March, 1974), Vol. 22, No. 2, pp. 65-68.

Early History of the Earth — Astronomical Aspects of Geology

18

The kindest words regarding these entertainments are those of H. H. Read, quoted by L. R. Wager (1958) in his address at the 1958 meeting of the British Association for the Advancement of Science: ". . . just as things too absurd to be said can yet with perfect propriety be sung, so views too tenuous, unsubstantial and generalized for ordinary scientific papers can yet appear with some measure of dignity in presidential addresses." This is then an appropriate occasion to review some speculations, by no means original, on the large-scale evolution of the Earth. Views on the early history of the Earth and on happenings in the deep interior can hardly fail to be tenuous and generalized, so I think they may qualify under Read's dictum. Nevertheless, we cannot hope to account for the development of the surface film which naturally engages most of our attention without considering the enormously larger interior from which this film was derived and which continues to determine its evolution . . . .

Introduction to Francis Birch's presidential address to The Geological Society of America, "Speculations on the Earth's Thermal History," *Bulletin*, Geological Society of America, Vol. 76, pp. 133-154, 1965.

The Earth[1] is part of the solar system, so we must look to astronomy for clues to the Earth's origin. Our main interest is not astronomy, so this chapter will be descriptive and much information will be presented without proof. Details will be discussed only where they have important application to geology.

Our sun is not the only star thought to have planets. Although the other stars are too far away to see the other planets, even with our largest telescopes, their existence is inferred from the wavering motion of three stars and the as yet unconfirmed wavering motions of five more stars.

Solar System

The solar system is composed of the sun and the planets that orbit around the sun, the satellites of the planets, and comets, meteors, and asteroids. All of the orbital motions are explained by the law of gravity.

The sun is an ordinary, average-size star, composed mainly of hydrogen and helium, but containing most of the elements found on Earth. The sun's energy, which makes life possible on Earth and ultimately causes most geologic processes, comes from nuclear reactions discussed in Chapter 2. The sun has 99.86 per cent of the mass of the solar system, and the Earth has about 1 per cent of the remaining 0.14 per cent.

The nine planets all revolve in the same direction around the sun. The orbits are all elliptical, very close to circular, and lie close to a single plane, except Pluto whose orbit is inclined. (See Fig. 18-1.) Seen

---

[1]In this section the astronomical convention of capitalizing the names of all planets will be followed.

Scientist-astronaut Harrison Schmitt standing next to a huge split lunar boulder. Photo from NASA.

343

from the Earth, the sun and the planets follow the same path across the sky. The planetary data are summarized in Table 18-1. The inner planets, Mercury, Venus, Earth, and Mars, form a group in which the individual members are more or less similar in size. The next four planets, Jupiter, Saturn, Uranus, and Neptune, form a similar grouping of much larger planets. Pluto is much smaller than the other outer planets and has a tilted, eccentric orbit, all of which suggest that it may not be a true planet. It may once have been a satellite of Neptune and escaped into its own orbit. Between Mars and Jupiter are several thousand small bodies, the asteroids, that are probably the origin of meteorites.

Most of the information that we have about the planets is summarized in Table 18-1. Only a few important points will be described in the text.

Inner Planets

The inner planets are Mercury, Venus, Earth, and Mars. Mercury is small and too close to the sun for detailed telescopic study; however, Mariner 10 revealed a cratered surface. Earth is the main subject of this book. Venus and Mars, our closest neighbors, are of great interest to us and have been studied by telescope and with space probes. Because they are our closest neighbors, they are the planets on which life might occur. For life, such as we know it, to exist, liquid water is probably necessary, and only Mars has temperatures in this range.

Mars has a reddish color and so was named for a god of war. It is farther from the sun than is Earth, and so has lower temperatures. Its rotational axis is tilted with respect to its orbital plane, and so it has seasons like the Earth. Telescopic observation reveals that polar ice caps form and disappear each year, and the color of the planet changes from bluish-green in spring to brown in the fall, suggesting seasonal growth to some observers. (See Fig. 18-2.) Mariner IX photographs reveal dust clouds, so the observed color changes may be caused by dust storms and clouds. Mars' atmosphere is composed mainly of carbon dioxide and a little water vapor. Mariner spacecraft revealed that this inhospitable atmosphere is less than one per cent as dense as the Earth's. The average

Table 18-1. Summary of solar system data. These statistics are approximations and vary somewhat in the different sources from which they were gathered.

Body	Distance from Sun (average) (millions of miles)	Radius[1] (miles)	Mass (Earth = 1)	Specific Gravity	Eccentricity of Orbit	Inclination of Spin Axis	Period of Revolution around Sun	Period of Rotation
SUN	—	432,000	332,000	1.4	—	—		25 days at equator 34 days at poles
Mercury	36	1,500	0.05	5.1	0.206	10° (?)	88 days	58.6 days
Venus	67	3,800	0.81	5.0	0.007	6°	225 days	242.9 days[2]
Earth	93	3,963	1.00	5.5	0.017	23.5°	365.25 days	23 hrs. 56 min.
Mars	141	2,100	0.11	4.0	0.093	25.2°	687 days	24.5 hours
Jupiter	483	43,500	318	1.3	0.048	3.1°	11.9 years	9.9 hours
Saturn	886	37,000	95	0.7	0.056	26.7°	29.5 years	10.2 hours
Uranus	1,783	15,750	14.6	1.6	0.047	98°	84 years	10.8 hours[2]
Neptune	2,793	16,500	17.3	2.3	0.009	29°	165 years	15.7 hours
Pluto	3,666	1,800 (?)	0.1 (?)	?	0.249	?	248 years	6.4 days(?)

[1]Data from various sources varies up to ten per cent.
[2]Venus rotates clockwise, opposite to other planets.
 (Uranus does also because of its 98° tilt of its spin axis.)

temperature at the equator at noon is probably 40°F, and the temperature may fall to –200°F at night. Thus the conditions are not very favorable for life on Mars, but some may exist. The pictures returned by Mariner space probes show evidence of both internal and surface activity. (See Fig. 18-3.)

Dust storms have been observed on Mars. As the storm subsides, first the higher elevations come into view and finally, the lowlands. In spite of the thin atmosphere, winds up to 300 miles per hour are possible and 170-mile-per-hour winds are not uncommon. The dust is probably the size of sand. The dust may result from meteorite impacts or other processes. The wind may move the dust to lower elevations, and in this way, create the featureless plains of Mars. The pits and basins of Fig. 18-4 could also be caused by wind erosion. The cliff, up to a mile high, that

surrounds Nix Olympia (see Fig. 18-5) may also have been produced by wind erosion.

The volcanic area has Nix Olympia (see Fig. 18-5), the largest volcano ever seen. It is about five miles

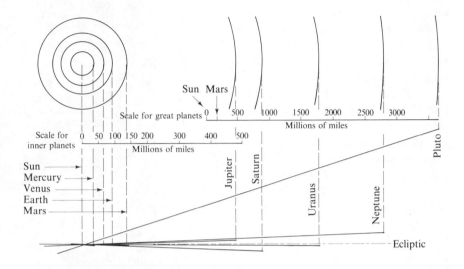

Figure 18-1. The orbits of the planets. Because Pluto is almost 40 times further away from the sun than is the Earth, it is not practical to draw this diagram to scale. The inner planets, the circles, are drawn to scale, and the distances among the great planets are drawn to a different scale. Eccentricity of the orbits is neglected. The edge view shows that, with the exception of Pluto, the orbits lie nearly in the same plane (ecliptic).

SOLAR RADIATION RECEIVED (EARTH = 1)	SOLAR RADIATION REFLECTED (ALBEDO) %	TEMPERATURE (°C)			ATMOSPHERIC PRESSURE AT SURFACE (EARTH = 1)	COMPOSITION OF ATMOSPHERE
		Theo-retical[3]	Actual			
			Top of atmosphere	Surface		
——	——	——	——	5,500°	——	——
7	6	Similar to actual	——	425° sun; 17° dark	0.003	Carbon dioxide
2	73	–35°	–38°	425°	100	Mainly carbon dioxide, minor water
1	34	–20°	——	15°	1	78% nitrogen 21% oxygen
0.4	30	–65°	–50°	–42°[4]	.006	Mainly carbon dioxide, possibly minor water
0.04	45	–168°	–123°	2,000° est.	200,000	60% hydrogen, 36% helium, 3% neon, 1% methane and ammonia
0.01	50	–195°	–167°	2,000° est.	200,000	Same as Jupiter
0.003	66	–185°	–185° approx.	?	8	Similar to Jupiter, with more methane, no ammonia
0.001	62	–185°	–185° approx.	?	?	Same as Uranus
0.0006	15 (?)	–212°	Large variation because of eccentricity		?	?

[3]Theoretical temperatures are calculated from the energy input from the sun.
    Actual temperatures are generally higher because of absorption in the atmospheres.
[4]Average varies from 20° to –95° depending on latitude.
    Actual measurements range from –123° to 17°.

high and 300 miles across at the base; it has a broad complex crater, and its flanks are covered with what appears to be flow material. Nearby are three big volcanoes in a row, one of which is over 15 miles high.

The equatorial region near the volcanic area is a plateau cut by what appear to be faults. (See Fig. 18-6.) The valleys shown in Fig. 18-6A are part of a parallel system that extends more than 3000 miles. They are interpreted as tensional features, perhaps associated with uplift. Many years ago, some astronomers reported a network of faint lines that were incorrectly called canals. The report of these "canals" caused speculation as to the existence of intelligent life on Mars. Perhaps these valleys are the "canals" seen by some astronomers. Fig. 18-6B shows another valley that also may be of fault origin. Note the lines of craters that parallel the main valley. This same part of Mars also has features that suggest erosion by running water, although very little water is present in the atmosphere. Fig. 18-7 shows an example of a long, sinuous gully that resembles a river course. The water might have come from melting part of the polar caps; that will be discussed later. It is also possible that water in the form of ice is present in the "soil." The pits and basins shown in Fig. 18-4 are near the south pole, and so could have formed by the melting of subsurface ice. The chaotic terrain of short ridges and small valleys shown in Fig. 18-8 is similar to landslide and collapsed areas on Earth caused by melting subsurface ice. Clouds on some of the big volcanoes suggest the possibility that they may be emitting gases, possibly water vapor, that

Figure 18-2. The northern hemisphere of Mars. The shrinking north polar ice cap is at the top, and the huge volcanoes are near the bottom. Photo from NASA.

might at times be abundant enough to cause rain and so erosion.

Craters are common in many areas, and, although some are of volcanic origin, many are caused by meteorite impact. This is not unexpected because Mars is near the asteroid belt and has a thin atmosphere. The effects of meteorite impacts are clearly shown on Mars' moon Phobos. (See Fig. 18-9.)

The polar ice caps appear to be dual ice caps. Much of the ice is frozen carbon dioxide, and this is the part that disappears and reforms seasonally. The rest may be composed of water ice. The laminated structure near the ice caps suggests that wind-blown dust accumulates in

the ice at times. These deposits seem to be up to 300 feet thick.

Venus is named for the goddess of love because it is visible only near sunrise and sunset. It is closer to the sun than is the Earth and thus hotter. It has a thick atmosphere, composed almost entirely of carbon dioxide and, as discussed in Chapter 10, the carbon dioxide causes a greenhouse effect, which would account for Venus' high temperature. Also present in minor amounts are hydrogen, water vapor, carbon monoxide, and hydrofluoric and hydrochloric acids. The surface temperature is near 536°C, making life a very remote possibility. The atmosphere has a

Figure 18-3. A mosaic of the equatorial region of Mars. Many of the features shown in the other photographs can be seen here. Big volcanoes are in the upper left, and large canyons parallel the equator. Photo from NASA.

Figure 18-4. A dune field within the 93-mile-wide crater on Mars. Photo from NASA.

solid cloud cover 45 to 60 miles thick so the surface is not visible. In spite of its slow rotation, the surface temperature does not change as much as expected from day to night, suggesting a circulation in the atmosphere that transfers the heat. Venus is the only planet that rotates on its axis in a clockwise direction (see discussion of Uranus below). No magnetic field has been detected, suggesting that it, too, lacks a liquid core, although its rate of rotation may be too slow to develop a magnetic field. Most of this recent information came from space probes that passed near Venus and from the Russian soft-lander in 1967. Radar observations suggest that Venus may have mountain ranges and craters, but little can be said about their type or origin.

Great Planets

The great planets, Jupiter, Saturn, Uranus, and Neptune, are so called because they are much larger than the other planets. Although larger in diameter and mass, they are much less dense than the other planets. They all have thick atmospheres, which are, of course, all that we can see. Their rapid rotation causes pronounced bulges of the gaseous exteriors. The upper surfaces of their atmospheres are very cold because of their distances from the sun. They all have methane in their atmospheres, with ammonia, also, in Jupiter and Saturn; but most of their atmospheres are probably hydrogen and helium. Their low overall densities suggest that they, like the sun, are composed mainly of hydrogen and helium. Their cores are denser than the Earth's; Uranus and Neptune may have rock cores, and the others may be liquid. The temperatures of the surfaces of the great planets are not known, and they may be cold; gravitational compression of the thick atmospheres may have caused the surface temperatures of Jupiter and perhaps Saturn to be as high as 2,000°C, making their structure more like that of the sun than a planet. Saturn's rings are unique and probably are dust similar in composition to a moon. (See Fig. 18-10.) This material is too close to the planet to consolidate and form a moon. The rings may be left over from **the** formation of the planet, or they might be satellites that approached the planet too closely and were broken up when the gravitational attraction overcame their strength. Uranus is unique in having its rotational axis tilted 98 degrees to its orbital plane; alternately, it could be considered tilted 82 degrees and revolving in the opposite direction to the majority of the planets. Its moon's orbit is in the same direction as its rotation.

Moon

Just as the planets orbit the sun, most of the planets have moons, or satellites, that revolve around them. (See Table 18-2.) Most of the moons revolve in the same direction as the planets, but some move in the opposite direction (retrograde). Most of the moons are much smaller than their planets. Ours is much closer in size to the Earth so that the Earth-moon system should probably be considered a double planet. The moons that orbit in the opposite direction and have very inclined orbits may have been captured by the planets instead of having been formed

Figure 18-5. Nix Olympia, a gigantic volcanic mountain on Mars. A. The mountain is 310 miles in diameter across the base. The complex crater is 40 miles in diameter. B. Closer view of the summit. C. Telephoto view of the flank showing features that resemble flow material on earth. Part C shows an area 27 × 34½ miles. Photos from NASA.

with the planets. Origin of satellites is a problem discussed later.

The moon orbits the Earth and rotates on its axis at the same rate

Table 18-2. Satellites of the planets. The planets that are not listed do not have known moons. Data are mainly from Werner Sandner, *Satellites of the Solar System*, American Elsevier Publishing Co., Inc., New York, 1965, 151 pp.

SATELLITE OF	DISTANCE FROM PLANET (THOUSANDS OF MILES)	DIAMETER (MILES)	ORBITAL ECCENTRICITY	INCLINATION OF SATELLITE'S ORBIT TO ORBIT OF PLANET (DEGREES)	REMARKS
Earth	238.9	2,160	0.055	5	—
Mars	5.8	9-12	0.017	2	—
	14.6	6-9	0.003	2	
Jupiter	113	100	0.003	3	
	262	2,450	0.000	3	
	417	2,040	0.000	3	
	666	3,560	0.001	3	
	1,170	3,340	0.007	3	
	7,120	100	0.152	28	
	7,290	40	0.207	28	
	7,300	16	0.132	28	
	13,000	16	0.200	160	Retrograde
	14,000	19	0.207	163	Retrograde
	14,600	40	0.378	147	Retrograde
	14,700	20	0.222	155	Retrograde
Saturn	52	100-200	Unknown	Unknown	Discovered Dec. 1966
	115	400	0.019	27	
	148	500	0.000	27	
	183	800	0.000	27	
	234	750	0.002	27	
	327	1,100	0.001	27	
	760	2,600	0.029	27	
	920	310	0.104	26	
	2,200	990	0.028	16	
	8,030	200	0.166	175	Retrograde
Uranus	80.8	Unknown	Unknown	98	Rotation axis at almost right angle to orbit, so can also be considered to be inclined at 82° and have retrograde movement.
	119	370	0.007	98	
	166	250	0.008	98	
	272.6	620	0.002	98	
	364.5	500	0.001	98	
Neptune	220	2,500	0.000	139	Retrograde
	3,700	180	0.760	Small	

and so only one side of the moon is visible from the earth. (See Fig. 18-11.) Our only knowledge of the far side of the moon comes from space probes that have photographed the far side. They have revealed a cratered surface with almost no smooth areas. (See Fig. 18-12.)

The visible face of the moon has large, dark, smooth areas and lighter colored, more rugged areas. The early astronomers called the dark areas seas [maria (pl.), mare (sing.)] and the lighter areas, lands, in a fanciful analogy to the earth. Both have craters but the lighter uplands have more craters than the lower maria. The craters range from hundreds of miles across down to a few inches in diameter. Most of these craters probably were formed by meteorite impacts. The moon, unlike the Earth, is not protected by an atmosphere, so more meteorites hit the moon than the Earth. The relative ages of various parts of the moon's surface can be determined by the density of the craters, the older surfaces having more craters. The present estimated rate of meteorite impact is not enough to have formed all of the moon's craters.

The craters, however, are of several types and may have more than one origin. The rayed craters are the most spectacular. (See Fig. 18-13.) The rays are believed to be material splashed from the crater by impact. They are visible from Earth at full moon and appear lighter colored in such illumination. The rays cover older craters, and the rays have more small, sharp, young craters than their surroundings. At places these young craters form in rows radial to the main crater, suggesting that their origin is from impact of material thrown out of the main crater.

Evidence of volcanic activity on the moon is abundant. Some of the craters may be volcanic in origin, and volcanic activity probably modifies many of the craters. Figures 18-13 and 18-14 show domes; and these domes could be volcanoes, laccoliths, or hills of other origins, partially buried by volcanic flows or other surface layers. Probably all of these features exist on the moon, but Figs. 18-13 and 18-14 show a number of domes with pits or craters at the top, strongly suggesting volcanoes.

Other probable volcanic features are the ridges shown in Fig. 18-14. Ridges like these are several hundred feet high and extend for hundreds of miles. They are almost entirely confined to the maria. Their origin is unknown, and it has been suggested that they may be caused by downsinking of the maria. More likely, they may be the location of dikes that fed the surface flows.

The nature of the moon's surface layer and surface processes is very different from the Earth's because

A.

B.

9500 ft

Figure 18-6. Valleys of probable fault origin on Mars. A. These structures are part of a system of parallel fissures extending more than 3000 miles. The widest valley is about one mile across and has a smaller valley inside it. The photo covers an area 21 by 26 miles. B. A valley on Mars much deeper than the Grand Canyon. This valley is 75 miles across. Photos from NASA.

A.

B.

C.

Figure 18-7. Erosional features on Mars. A. River-like gully suggests erosion by running water. B. A possible braided river channel. C. An intricate network of big canyons. Photos from NASA.

of the lack of water and atmosphere. Downslope movements occur on steep crater walls. (See Fig. 18-15.) Some very slow erosion does occur because the older craters are more rounded than younger ones. This rounding is believed to be caused by the impact of micrometeorites and atomic particles from the sun.

The moon has a rugged surface with many mountains; however, these mountains have apparently been formed by impacts and volcanic activity. Thus, they are very different from the compressive folded and faulted mountains of the Earth.

Several types of rocks have been returned from the moon. Those from the maria resemble earth basalt, but have much more titanium and much less of the more volatile elements such as sodium and potassium. Gas holes are common. The main minerals are plagioclase feldspar and titanium-rich pyroxene, so that the mineralogy and texture are much like earth rocks. (See Fig. 18-16.) The plagioclase feldspar, unlike in most earth rocks, is almost pure calcium plagioclase. One unusual aspect is the presence of small blebs of glass of two different compositions inside some minerals. This suggests that at some stage in crystallization, the melt may have separated into two immiscible liquids like oil and water. Many of these rocks are breccias and were formed by meteorite impact. The fragments in these breccias are basalt, gabbro, breccia, glass, anorthosite (feldspar rock) and peridotite. Much of this material is dust size. (See Fig. 18-17.) Melting and fracturing from impact shocks are visible in most samples, and spherical glass particles and iron-nickel spheres of impact origin are common. (See Figs. 18-18, 18-19 and 18-20.) Layering suggests that the surface formed in stages. (See Fig. 18-21.) The maria basaltic rocks all have radioactive dates between about 3900 and 3100 million years, and are the youngest rocks so far found on the moon.

The rocks from the rugged uplands of the moon are different. In composition many are anorthosite, a rock composed of between 70 and 90 per cent plagioclase, and in texture they are breccias. This suggests that the uplands may be underlain by anorthosite that has been brecciated by meteorite impacts. Many of these breccias were partially melted. Some of the highland rocks are lavas, and a few have textures similar to those of hornfels and so may be thermally metamorphosed

Figure 18-8. Chaotic terrain on Mars is shown on the right. The chaotic terrain is a jumble of ridges, each of which is one half to two miles wide and one to five miles long. Mariner 6 photo from NASA.

Figure 18-9. Mariner IX photograph of Mars' moon Phobos. Phobos is about 13 by 16 miles. The age and strength of Mars' innermost moon is shown by the impact crater. Photo from NASA.

Figure 18-10. Saturn and its ring system. Photo from Mount Wilson and Palomar Observatories.

rocks. The highland rocks all have radioactive ages of about 4000 million years.

Carbon compounds have been found in the moon rocks, but no trace of life has been found.

Study of the orbits of spacecraft around the moon has shown that some parts of the moon are much denser than others. The areas of high mass concentrations (called mascons) are the large circular maria. The origin of these mascons is not known. They could be caused by large iron-nickel meteorites that were buried by impact, or they may be buried concentrations of mare basalt. Larger gravity anomalies exist on the Earth.

The seismographs left on the moon reveal that most of the moon's "earthquakes" are caused by tidal attraction and occur when the moon is closest to the Earth. In comparison to the Earth, the moon is an almost dead planet. Most of the moonquakes have magnitudes less than two on the Richter Scale and most appear to originate from two active zones about 600 miles long, one running north-south and the other northeast-

southwest. The foci of the moon-quakes are between about 500 and 700 miles deep, which is much deeper than earthquakes. Moon-quakes build up and decay much slower than earthquakes, probably because of the thick surface layer of breccia on the moon. Meteorite falls have also been recorded by the seismographs.

Seismic studies reveal that the moon is layered. The surface to about 12 miles is apparently rubble. The base of the crust is at about 40 miles, and there the velocity increases. The mantle extends to about 600 miles. At that depth the shear waves are somewhat reduced in amplitude, suggesting that the core may be partly melted.

The history of the moon begins with its formation, probably about 4550 million years ago. This is the same date as the origin of the Earth and the solar system. The moon probably was formed by accumulation of material in the same manner as the rest of the solar system — a topic that will be covered later. The outer part of the moon probably melted at this time. The infall of

material and the resulting compaction would cause heating. This melting would produce the anorthosite crust of the moon which crystallized about 4000 million years ago. After the crust formed, the interior of the moon would begin to heat. This heating would be caused by radioactive elements. It would first melt the layer below the crust and could be the origin of the mare basalt that formed between 3900 and 3100 million years ago. Melting would then progress downward as the upper layers cooled. This process could produce the partially melted core predicted by the seismic data. After 3100 million years, the only process acting on the surface has been meteorite impacts.

This history agrees well with all of the data except the remanent magnetism of some moon rocks. At present the moon has almost no magnetic field, but some of the moon rocks have remanent magnetism that could only have formed in a much stronger magnetic field. If the moon once had a magnetic field, that field might have originated in a liquid iron core, just as the earth's magnetic field does. This would imply that in the past the moon had a hot liquid core that has since cooled, and this is the opposite of the history outlined above. The other possibility is that the moon's magnetic field originated in some other way that we do not understand.

Meteorites

Meteorites are objects from space that hit the Earth. Their composition and some other features were discussed earlier in relation to the layered structure of the Earth. The iron-nickel meteorites are believed to be similar in composition to the core of the Earth, and the stony meteorites similar to the mantle. One interesting aspect of meteorite recognition is that over 90 per cent of those seen to fall are stony meteorites, but only a third of the finds not seen to fall are stony. Only a few aspects of meteorites will be discussed here; much more information can be found in the references at the end of this chapter.

Meteorite specimens range from dust to 60 tons. Those that have produced the large impact craters on the Earth were much larger, weighing perhaps hundreds or thousands of tons. (See Figs. 18-22 and 18-23.) The shock wave caused by such impacts can cause metamorphism of the rocks as well as shattering. A few

Figure 18-11. Map of the visible side of the moon. U.S. Air Force, 1962.

very high pressure minerals form in this way and are found only at sus- pected impact craters. Meteorite craters on the Earth are less than 10 miles across; however, on the moon some craters are more than

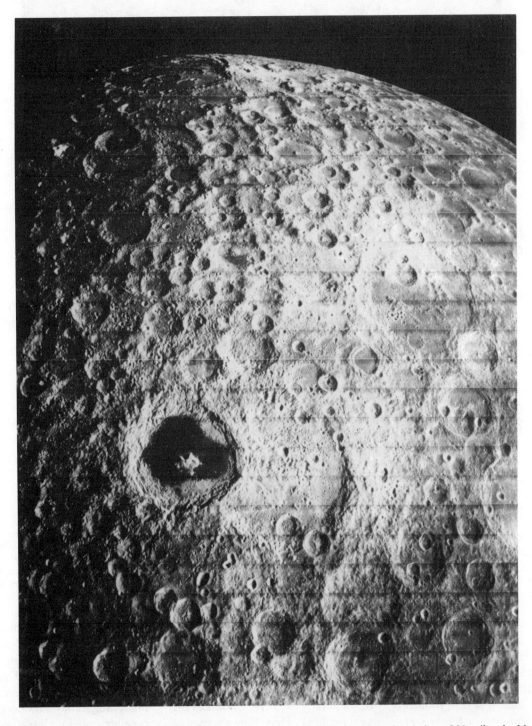

Figure 18-12. Hidden side of the moon photographed by Lunar Orbiter III from an altitude of about 900 miles, looking south from a point about 250 miles south of the lunar equator. The prominent crater is about 150 miles in diameter, and its rim shows evidence of inward slumping like many of moon's craters. Photo from NASA.

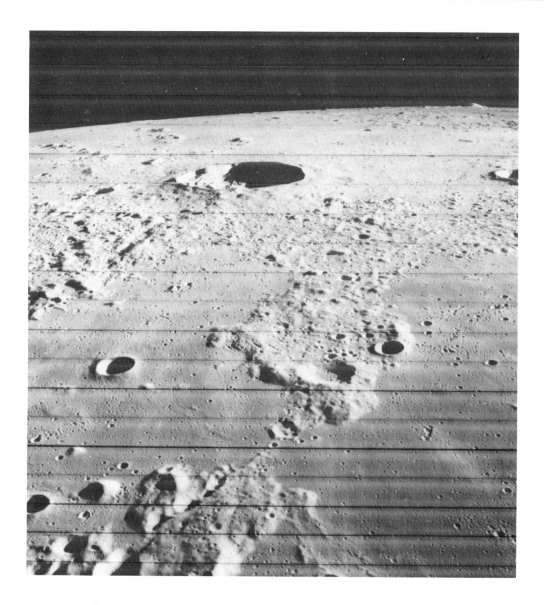

Figure 18-13. Oblique view of the crater Kepler taken by Orbiter III from a height of 32.7 miles. Kepler has rays extending radially in all directions. The rays are believed to be the material splashed out by the impact and are visible from earth only near full moon. The light color streaks in the lower right may be ray material, especially as the associated smaller craters form lines radial from Kepler. Many domes are visible; and one in the foreground has a crater at its summit, suggesting that the domes may be volcanic. The line in the foreground may be a fault. Photo from NASA.

150 miles in diameter. Perhaps such large craters are not recognizable on the Earth. Many of the smaller meteorites do not survive the heat generated by friction in passing through the Earth's atmosphere, producing meteors, or shooting or falling stars.

It is estimated that between 1,000 and 1,000,000 tons of meteorites fall on the Earth each year.

Meteorites are believed to have formed at the same time as the rest of the solar system (discussed in the next section). Thus, they are used to

date the formation of the solar system and as models of the Earth's structure. Meteorites are fragments of much larger bodies. This is shown by the coarse crystalline nature of many, suggesting slow cooling that is possible only in a large body. Many,

Figure 18-14. An array of domes near the older impact crater Marius. The domes are from two to 10 miles in diameter and from 1,000 to 1,500 feet high. Many have pits on the summit and so resemble volcanoes. The prominent low ridges are also believed to be volcanic features, probably feeder dikes. Photo from NASA.

however, have complex crystallization histories. As shown in the next section, the Earth developed its layered structure by melting after initial formation. Some meteorites are also differentiated bodies, and this melting can only occur in bodies large enough to generate the necessary heat, again suggesting that meteorites are fragments of large bodies. Meteorites are generally believed to be fragments from the asteroid belt. The asteroids are a group of about 30,000 small bodies ranging from about 500 miles to about one mile in diameter. Billions more down to dust size are also thought to exist. The asteroids are believed to have formed by collisions between two or more small planets. The orbits of many meteors have been recorded by radar and are consistent with an asteroid origin.

However, some of the meteor swarms that the Earth passes through each year are very different in that they are predictable and are not associated with large meteorite falls. They are in the orbits of now dead comets. Small particles of meteorite material, called micrometeorites, probably originate from such meteor swarms and so have their origin in comets. Micrometeorites of this or other origin are believed to cause the erosion of the moon. Comets are described next.

Cosmic ray bombardment in space can cause the formation of new elements, and the amounts of such elements record the time that the object has been in space. Large nuclei cosmic rays can also produce tracks like those shown in Fig. 2-13, and the number of such tracks is a measure of the time of exposure. The various types of meteorites have different exposure ages, again suggesting that meteorites have complex histories.

Comets

Comets are among the most awe-inspiring spectacles in nature. (See Fig. 18-24.) In the past they were considered harbingers of evil times. We now know their make-up in general terms and can predict their reappearance, but we know almost nothing about how they form. Comets have very small masses. One passed through Jupiter's moons without noticeable effect on the moons, and the Earth has passed through the tail of a comet in historic time. However, in 1908 a great explosion, heard

500 miles away, occurred in Siberia that was believed to have been caused by a meteorite. When the area was visited in 1927, although the forest was knocked down for over ten miles, no large crater was found. It has been postulated that the explosion might have been caused by a gaseous comet rather than a solid meteorite. Comets

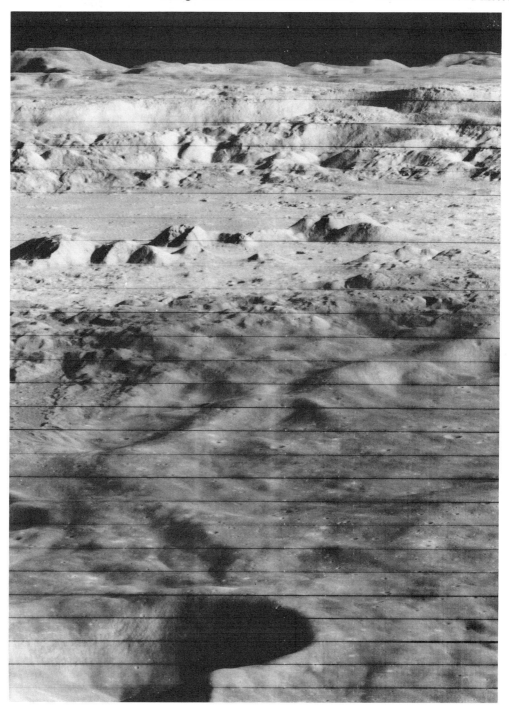

Figure 18-15. Closer view of Copernicus from Orbiter II. Hummocky terrain in foreground is ejecta from the crater. Details of the sliding of the rim can be seen. The central peaks within the crater are about 1,000 feet high. The low area on the left foreground is apparently a fault trough filled by younger volcanic rocks. Photo by NASA.

are believed to be masses of frozen gas with some very small particles similar in composition to meteorites. This is sometimes called the "dirty snowball" theory. In the case of the Siberian explosion the ice of the comet is believed to have turned to gas very rapidly by frictional heating in the atmosphere and the resulting shock wave caused the explosion and damage.

When distant from the sun, a comet is very small, perhaps a half mile in diameter. At this distance it is seen only by the sunlight that it reflects. When closer to the sun, the sun's radiation melts some of the ice and ionizes some of the resulting gases, making the comet more visible. Thus, comets disintegrate by passing near the sun and die after a

Figure 18-16. Thin section of basalt from the moon. The texture closely resembles terrestrial basalt. White is plagioclase, gray is pyroxene, and black is ilmenite, a titanium mineral. Photo from NASA.

number of passes near the sun. During the process of disintegration, the spectra of water, carbon dioxide, methane, and ammonia can be seen.

At this stage an average comet may have a head 80,000 miles in diameter, made mostly of very thin gas. The solar wind, to be discussed in the next section, pushes this gas outward from the head to form a tail up to 50 million miles long. This is the spectacular stage of a comet and occurs only when the comet is near the sun. One interesting point is that the solar wind pushes the tail away from the sun without regard to the direction in which the comet is moving. Most comets have very elliptical orbits, which explains why they reappear at intervals. As with the planets, the sun is at one focus of the ellipse, but when distant from the sun, comets are too small to see. Some comets have orbits that bring them back only after thousands of years; some may be parabolic and so never return. (See Fig. 18-25.)

The origin of comets is not known; they may form at the fringe of the solar system or may come from outside the solar system. Their composition is not too unlike the material from which the solar system is believed to have formed (discussed in the next section). Thus they may be fragments "left over" from the origin of the solar system.

Figure 18-17. The surface of the moon. In the foreground is the orange "soil" found by the crew of Apollo 17. The tripod-like object is used as a photographic reference and to establish sun angle. Note the footprints in the fine surface material. Photo from NASA.

Tektites

Tektites are pieces of glass found on the Earth, both on the surface and in young sediments. They are foreign to the places where they are found, occurring as rare objects scattered in several distinct fields. (See Figs. 18-26 and 18-27.) Some have unusual shapes that can be shown to be aerodynamic shapes that a melted object would assume traveling at high speed in the atmosphere. Their chemical composition is similar to that of the continental crust. Their origin is a problem, and several theories have been proposed. One theory suggests that they are fragments thrown out of the moon by meteorite impacts. The distribution pattern and other features of the 700,000-year-old tektites are consistent with an origin from one of the rays of the moon's crater Tycho. Another theory is that they are earth material dislodged by meteorite impacts. A third theory is that they may originate from comets because microtektites have been found near the site of the 1908 impact mentioned in the preceding section. Their shapes and distribution are consistent with any of these theories.

Recent studies suggest that at least some tektites may be of terrestrial origin. In each field the potassium 40-argon 40 dates are similar. Because argon, a gas, would be lost in melting, this is the date of melting. Rubidium-strontium dates for the fields studied have dates similar to the age of the bedrock at nearby meteorite-impact craters. These latter elements, unlike argon, would not be lost during melting. However, the distribution is not radial to the possible origin crater.

Figure 18-18. Glass spheres from the moon's surface. The largest is .5 mm in diameter. Colors are clear, yellow, red, and black. Their origin is probably from melting during meteorite impact. Photo from NASA.

Recently microtektites were found in sediments in many parts of the Pacific and Indian Oceans by core sampling. Interestingly, the microtektites were found at the point in the cores where the Earth's magnetic field reversed. This suggests the possibility that whatever event caused the tektites might also have caused the magnetic reversal. The total amount of microtektites in the Indian and Pacific Oceans is estimated at a quarter of a million tons. The level at which the microtektites and the magnetic reversal are found also corresponds to the level at which a change in fossil life occurs.

Origin of the Solar System

The origin of the solar system is a problem that man has attempted to solve for many years. No completely satisfactory theory has yet been offered, and objections exist to all theories, including the one presented here. The best theory offered to date states that the planets and the other orbiting objects formed at the same time the sun formed. Other theories assumed that the planets formed from material pulled from the existing sun, generally by the gravitation of another star that happened to pass nearby. The many physical objections to this latter theory suggest that it is impossible. Only a brief outline of the newer theory can be given here.

Stars are believed to form by the condensation of a cosmic dust or gas cloud called a *nebula* (cloud), composed mainly of hydrogen and helium and perhaps frozen like snow. This condensation is caused largely by gravitational attraction. Condensation causes the temperature to rise; and when the temperature rises high enough, nuclear reactions begin and the new star begins to radiate energy.

The planets are believed to have formed in a similar way during the origin of the sun. During condensa-

A.

B.

C.

Figure 18-19. Moon rocks. A. Basalt sample returned by Apollo 12. The cavity on the lower right was probably caused by a gas bubble. Such features are common in volcanic rocks. The specimen is 9 cm in diameter. B. This rock, also returned by Apollo 12, was collected from the bottom of a one-meter-deep crater. The bottom part is covered by a thin layer of glass that was probably produced by melting caused by impact. The rest of the rock also appears to be shattered by impact. C. Rock returned by Apollo 16. This fragment was broken from a larger rock. Gas holes are prominent. Photos from NASA.

tion, such a gas cloud would almost certainly rotate because of the initial velocities of the particles brought together. As condensation continued, the rotation became more uniform as the random velocities cancelled out. Ultimately, the cloud flattened and became disk-shaped with most of the mass concentrated near the center. Further contraction and flattening made the disk unstable, and it broke into a number of smaller clouds. These smaller clouds were all in the same plane. Condensation occurred in each of these clouds, concentrating the heavier elements at the center. Because the nebula had about the same composition as the sun, the ancestral planets were surrounded by envelopes of hydrogen and helium. At about this stage, condensation had heated the ancestral sun to the point that nuclear reactions began. The sun's radiation ionized the remaining part of its nebula; and these ionized particles interacted with the sun's magnetic field, reducing the sun's angular momentum (spin).

Thus, the sun was slowed to its present slow rate of rotation, and the particles in the dust cloud gained in energy. Ionized particles radiated from the sun swept away the remaining gaseous dust surrounding the sun and the planets in the same way that the sun's radiation makes the tail of a comet point away from the sun, regardless of the direction of motion of the comet. In this process the planets nearest the sun lost the most material. They are much smaller and denser than the more distant planets. As each planet condensed, it rotated faster, just as a spinning skater spins faster when she brings her arms in to her body. This theory explains the size and density of the planets, the distribution of momentum in the solar system, and why most of the planets revolve in the same direction.

The asteroid belt between Mars and Jupiter may have originated from a number of small centers of accumulation rather than from a single center, as did the other planets. Collisions among these bodies have increased their total number, and it is the resulting fragments from larger bodies that are the meteorites.

The moons formed close to the planets in the same way. When the planets lost much of their original masses because of the sun's radiation, the gravitational attraction between the planets and their moons was reduced. As a result, the moons moved further away from the planets. Some of these satellites may have escaped from their original planets to be captured by other planets. Such a transposition could account for the moons that revolve in the opposite direction. Pluto may once have been a moon of Neptune. Some of these erratic moons may even be captured asteroids.

This theory accounts for many other features of the solar system. The composition of the crust of the earth was our starting point in Chapter 2, and this can now be discussed further. The estimated composition of the universe is shown in Table 18-3; this is probably representative of the composition of the nebula from which the solar system condensed. Note that hydrogen predominates, that helium forms most of

A.

B.

Figure 18-20. Breccia from the moon. A. Two centimeters in length. B. Closeup photo showing glass-lined pits, probably caused by impact. Large pits are about 1.5 mm in diameter. Photos from NASA.

the rest, and that all the other elements constitute less than one per cent of the total. Most astronomers believe that originally the universe consisted entirely of hydrogen, the simplest atom, and that all other elements have been made by combining hydrogen atoms by a process called fusion. Such fusion can occur at the high temperatures, millions of degrees, inside stars. The develop-

ment of the elements is believed to begin with the condensation of a gas cloud, or nebula, composed of hydrogen. The resulting star produces energy by atomic fusion of hydrogen to helium. Some of the possible reactions were described in Chapter 2. As the star ages, a stage is reached when the atomic "fuel" in the core has been nearly used up. At this time the temperature increases greatly and energy is produced by the formation of heavier elements from helium. At this stage, energy may be produced in the core faster than it can be radiated from the surface and an explosion results. The explosion may disintegrate the star, making it a *supernova* (a very bright, short-lived "star"); or the star may lose its material in less spectacular ways. The complete history of stars is complex and not entirely known; however, it is known that the history of a star depends greatly on its mass, and that larger stars move through this cycle much more rapidly than do average-size stars, such as our sun. In any case, a star formed initially of hydrogen produces heavier elements and scatters these elements into the universe. The solar system formed from a dust cloud that contained heavy elements as well as hydrogen.

The correspondence between the composition of the universe and the composition of the earth's crust is shown in Table 18-4. The light elements, hydrogen and helium, are most abundant but were lost during the formation of the earth, as de-

Figure 18-21. Core sample of moon's surface collected by the crew of Apollo 12. The fine-grained nature of the material is well shown. Other cores from this area show distinct layering. Photo from NASA.

scribed above. The next most abundant element in the universe is oxygen, which is most abundant in the crust.

Earth-Moon System

Interaction between the earth and the moon is believed to be slowing the earth's initially rapid rotation. This process is a continuing one, and the length of day is increasing. The year, the period of the earth's trip around the sun, is not affected because it is determined by the distance between the sun and the earth. Thus the number of days in the year is decreasing. The present average rate of slowing of the earth's rotation is estimated at between one second per day per 120,000 years to two seconds per day per 100,000 years. This is enough to produce apparent errors if the times of eclipses computed back in terms of our present day are

Figure 18-22. Meteor Crater (Barringer Crater), Arizona. A similar crater would not be as well preserved in a more humid climate. Photo from Yerkes Observatory.

compared with the ancient records. An ingenious check on this lengthening of the day in the geologic past has been devised. Some corals grow by adding layers much like tree rings. Thinner layers, apparently daily, occur within the annual layers. These daily layers are commonly not well preserved, but the few that can be counted suggest that in Middle Devonian time, about 350,000,000 years ago, the year had 400-410 days and in Pennsylvanian time, about 280,000,000 years ago, 390 days. (See Fig. 18-28.) The figures are in good agreement with the estimated slowing of the earth's rotation. Other types of fossils also show daily, monthly, tidal, and annual cycles, and it may soon be possible to determine the early history of the earth-moon system from such studies.

The slowing of the earth's rotation

Moenkopi formation

Kaibab limestone

Coconino and Toroweap formations

1. Meteorite approaches ground at 15 km/sec.

fused rock

2. Meteorite enters ground, compressing and fusing rocks ahead and flattening by compression and by lateral flow. Shock into meteorite reaches back side of meteorite.

shock front   fused rock

3. Rarefaction wave is reflected back through meteorite, and meteorite is decompressed, but still moves at about 5 km/sec into ground. Most of energy has been transferred to compressed fused rock ahead of meteorite.

shock front

4. Compressed slug of fused rock and trailing meteorite are deflected laterally along the path of penetration. Meteorite becomes liner of transient cavity.

shock front

5. Shock propagates away from cavity, cavity expands, and fused and strongly shocked rocks and meteoritic material are shot out in the moving mass behind the shock front.

reflected rarefaction

shock front

6. Shell of broken rock with mixed fragments and dispersed fused rock and meteoritic material is formed around cavity. Shock is reflected as rarefaction wave from surface of ground and momentum is trapped in material above cavity.

reflected

shock front   rarefaction

7. Shock and reflected rarefaction reach limit at which beds will be overturned. Material behind rarefaction is thrown out along ballistic trajectories.

8. Fragments thrown out of crater maintain approximate relative positions except for material thrown to great height. Shell of broken rock with mixed meteoritic material and fused rock is sheared out along walls of crater; upper part of broken rock is ejected.

9. Fragments thrown out along low trajectories land and become stacked in an order inverted from the order in which they were ejected. Mixed broken rock along walls of crater slumps back toward center of crater. Fragments thrown to great height shower down to form layer of mixed debris.

0   1000   2000 feet

Figure 18-23. Diagrammatic sketches showing sequence of events in the formation of Meteor Crater, Arizona. From E. M. Shoemaker in Part 18, Report of XXI International Geological Congess, 1960.

Figure 18-24. Photo of Halley's comet. Photo from Mount Wilson and Palomar Observatories.

is caused by the tidal drag of the moon. The total rotational energy of the earth-moon system is conserved, so the energy lost by the earth is gained by the moon. The moon moves farther from the earth, and the length of the month, the moon's period around the earth, increases. Calculations show that ultimately the earth's rotation and the moon's orbital period will be the same, so that one part of the earth will always face the moon just as only one part of the moon now faces the earth. Similar calculations can be made to find the moon's past position and orbit. The results of such calculations depend to a large extent on the assumptions made. This is unfortunate, because otherwise the problem of the moon's origin and early history probably could be solved.

The origin of the moon cannot be determined, and there are serious objections to all theories so far proposed. The theory that the earth and the moon formed at about the same time at different centers of accumulations in a dust cloud was described above. The main objection to this is that both earth and moon should have the same composition if they formed in this way, but their average specific gravities are different. The average specific gravity of the earth is 5.5, and that of the moon is 3.3. Capture hypotheses were also mentioned in that earlier section, especially in regard to moons of other planets. Some of the recent calculations of the moon's early orbit strongly suggest a capture origin. However, if the earth captured the moon, the orbital energy (kinetic energy) of the moon would have heated the earth, probably to near its melting point, an event not suggested by the earth's thermal history.

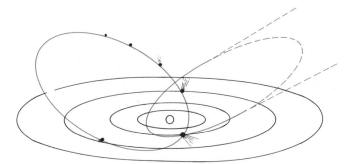

Figure 18-25. The solar system showing some possible comet orbits. The comet is shown on one orbit. Note that the tail always points away from the sun and that the comet gets larger where it is heated by the sun. The other orbit shows possible elliptical and parabolic orbits.

Table 18-3. Cosmic abundance of the chemical elements.

ELEMENT	ATOMIC NUMBER	APPROX. ATOMIC WEIGHT	RELATIVE NUMBER OF ATOMS*	RELATIVE PER CENT OF ATOMS
Hydrogen	1	1	25,000,000.0	86.64
Helium	2	4	3,800,000.0	13.16
Oxygen	8	16	25,000.0	0.09
Neon	10	20	14,000.0	0.05
Carbon	6	12	9,300.0	0.03
Nitrogen	7	14	2,400.0	0.008
Silicon	14	28	1,000.0	0.003
Magnesium	12	24	910.0	0.003
Sulfur	16	32	380.0	0.001
Argon	18	40	150.0	0.0005
Iron	26	56	150.0	0.0005
Aluminum	13	27	95.0	0.0003
Calcium	20	40	49.0	0.0002
Sodium	11	23	44.0	0.0002
Nickel	28	59	27.0	0.00009
Phosphorus	15	31	10.0	0.00003
Chromium	24	52	7.8	0.00003
Manganese	25	55	6.8	0.00002
Potassium	19	39	3.2	0.00001
Chlorine	17	35	2.6	0.000009
Cobalt	27	59	1.8	0.000006
Titanium	22	48	1.7	0.000006
Fluorine	9	19	1.6	0.000006
Vanadium	23	51	0.2	0.0000007
Copper	29	64	0.2	0.0000007
Zinc	30	65	0.2	0.0000007
Lithium	3	7	0.1	0.0000003
Strontium	38	88	0.06	0.0000002

*The number of atoms, or the cosmic abundance, in the fourth column is relative to a value of 1,000 for silicon, atomic number 14. The figures in this column are from A. G. W. Cameron's revision of an earlier list by H. E. Suess and H. C. Urey.

Many other ideas have been suggested, but all have physical objections.

Crystallization of the Interior of the Earth

The oldest crustal rocks so far dated by radioactive methods are 3,980 million years old. This is probably close to the time that the continental crust formed. The earth is probably appreciably older than this. Meteorites yield dates of 4,550 million years; this is the time that they crystallized. As noted earlier, meteorites are dated using the uranium-lead and thorium-lead methods, and the correction for original lead is a problem. Meteorites are believed to have the same origin as the earth and to have the same composition as the interior of the earth. Thus the 4,550 million year date is perhaps the time that the interior of the earth crystallized. As will be seen shortly, this event is believed to have occurred sometime after the formation of the initially homogeneous earth. How much after is the problem because the size of the planet probably is important in determining the rate at which recrystallization occurred. At any rate, 4,550 million years is the time at which the bodies that became meteorites crystallized, and the solar system, including the earth, formed sometime earlier.

The earth, as originally formed, was probably nearly homogeneous and relatively cool. The contraction caused some heating, and radioactivity also must have caused heating. The radioactive heating must have been much greater than at present

Table 18-3. Cosmic abundance of the chemical elements (continued).

Element	Atomic Number	Approx. Atomic Weight	Relative Number of Atoms*	Relative Per Cent of Atoms
Krypton	36	84	0.04	0.0000001
Scandium	21	45	0.03	0.0000001
Germanium	32	73	0.03	0.0000001
Beryllium	4	9	0.02	0.00000007
Boron	5	11	0.02	0.00000007
Selenium	34	79	0.02	0.00000007
Lead	82	207	0.02	0.00000007
Zirconium	40	91	0.014	
Gallium	31	70	0.009	
Yttrium	39	89	0.009	
Rubidium	37	85	0.006	
Bromine	35	80	0.004	
Barium	56	137	0.004	
Tellurium	52	128	0.003	
Xenon	54	131	0.003	
Arsenic	33	75	0.002	
Molybdenum	42	96	0.002	
Tin	50	119	0.001	
Platinum	78	195	0.001	
Ruthenium	44	102	0.0009	
Cadmium	48	112	0.0009	
Neodymium	60	144	0.0009	
Niobium	41	93	0.0008	
Palladium	46	107	0.0007	
Dysprosium	66	162	0.0007	
Iodine	53	127	0.0006	
Cerium	58	140	0.0006	
Erbium	68	167	0.0006	
Osmium	76	190	0.0006	
Cesium	55	133	0.0005	
Lanthanum	57	139	0.0005	
Gadolinium	64	157	0.0005	
Iridium	77	193	0.0005	
Ytterbium	70	173	0.0004	
Mercury	80	201	0.0004	
Silver	47	108	0.0003	
Thallium	81	204	0.0003	
Bismuth	83	209	0.0003	
Rhodium	45	103	0.0002	
Antimony	51	122	0.0002	
Praseodymium	59	141	0.0002	
Samarium	62	150	0.0002	
Holmium	67	165	0.0002	
Indium	49	115	0.0001	

*The number of atoms, or the cosmic abundance, in the fourth column is relative to a value of 1,000 for silicon, atomic number 14. The figures in this column are from A. G. W. Cameron's revision of an earlier list by H. E. Suess and H. C. Urey.

Table 18-3. Cosmic abundance of the chemical elements (continued).

ELEMENT	ATOMIC NUMBER	APPROX. ATOMIC WEIGHT	RELATIVE NUMBER OF ATOMS*	RELATIVE PER CENT OF ATOMS
Europium	63	152	0.0001	
Hafnium	72	179	0.0001	
Tungsten	74	184	0.0001	
Gold	79	197	0.0001	
Terbium	65	159	0.00009	
Thulium	69	169	0.00009	
Rhenium	75	186	0.00005	
Lutetium	71	175	0.00004	
Thorium	90	232	0.00003	
Tantalum	73	181	0.00002	
Uranium	92	238	0.000008	
Technetium	43	99		
Promethium	61	147		
Polonium	84	210		
Astatine	85	211		
Radon	86	222		
Francium	87	223		
Actinium	89	227		
Protactinium	91	231		
Neptunium	93	237		
Plutonium	94	242		
Americium	95	243		
Curium	96	243		
Berkelium	97	245		
Californium	98	246		
Einsteinium	99	253		
Fermium	100	254		
Mendelevium	101	254		
Nobelium	102			
Lawrencium	103			

*The number of atoms, or the cosmic abundance, in the fourth column is relative to a value of 1,000 for silicon, atomic number 14. The figures in this column are from A. G. W. Cameron's revision of an earlier list by H. E. Suess and H. C. Urey.

because much more radioactive material was present on the earth. (See Table 2-7.) This follows because after each half-life, half of the original radioactive element is gone. Meteorite-like material hitting the earth probably also caused local heating. The heating of the earth caused melting, and, under the influence of gravity, differentiation of the earth into core, mantle, and crust began. The heavier materials such as iron and nickel melted and moved toward the center, and the lighter silicates moved outward. The silicate mantle solidified from the bottom outward. The core, unable to lose its heat, remains liquid. The inner core may be crystallized. The radioactive elements that caused most of the heating — uranium, thorium, and potassium—all moved outward. These are all large atoms that do not fit into the iron-magnesium silicates, so they stayed with the melt as crystallization progressed from the core outward. The concentration of the radioactive elements near the surface enables their heat to be radiated at the surface. If they were still deep in the earth, their heat would be able to cause remelting.

The resulting layered structure of the earth has great geological significance. The fluid iron core makes the earth's magnetic field possible. The heat of the core may be responsible for convection currents in the mantle. As noted above, the other inner planets that have been studied do not have magnetic fields, and may not have mountain ranges either. The moon also does not have folded mountains. Perhaps it is significant that the earth is the largest of the inner planets, and for this reason may be the only one with a liquid core.

The actual formation of the earth is believed to have occurred 5,000 to 6,000 million years ago, perhaps closer to 6,000 million years. This estimate is reached from at least two lines of reasoning, but neither method gives more precise figures. The age of the sun is estimated to be about 6,000 million years. This figure is based on knowledge of stellar structure and rate of energy production, but it is not possible to discuss the complex physical reasoning further here. This does put an upper limit on the age of the earth, however, because the whole solar system is believed to have originated at the same time. The second line of reasoning is based on estimates of the time necessary for melting and recrystallization to occur. About 500 to 1,000 million years is a reasonable time for these processes, so this figure should be added to the 4,550 million years.

Origin of Oceans and Atmosphere

So far we have accounted for the main events between the origin of the earth, 5,000 to 6,000 million years ago, and 4,550 million years ago when the interior crystallized. Conventional geology begins with the oldest crustal rocks, 3,980 million years old. We know very little of the events in this interval between 4,550 and 3,980 million years ago. The continents had at least begun to form. After the initial continents formed in some unknown way, the crust was probably similar to that of the later times, but no details can yet be deciphered. Other events occurred during this time, however; the earth's initial rapid rotation was being slowed, as described earlier, and the atmosphere and the oceans began to evolve.

The present atmosphere and oceans have developed since the earth solidified, but the development is known only in general terms. Although the earth and the sun formed from dust clouds of probably the same composition, the earth's overall composition now differs from that of the sun. Therefore, it is clear that sometime after the initial formation of the earth by accumulation from a dust cloud, the earth lost much of its original thick gaseous envelope of mainly hydrogen and helium. We must account for the loss of these gases by processes not now operating because the earth is at present losing very little of its atmosphere. Much of the original dust cloud was lost when it was ionized by the sun's radiation and then swept away by moving ionized particles from the sun, as described above. The remain-

Table 18-4. Comparison of the estimated composition of the crust, total solid earth, and universe. The figures for the total earth and the universe are much more uncertain than those for the crust. The compositions of the whole solid earth and the universe are discussed in Chapters 13 and 18 respectively.

Element	Atomic Number	Approx. Atomic Weight	Abundance (weight per cent) Crustal rocks[1]	Abundance (weight per cent) Total solid earth[2]	Abundance (weight per cent) Universe[3]
Hydrogen	1	1	0.14	—	61.0
Helium	2	4	0.0000003	—	36.8
Carbon	6	12	0.02	—	0.25
Nitrogen	7	14	0.002	—	0.08
Oxygen	8	16	46.60	30.0	1.0
Silicon	14	28	27.72	15.0	0.06
Aluminum	13	27	8.13	1.1	0.006
Iron	26	56	5.00	35.0	0.02
Calcium	20	40	3.63	1.1	0.006
Sodium	11	23	2.83	0.57	0.003
Potassium	19	39	2.59	0.07	0.0003
Magnesium	12	24	2.09	13.0	0.05
Titanium	22	48	0.44	0.05	0.0002
Phosphorus	15	31	0.105	0.10	0.0006
Manganese	25	55	0.095	0.22	0.0008
Nickel	28	59	0.0075	2.4	0.004
Sulfur	16	32	0.026	1.9	0.02
Chromium	24	52	0.010	0.26	0.001
Cobalt	27	59	0.0025	0.13	0.0002

[1]From Table 2-1.

[2]From Brian Mason, *Principles of Geochemistry*, John Wiley & Sons, 3rd ed., 1966, p. 53. (These data are estimates, and because only two significant figures are used the total is 101 per cent.)

[3]Calculated from Table 18-3.

ing gas nearer the earth was probably lost as a result of the rapid rotation of the earth, mentioned earlier, and as a result of the heating that melted the earth and formed the core and mantle.

It is difficult to reconstruct the next stage in the development of the early atmosphere; it may have consisted of such gases as ammonia, methane, and steam. This is the composition of the atmospheres of the great planets such as Jupiter (where the water is ice and hydrogen and helium are also present). In the case of the great planets, the source is

believed to be the initial dust cloud. The earth, being nearer the sun, had lost by this stage the gases of the original dust cloud; however, frozen chunks of these gases left in the earth would have melted and been released during the later remelting of the earth to form this atmosphere. Evidence for the existence of this atmosphere is very weak. If such an atmosphere ever existed on earth, it probably was lost quickly because of the prevailing high temperature. Ammonia would be quickly decomposed by the sun's ultraviolet energy. Methane would cause fixation of carbon in

A.

B.

Figure 18-26. Tektites. A. Typical tektites from Indochina. B. An actual tektite (right) compared with a synthetic tektite formed in a high speed wind tunnel that simulated the conditions of entry into the earth's atmosphere. The map in the background shows the area where the tektite was found. A. from Ward's Natural Science Establishment, Inc., Rochester, N.Y., B. from NASA.

life will be discussed in more detail later. Returning to our main theme, volcanoes today emit mainly steam, carbon dioxide, and nitrogen, together with some sulfur gases; and there is no reason to believe that earlier volcanoes were different. Thus the atmosphere became steam, carbon dioxide, and nitrogen. With the cooling of the earth some of the steam condensed, and some oxygen may also have formed from the breakdown of steam. Such oxygen would recombine with hydrogen to form steam. The oxygen now in the atmosphere came from photosynthesis by plants after life began, and life probably could not form until the temperature was below 200°F. At the present time the atmosphere is 78 per cent nitrogen and 21 per cent oxygen, with all other gases forming about one per cent. The disposition of the volcanic gases probably follows this pattern: the steam has become water and is mainly in the oceans; the nitrogen is in the atmosphere; and the carbon dioxide has been partly broken down by photosynthesis to carbon in life and to oxygen in the atmosphere, and is partly in carbonate rocks.

This development of the atmosphere seems reasonable, and it is possible to work out a crude timetable. By about 4,000 million years, the crust had formed. By 3,375 million years, water was present because sedimentary rocks of this age have been found in southern Africa. Prior to that time, anaerobic life probably existed. Oxygen, however, was not present in the atmosphere because up to about 1,800 million years, clastic fragments of pyrite and uraninite are found in sedimentary rocks. If oxygen had been present, these minerals would have been oxidized.

the early sedimentary rocks, but there is no evidence that this occurred. The present atmosphere developed mainly from volcanic emanations; and the cooler, more slowly rotating earth retained these gases. Methane, ammonia, hydrogen, and water would form an atmosphere in which life could have begun; but as outlined here, the temperature was probably too high for life at the time the atmosphere had this composition. Life could also form in a volcanic-derived atmosphere. The origin of

Another problem is that unless oxygen-mediating enzymes had preceded photosynthesizers, the oxygen would destroy the photosynthesizers. Banded iron formation may have kept the oxygen out of the atmosphere. Banded iron formation is found on every continent, but only in rocks 3,200 to 2,000 million years old and never in younger rocks. The iron could not be transported from the weathering site to the place of deposition in the oxidized state. Thus the iron must have been transported in solution in an unoxidized state, and oxidized at the place of deposition. It is believed that the photosynthesizers lived in the ocean, and that the iron combined with the biological oxygen and was precipitated to form banded iron formation. The banding or layering suggests a fluctuation between iron and oxygen.

Free oxygen appeared in the atmosphere at about 2,000 million years because from that time on, red beds—that is, clastic rocks in which the iron is oxidized—are common. Thick extensive red beds appear in the range 1,800 to 2,000 million years and so slightly overlap the last of the banded iron formation. Apparently at this time, oxygen-mediating enzymes formed, allowing life to expand rapidly. The life, probably algae, produced much oxygen; and as the oxygen accumulated in the atmosphere, iron was oxidized in weathering, preventing further development of banded iron formation.

At this stage, ultraviolet from the sun was lethal to life at the surface, so the organisms must have lived in water deep enough to shield them.

Some of the oxygen in the atmosphere was changed into ozone by ultraviolet radiation from the sun. Ordinary atmospheric oxygen is in

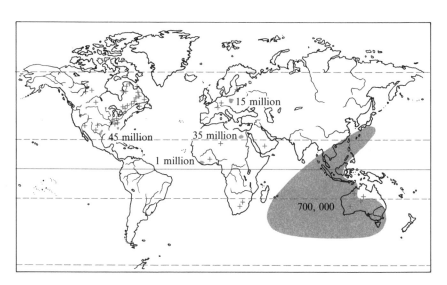

Figure 18-27. Map showing locations of impact craters and possible impact craters (crosses). Tektite fields are shown in color, and their approximate ages in years are given.

the diatomic form, O_2, in which two oxygen atoms share electrons to form stable outer electron shells. In the ozone molecule, O_3, three oxygen atoms share electrons. Ozone is not stable and ultimately returns to ordinary diatomic oxygen. The formation of ozone absorbs much ultraviolet radiation that would otherwise reach the surface of the earth. Higher forms of life could not originate until the ozone layer had formed.

Origin of Life

All discussions on the origin of life are necessarily speculations. Life must have begun very early in the history of the earth, probably almost as soon as it became cool enough, perhaps 4,000 million years ago.

The steps involved in the formation of life are believed to start with the formation of amino acids. Next, these amino acids are brought together to form proteins, nucleic acids, and other compounds. Finally,

the most difficult step is the transformation into living, self-replicating structures. The first of these steps is relatively easy. In a number of experiments, amino acids have been produced by passing an electrical discharge through a container of water, methane, ammonia, and hydrogen. This mixture was chosen because it contains the necessary elements and it is the composition of the atmospheres of some planets such as Jupiter. As discussed earlier, this mixture is believed to be one of the possibilities for the composition of the early atmosphere on the earth.

At the time life is believed to have formed on the earth, the atmosphere is believed to have consisted of nitrogen, carbon dioxide, carbon monoxide, and perhaps other volcanic gases. These gases dissolved in the ocean could produce amino acids using energy from ultraviolet radiation from the sun. Thus the early ocean would have been a very thin soup. The other steps necessary to form life could also take place in

Figure 18-28. Growth lines in coral. The fine bands are daily growth layers. By counting the number of such bands within the yearly growth bands, it is possible to determine the number of days in the year at the time the coral lived. Photo courtesy of S. K. Runcorn.

the ocean. The sun's ultraviolet radiation would probably destroy life without the protection of the ocean to absorb this energy.

Life could only have formed in an oxygen-free atmosphere or environment. If oxygen had been present, it would have destroyed the amino acids by oxidation. After life formed and oxygen was produced by photosynthesis, no new life could be formed. Thus all of the present life on the earth must have evolved from that first population of living organisms.

Clearly, if life formed in the manner described, it could form anywhere where these conditions exist. That is, it could have formed on another planet in the Solar System or elsewhere in the universe. The conditions needed are the elements carbon, nitrogen, and hydrogen, plus water in the liquid state. Such conditions must occur at many places in the vast universe, and so there must be other life in the universe. In the current space program, one of the goals is to search for life on Mars, the only planet with anything like the right temperature range. The materials returned from the moon have been carefully searched for evidence of life. Amino acids and other hydrocarbons were found in small amounts on moon rocks. They are believed to be of inorganic origin. On September 28, 1969, a carbonaceous meteorite fell at Murchison, Victoria, Australia. This meteorite that was dated as 4,500 million years old contains amino acids of non-biologic origin. These discoveries suggest that the conditions necessary to form life are not uncommon and that amino acids may have formed in pre-geologic time in the Solar System.

The Oldest Fossils

Until the late 1950's, very few Precambrian fossils were found; and many of the reported fossils found before that time are probably of inorganic origin. In recent years, microfossils have been found at several localities around the world, and larger fossils have been found in upper Precambrian rocks in a dozen or so places.

The first living organisms probably could not produce their own food but lived on material dissolved in the ocean. This situation could only exist for a short time; otherwise, the life would use all of the available nutrients. The development of photosynthesis probably made the first self-sustaining organism. Some of the oldest sedimentary rocks so far found seem to confirm this theory. These rocks at Barberton Mountain Land in southern Africa are about 3,400 million years old. They contain tiny blue-green algae and bacteria, and chemical studies suggest that photosynthesis had developed.

Abundant Precambrian fossils have been studied from the Gunflint Chert on the north shore of Lake Superior in Canada. (See Fig. 18-29.) Again, blue-green algae and bacteria are found, and these organisms were clearly photosynthetic. These rocks are about 1,700 million years old.

The organisms found at these two localities are primitive. Blue-green

A.

B.

C.

D.

E.

F.

G.

H.

I.

Figure 18-29. Microfossils from the Precambrian Gunflint Chert. A. Tangled filaments mainly *Gunflintia minuta,* and spore-like bodies *Huroniospora.* Diameter of field 0.1 mm. B. *Entosphaeroides amplus* with sporelike bodies within the filament. Diameter of field 40μ. C. *Animikiea septata* probable algal filament with transverse septae. Diameter of field 50μ. D. and E. Filaments with well-defined septae similar to iron-precipitating bacteria and blue-green algae. F. *Eosphaera tyleri.* Diameter of field 32μ. G. and H. *Kakabekia umbellata.* Diameter of field of G. is 20μ and H. is 12μ. I. Radiating structure similar to manganese- and iron-oxidizing colonial bacteria. One micron (μ) is one one-thousandth of a millimeter. Photos A, B, C, F, G, and H courtesy of E. S. Barghoorn; D, E, and I courtesy of P. E. Cloud, Jr.

algae and bacteria are the only living things whose reproduction does not involve genetic material from two parents. They are essentially sexless, and so their form does not change. Thus these organisms in rocks over 3,000 million years old are much like living species. Because genes are not involved, any mutation that occurs, in general, dies out in a few generations.

The first organisms that have genes are found in rocks about 1,000 million years old at Bitter Springs in the Northern Territory of Australia. The organisms are green algae, and all green algae differ from blue-green algae in having sexual reproduction. Some of the fossils recovered from this chert show cell division. The development of these early forms of life is summarized in Fig. 18-30.

These discoveries suggest that true evolution could not begin until about 1,000 million years ago. Up to that time, the organisms were too primitive and evolutionarily too conservative to change very much. After that time, evolution could proceed, and apparently did, because about 600 million years ago, life became abundant. Thus, in less than 500 million years, all of the many types of life found in Cambrian rocks may have evolved from green algae. Representatives of every phylum capable of fossilization are found in Cambrian or early Ordovician rocks.

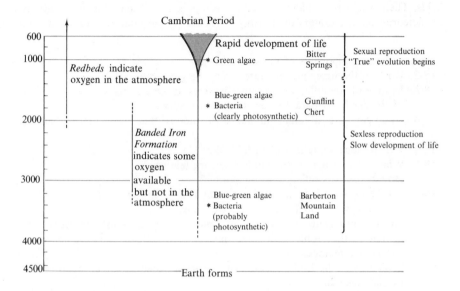

Figure 18-30. Development of Precambrian life and oxygen in the atmosphere.

Summary

The *solar system* is composed of the sun and the planets, comets, meteors, asteroids, and the satellites of the planets.

The *inner planets*, Mercury, Venus, Earth, and Mars, are smaller and denser than the more distant great planets.

Mars has dust storms, seasonal ice caps, huge volcanoes, evidence of faulting, apparent erosion by running water, and impact craters, as well as featureless and chaotic terrains. It is the other planet most likely to have life.

Venus has a thick atmosphere and is very hot.

None of the inner planets have mountain ranges like the Earth's.

The *great planets*, Jupiter, Saturn, Uranus, and Neptune, are composed mainly of hydrogen and helium.

The *moon* orbits the Earth and rotates on its axis at the same rate, so only one side is visible from Earth. The dark, smooth areas are called *maria* (*mare,* singular) and the lighter areas are uplands.

Most of the moon rocks are breccias formed by the meteorite impacts that also cause the many craters. The mare rocks are basaltic in composition, and those of the uplands are feldspar-rich rocks.

Mascons are areas of concentration of mass under the large circular maria.

Most *moonquakes* are low magnitude and are caused by tidal effects. They originate from a single zone and reveal a layered moon with about 12 miles of surface rubble with velocity increases at about 12 and 40 miles deep. The core, starting at about 600 miles, may be partially melted because seismic waves are somewhat attenuated there.

The moon probably formed about 4,550 million years ago and probably melted soon after this. The oldest rocks so far found are from the highlands and are about 4,000 million years. The maria basalts were extruded between 3,900 and 3,100 million years. Since then the moon has been quiet, with only meteorite impacts and perhaps some limited local volcanic activity.

Meteorites are objects from space that hit the Earth. Iron-nickel meteorites are believed to be similar in composition to the earth's core. Over 90 per cent of the meteorites seen to fall are stony. Meteorites range from dust to 60 tons. Large meteorites cause craters. Their origin is probably the asteroids.

Comets are masses of frozen gas with some small particles. They orbit the sun and are visible when they are close enough to the sun that their gases are ionized by the sun's energy.

Tektites are glass fragments whose shape indicates that they have been melted by high speed travel in the atmosphere. They are found in distinct fields. Their origin is probably related to meteorite impact craters on Earth, but there is some evidence that they come from such impacts on the moon.

The solar system probably formed when the sun formed by condensation of a cosmic dust cloud or *nebula*.

Growth rings on corals and other marine animals show that the earth's rotation has been slowing. This is caused by the tidal drag of the moon.

Questions

18-1. List the components of the solar system.

18-2. How do the inner planets differ from the great planets?

18-3. In spite of the different sizes and densities of the planets, why is the origin of the earth inseparable from that of the solar system?

18-4. How can the different numbers of moons that each planet has be explained?

18-5. Which planets might contain life? Why?

18-6. What is the origin of the moon's surface features?

18-7. How do comets differ from meteorites?

18-8. How long would it take for meteorite falls to produce the earth? Is this a reasonable theory for the origin of the earth?

18-9. The uranium-lead and thorium-lead methods are used to date meteorites in spite of the problems in correcting for original lead. Why are not other radioactive methods used?

18-10. Explain why the initial melting of the earth occurred sometime after the formation of the earth.

18-11. How did the earth's layered structure form?

18-12. When did the first life form?

18-13. Outline the steps by which the atmosphere is believed to have formed.

18-14. What is the significance of banded iron formation?

18-15. How is life believed to have originated on the earth? Can this process account for the development of new types of life?

Supplementary Reading

Planets

Ebbighausen, E. G., *Astronomy* (2nd Ed.) Columbus, Ohio: Charles E. Merrill Publishing Co., 1971, 160 pp. (paperback).

Faul, Henry, *Ages of Rocks, Planets, and Stars*. New York: McGraw-Hill Book Co., 1966, 109 pp. (paperback).

Lewis, J. S., "The Chemistry of the Solar System," *Scientific American* (March, 1974), Vol. 230, No. 3, pp. 51-65.

Ohring, George, *Weather on the Planets*. Garden City, New York: Doubleday and Co., Science Study Series, 1966, 144 pp. (paperback).

Sagan, Carl, and J. N. Leonard, *Planets*. New York: Life Science Library, Time, Inc., 1966, 200 pp.

Mars

Hammond, A. L., "The New Mars: Volcanism, Water, and a Debate over its History," *Science* (February 2, 1973), Vol. 179, No. 4072, pp. 463-465.

Murray, B. C., "Mars from Mariner 9," *Scientific American* (January, 1973), Vol. 228, No. 1, pp. 48-69.

Venus

James, J. N., "The Voyage of Mariner II," *Scientific American* (July, 1963), Vol. 209, No. 1, pp. 70-84.

Moon

Alfvén, Hannes, "Origin of the Moon," *Science* (April 23, 1965), Vol. 148, No. 3669, pp. 476-477.

Baldwin, R. B., *A Fundamental Survey of the Moon*. New York: McGraw-Hill Book Co., 1965, 149 pp. (paperback).

Dyal, Palmer, and C. W. Parkin, "The Magnetism of the Moon," *Scientific American* (August, 1971), Vol. 225, No. 2, pp. 63-73.

Goldreich, Peter, "Tides and the Earth-Moon System," *Scientific American* (April, 1972), Vol. 226, No. 4, pp. 43-52.

Hammond, A. L., "Lunar Science: Analyzing the Apollo Legacy," *Science* (March 30, 1973), Vol. 179, No. 4080, pp. 1313-1315.

Hess, Wilmot, and others, "The Exploration of the Moon," *Scientific American* (October, 1969), Vol. 221, No. 4, pp. 54-72.

Hibbs, A. R., "The Surface of the Moon," *Scientific American* (March, 1967), Vol. 216, No. 3, pp. 60-74.

Mason, Brian, "The Lunar Rocks," *Scientific American* (October, 1971), Vol. 225, No. 4, pp. 48-58.

Runcorn, S. K., "Corals as Paleontological Clocks," *Scientific American* (October, 1966), Vol. 215, No. 4, pp. 26-33.

Schmitt, H. H., "Apollo 17 Report on the Valley of Taurus-Littrow," *Science* (November 16, 1973), Vol. 182, No. 4113, pp. 681-690. A report on the geology of the moon, written by a geologist who was there.

Schurmeier, H. M., *et al.*, "The Ranger Missions to the Moon," *Scientific American* (January, 1966), Vol. 214, No. 1, pp. 52-67.

Science, Apollo II issue, (30 January, 1970), Vol. 167, No. 3918.

Shoemaker, E. M., "The Geology of the Moon," *Scientific American* (December, 1964), Vol. 211, No. 6, pp. 38-47.

Wood, J. A., "The Lunar Soil," *Scientific American* (August, 1970), Vol. 223, No. 2, pp. 14-23.

Meteorites

Dietz, R. S., "Astroblemes," *Scientific American* (August, 1961), Vol. 205, No. 2, pp. 50-58. Reprint 801, W. H. Freeman and Co., San Francisco.

Heide, Fritz, *Meteorites*. Chicago: Phoenix Science Series, The University of Chicago Press, 1963, 144 pp. (paperback).

Lawless, J. G., and others, "Organic Matter in Meteorites," *Scientific American* (June, 1972), Vol. 226, No. 6, pp. 38-46.

Mason, Brian, *Meteorites*. New York: John Wiley & Sons, Inc., 1962, 274 pp.

Urey, H. C., "Meteorites and the Moon," *Science* (March 12, 1965), Vol. 147, No. 3663, pp. 1262-1265.

Comets

Whipple, F. L., "The Nature of Comets," *Scientific American* (February, 1974), Vol. 230, No. 2, pp. 49-57.

Tektites

Barnes, V. E., "Tektites," *Scientific American* (November, 1961), Vol. 205, No. 5, pp. 58-64. Reprint 802, W. H. Freeman and Co., San Francisco.

Faul, Henry, "Tektites are Terrestrial," *Science* (June 3, 1966), Vol. 152, No. 3727, pp. 1341-1345.

Glass, B. P., and B. C. Heezen, "Tektites and Geomagnetic Reversals," *Scientific American* (July, 1967), Vol. 217, No. 1, pp. 32-38.

Origin of the Earth

Kuiper, G. P., "Origin, Age, and Possible Ultimate Fate of the Earth," in *The Earth and Its Atmosphere* (D. R. Bates, ed.), pp. 12-30. New York: Science Editions, Inc., 1961, 324 pp. (paperback). (This book was also published under the title, *The Planet Earth*. London: Pergamon Press, 1957).

Origin of Life

Abelson, P. H., "Chemical Events on the Primitive Earth," *Proceedings of the National Academy of Sciences* (June, 1966), Vol. 55, No. 6, pp. 1365-1372.

Barghoorn, E. S., "Origin of Life," in *Treatise on Marine Ecology and Paleoecology*, Part 2 (H. S. Ladd, ed.), pp. 75-86. Geological Society of America, Memoir 67, 1957. This paper reprinted in J. F. White, (ed.), *Study of the Earth*, pp. 335-347. Englewood Cliffs, N.J.: Prentice-Hall, Inc., 1962 (paperback).

Barghoorn, E. S., "The Oldest Fossils," *Scientific American* (May, 1971), Vol. 224, No. 5, pp. 30-42.

Cloud, P. E., Jr., "Significance of the Gunflint (Precambrian) Microflora," *Science* (April 2, 1965), Vol. 148, No. 3666, pp. 27-35.

Eglinton, Geoffrey, and Melvin Calvin, "Chemical Fossils," *Scientific American* (January, 1967), Vol. 216, No. 1, pp. 32-43.

Glaessner, M. F., "Pre-Cambrian Animals," *Scientific American* (March, 1961), Vol. 204, No. 3, pp. 72-78. Reprint 837, W. H. Freeman and Co., San Francisco.

Keosian, John, *The Origin of Life*. New York: Reinhold Publishing Corp., 1964, 118 pp. (paperback).

Lansberg, H. E., "The Origin of the Atmosphere," *Scientific American* (August, 1953), Vol. 189, No. 2, pp. 82-86. Reprint 824, W. H. Freeman and Co., San Francisco.

Woodford, A. O., "Origin and History of Oxygen in the Air," *Journal of Geological Education* (November, 1972), Vol. 20, No. 6, pp. 276-280.

Appendices

Appendix A
Minerals

Appendix B
Topographic Maps

Appendix C
Physical Methods of
Age Determination

Appendix D
Rock Textures

Appendix A

Minerals

Table A–1 on the following two pages helps to identify the more common minerals. To use the chart, first decide the sort of luster (metallic or non-metallic) the mineral has; next, its color (light or dark); third, its hardness (compared to a knife blade); fourth, whether it has cleavage; and finally, read the brief descriptions to identify it.

Table A-1. Mineral identification key.

Dark-colored nonmetallic luster	**Hard — not scratched by knife**	**Shows cleavage**	Black to dark green; cleavage 2 planes at nearly right angles; hardness, 5-6.	AUGITE
			Black to dark green; cleavage 2 planes about 60°; hardness, 5-6.	HORNBLENDE
		No cleavage	Red to red-brown; fracture resembles poor cleavage; brittle equidimensional crystals; hardness, 6.5-7.5.	GARNET
			Various shades of green and yellow; glassy luster; granular masses and crystals in rocks; hardness, 6.5-7 (apparent hardness may be much less).	OLIVINE
			White, clear, or any color; glassy luster; transparent to translucent; hexagonal (6-sided) crystals; hardness, 7; conchoidal fracture.	QUARTZ
			Any color or variegated; glassy luster; hardness, 5-6; conchoidal fracture.	OPAL
			Any color or variegated; waxy luster; hardness, 7; conchoidal fracture.	CHALCEDONY (AGATE)
			Red to brown; red streak; earthy appearance; hardness, 5.5-6.5 (apparent hardness may be less).	HEMATITE
			Yellow-brown to dark brown, may be almost black, streak yellow-brown; earthy; hardness, 5-5.5 (may have lower apparent hardness).	LIMONITE (GOETHITE)
	Soft — scratched by knife	**Shows cleavage**	Brown to black; cleavage, 1 direction; hardness, 2.5-3 (black mica).	BIOTITE
			Various shades of green; cleavage, 1 direction; hardness, 2-2.5 (green "mica").	CHLORITE
			Yellow-brown, dark brown, or black; streak white to pale yellow; resinous luster; cleavage, 6 directions; hardness, 3.5-4.	SPHALERITE
		No cleavage	Red to brown; red streak; earthy appearance; hardness, 5.5-6.5 (apparent hardness may be less).	HEMATITE
			Scarlet to red-brown; scarlet streak; hardness, 2-2.5; high specific gravity.	CINNABAR
			Lead-pencil black, smudges fingers; hardness, 1; (one cleavage that is apparent only in large crystals).	GRAPHITE
			Yellow-brown to dark brown, may be almost black; streak yellow-brown; earthy; hardness, 5-5.5 (may have lower apparent hardness).	LIMONITE (GOETHITE)
			Dark to light green; greasy or waxy luster; some varieties are fibrous; hardness, 2-5, generally 4.	SERPENTINE
Metallic luster			Black; strongly magnetic; hardness, 6.	MAGNETITE
			Lead-pencil black; smudges fingers; hardness, 1; one cleavage that is apparent only in large crystals.	GRAPHITE
			Brass yellow; black streak; cubic crystals, commonly with striations; hardness, 6-6.5.	PYRITE
			Brass-yellow; may be tarnished; black streak; hardness, 3.5-4; massive.	CHALCO-PYRITE
			Shiny gray; black streak; very heavy; cubic cleavage; hardness, 2.5.	GALENA

Table A-1. Mineral identification key (cont.).

Light-colored nonmetallic luster	Hard—not scratched by knife	Shows cleavage	White or flesh-colored; 2 cleavage planes at nearly right angles; hardness, 6. Large crystals which show irregular veining are PERTHITE.	ORTHOCLASE (POTASSIUM FELDSPAR)
			White or green-gray; 2 cleavage planes at nearly right angles; hardness, 6; striations on one cleavage.	PLAGIOCLASE
		No cleavage	White, clear, or any color; glassy luster; transparent to translucent; hexagonal (6-sided) crystals; hardness, 7; conchoidal fracture.	QUARTZ
			Various shades of green and yellow; glassy luster; granular masses and crystals in rocks; hardness, 6.5-7 (apparent hardness may be much less).	OLIVINE
			Any color or variegated; glassy luster; hardness, 5-6; conchoidal fracture.	OPAL
			Any color or variegated; waxy luster; hardness, 7; conchoidal fracture.	CHALCEDONY (AGATE)
	Soft—scratched by knife	Shows cleavage	Colorless to white; salty taste; cubic cleavage; hardness, 2.5.	HALITE
			White, yellow to colorless; rhombohedral cleavage; hardness, 3; effervesces with dilute hydrochloric acid.	CALCITE
			Pink, colorless, white, or dark; rhombohedral cleavage; hardness, 3.5-4; effervesces with dilute hydrochloric acid only if powdered.	DOLOMITE
			White to transparent; 3 unequal cleavages; hardness, 2.	GYPSUM
			Green to white; feels soapy; 1 cleavage; hardness, 1.	TALC
			Colorless to light yellow or green; transparent in thin sheets which are very elastic; 1 cleavage; hardness, 2-2.5 (white mica).	MUSCOVITE
			Green to white; fibrous cleavage; may form veins.	ASBESTOS
		No cleavage	Green to white; feels soapy; hardness, 1.	TALC
			White to transparent; hardness, 2.	GYPSUM
			Yellow to greenish; resinous luster; hardness, 1.5-2.5.	SULFUR

Table A-2. Summary of common minerals.

Mineral	Composition	Color	Luster	Streak	Hardness	Cleavage	Other Properties—Uses
Amphibole—*see hornblende*							
Asbestos	Hydrous magnesium silicate	Shades of green	Silky	None	Low apparent hardness	Fibrous	Fireproofing & insulation against heat & electricity
Augite	$(Ca,Na)(Mg,Fe^{II},Fe^{III},Al)(Si,Al)_2O_6$ Calcium, sodium, magnesium, iron, aluminum silicate	Black–dark green	Vitreous	None	5–6	2 at nearly right angles	Rock-forming mineral
Biotite	$K(Mg,Fe)_3(AlSi_3O_{10})(OH)_2$ Hydrous potassium, magnesium, iron, aluminum silicate	Black–dark green	Vitreous	None	2.5–3	1 perfect	Rock-forming mineral
Calcite	$CaCO_3$ Calcium carbonate	Generally white or colorless	Vitreous to earthy	None	2.5–3	Rhombohedral, 3 not at right angles	Effervesces in cold dilute hydrochloric acid. Manufacture of cement
Chalcedony	SiO_2 Silicon dioxide	Any	Waxy	None	7	None	Conchoidal fracture
Chalcopyrite	$CuFeS_2$ Copper-iron sulfide	Brass-yellow—may be tarnished to bronze or iridescent	Metallic-dull	Greenish-black	3.5–4	None	Brittle
Chlorite	$(Mg,Fe,Al)_6(Al,Si)_4O_{10}(OH)_8$ Hydrous magnesium, iron, aluminum silicate	Green	Vitreous	None	2–2.5	1 perfect	Rock-forming mineral
Cinnabar	HgS Mercury sulfide	Vermillion-red to brownish-red	Adamantine to dull earthy	Scarlet	2.5	1	High specific gravity. Only important ore of mercury.
Clay	A family of hydrous aluminum silicates which may contain potassium, sodium, iron, magnesium, etc.	Generally light	Earthy	None	Very low apparent hardness	None	Rock-forming mineral
Dolomite	$CaMg(CO_3)_2$ Calcium-magnesium carbonate	Generally some shade of pink, flesh color, or white	Vitreous to pearly	None	3.5–4	Rhombohedral, 3 not at right angles	Powder effervesces in cold dilute hydrochloric acid
Feldspar—*see plagioclase and orthoclase (potassium feldspar)*							
Galena	PbS Lead sulfide	Lead-gray	Bright metallic	Lead-gray	2.5	Cubic, 3 at right angles	High specific gravity. Ore of lead
Garnet	A family of silicate minerals containing aluminum, iron, magnesium, calcium, etc.	Commonly red	Vitreous	None	6.5–7.5	None	Characteristic isometric crystals

Table A-2. Summary of common minerals (cont.).

Mineral	Composition	Color	Luster	Streak	Hardness	Cleavage	Other Properties—Uses
Graphite	C Carbon	Black to steel-gray	Metallic or dull earthy	Black	1–2	1	Greasy feel—marks paper
Gypsum	$CaSO_4 \cdot 2H_2O$ Hydrous calcium sulfate	Colorless, white, gray	Vitreous, pearly, silky to earthy	None	2	1 perfect. Others poorer	Plaster of Paris
Halite	NaCl Sodium chloride	Colorless to white	Vitreous	None	2.5	Cubic, 3 at right angles	Salty taste
Hematite	Fe_2O_3 Iron oxide	Reddish-brown to black or gray	Earthy to bright metallic	Light to dark Indian-red	5.5–6.5	None	Most important ore of iron
Hornblende	$Ca_2Na(Mg,Fe^{II})_4 (Al,Fe^{III},Ti)$ $(Al,Si)_8 O_{22}(O,OH)_2$ Hydrous calcium, sodium, magnesium, iron, aluminum silicate	Black–dark green	Vitreous	None	5–6	2 at 60°	Rock-forming mineral
Limonite (Goethite)	$FeO(OH) \cdot nH_2O$ Hydrous iron oxide	Yellow-brown to black	Generally earthy	Yellowish-brown	5–5.5	None	Low apparent hardness
Magnetite	Fe_3O_4 Iron oxide	Black	Metallic to submetallic	Black	6	None	Strongly magnetic. Important iron ore
Mica — see biotite and muscovite							
Muscovite	$KAl_2(AlSi_3 O_{10})(OH)_2$ Hydrous potassium-aluminum silicate	Clear–light green	Vitreous	None	2–2.5	1 perfect	Rock-forming mineral
Olivine	$(Mg,Fe)_2SiO_4$ Magnesium-iron silicate	Green	Vitreous	None	6.5–7	None	Rock-forming mineral
Opal	$SiO_2 \cdot nH_2O$ Hydrous silicon dioxide	Any	Vitreous or resinous	None	5–6	None	Conchoidal fracture
Orthoclase (Potassium feldspar)	$KAlSi_3O_8$ Potassium-aluminum silicate	White to pink	Vitreous	None	6	2 at right angles	Rock-forming mineral
Plagioclase	$CaAl_2Si_2O_8$ & $NaAlSi_3O_8$ Calcium & sodium aluminum silicate	White to green	Vitreous	None	6	2 at right angles–striations on one cleavage direction	Rock-forming mineral
Pyrite	FeS_2 Iron sulfide	Pale brass-yellow	Metallic bright	Greenish or brownish-black	6–6.5	None	Brittle. Cubic crystals
Pyroxene — see augite							
Quartz	SiO_2 Silicon dioxide	Any	Vitreous	None	7	None	Rock-forming mineral

Table A-2. Summary of common minerals (cont.).

Mineral	Composition	Color	Luster	Streak	Hardness	Cleavage	Other Properties—Uses
Serpentine	$Mg_6Si_4O_{10}(OH)_8$ Hydrous magnesium silicate	Dark to light green	Greasy to waxy	None	2–5	None	Metamorphic rocks. Fibrous variety is Asbestos
Sphalerite	ZnS Zinc sulfide	Commonly yellow-brown to black	Resinous to adamantine	White to yellow and brown	3.5–4	6	Zinc ore
Sulfur	S Sulfur	Yellow	Resinous	None	1.5–2.5	None	Fracture conchoidal to uneven. Brittle
Talc	$Mg_3(Si_4O_{10})(OH)_2$ Hydrous magnesium silicate	Apple-green, gray, white, or silver-white	Pearly to greasy	None	1	1	Cosmetics

Appendix B

Topographic Maps

Topographic Maps

Maps are simply scale drawings of a part of the earth's surface. Thus they are similar to a blueprint of an object or to a dress pattern. Most maps are drawn on sheets of paper and so show only the two horizontal dimensions. Geologists, together with civil engineers and many other users of maps, require that the third dimension, elevation, be shown on maps. Maps that indicate the shape of the land are called *topographic maps*. A number of different ways of showing topography are in current use, such as the familiar color tint that uses shades of green for low elevations and brown for higher elevations. The most accurate method, which will be described here, involves the use of contour lines.

Map Scales

Map scales are designated in several ways. (1) A scale can be stated as a specified unit of length on the map corresponding to a specified unit of length on the ground—for example, one inch equals one mile, meaning that one inch on the map represents one mile on the ground. (2) A scale can also be stated as a scale ratio or representative fraction, meaning that one unit (any unit) on the map equals a specified number of the same units on the ground. In the example used above, the scale would be given as 1:63,360, and one inch on the map would represent 63,360 inches on the ground (1 mile = $12 \times 5280 = 63,360$ inches). (3) Most maps contain a graphic scale, generally in the lower margin, for measuring distances.

Contour Lines

A contour line is an imaginary line, every part of which is at the same elevation. Thus a shore line is an example of a contour. Because a single point can only have one elevation, two contour lines can never cross (except on an overhanging cliff).

By convention, contour lines on each map are a fixed interval apart.

385

This is called the contour interval. Typical contour intervals are 5, 10, 20, 40, 80, or 100 feet. The contour lines are always at an elevation that is a whole number times the contour interval. That is, the contour lines are drawn at 20, 40, 60, 80, etc., feet, never at 23, 43, or 63 feet.

A little imagination will reveal that if the contour lines are close together the surface is steeper than if they are far apart. (See Fig. B-1.)

If we think of an island and remember that the shore line is the zero contour, we can then imagine the successive contours, or shore lines, if the water level rises. From this we learn that a hill is represented on a contour map by a series of concentric closed lines, and, of course, a ridge would be similar, but elongate.

If we think of a valley, we see that the contour lines form V's pointing upstream. Remember a stream flows downhill at the lowest point in a valley; hence, if we start on one of the valley sides and walk in the upstream direction at a given elevation, we will eventually reach the stream.

A few minutes study of Fig. B-2 will reveal much information about reading contour maps.

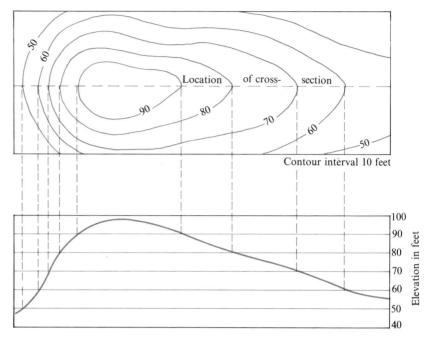

Figure B-1. Contour map of a simple hill. The cross-section shows that where the contour lines are closer together the hill is steeper.

Contour lines have the inherent limitation that the elevations of points between contours are not designated, although the approximate elevation can be determined by interpolation. In an effort to overcome this limitation, the elevations of hills, road junctions, and other features are commonly given on the maps.

Conventions Used on Topographic Maps

Because the United States Geological Survey produces most of the topographic maps in use in the United States, these are described here. Until recently these maps were made by actual field surveying, but are now

Table B-1. Sizes and scales of U.S.G.S. topographic maps.

Map Designation	Scale		Quadrangle Size (Lat.-Long.)	Quadrangle Area (sq. miles)	Paper Size (inches)
	Ratio	One inch equals			
7½ minute	1:24,000	exactly 2,000 feet	7½′ × 7½′	49-70	22 × 27
					23 × 27
	1:31,680	exactly ½ mile	7½′ × 7½′	49-68	17 × 21
15 minute	1:62,500	approximately 1 mile	15′ × 15′	197-282	17 × 21
					19 × 21
30 minute	1:125,000	approximately 2 miles	30′ × 30′	789-1,082	17 × 21
1 degree	1:250,000	approximately 4 miles	1° × 1°	3,173-4,335	17 × 21
1:250,000	1:250,000	approximately 4 miles	1° × 2°	6,349-8,669	24 × 34

made from aerial photographs with some field surveying to establish horizontal and vertical control points. The maps made since 1942 meet high accuracy standards. Ninety per cent of all well-defined features are within one-fiftieth of an inch of their true location on the published map, and the elevations of 90 per cent of the features are correct within one-half of the contour interval.

The maps, which are called quadrangles, are bounded by latitude and longitude lines. They are named for features within the map area, and because the names are not duplicated within a state, some are named for less important features. The sizes and scales used are shown in Table B-1. The number of square miles in a map of a given scale varies because of the convergence of longitude lines at the north pole.

Colors

Water features are in blue.
Works of man—roads, houses, etc. —are in black.
Contour lines are in brown.
In addition, some maps show:
    Important roads, urban areas, and public land subdivision lines in red.
    Woodlands, orchards, etc., in green.

If you would like to obtain topographic maps of any area, write to the addresses below for an index to topographic maps for the state desired. They are free on request and show all of the maps available.

For states east of the Mississippi River:

    United States Geological Survey Washington, D.C. 20242

For states west of the Mississippi River:

    United States Geological Survey Federal Center Denver, Colorado 80225

THE USE OF SYMBOLS IN MAPPING

These illustrations show how various features are depicted on a topographic map. The upper illustration is a perspective view of a river valley and the adjoining hills. The river flows into a bay which is partly enclosed by a hooked sandbar. On either side of the valley are terraces through which streams have cut gullies. The hill on the right has a smoothly eroded form and gradual slopes, whereas the one on the left rises abruptly in a sharp precipice from which it slopes gently, and forms an inclined tableland traversed by a few shallow gullies. A road provides access to a church and two houses situated across the river from a highway which follows the seacoast and curves up the river valley.

The lower illustration shows the same features represented by symbols on a topographic map. The contour interval (the vertical distance between adjacent contours) is 20 feet.

Figure B-2. Perspective view of an area and a contour map of the same area. After U.S. Geological Survey.

Appendix C

Physical Methods of Age Determination

Some of the methods are described in Chapter 2. Table C-1 on the following three pages after E. J. Zeller, "Modern Methods for Measurement of Geologic Time," Mineral Information Service, California Division of Mines and Geology, January, 1965, Vol. 18, No. 1, pp. 9–15.

Table C-1. Physical methods of age determination.

Method	Kind of Sample Used	Parameter Measured	Range and Accuracy	Remarks
		Isotopic or radioactive yield methods		
Uranium–Lead	Zircon, xenotime, and other igneous and metamorphic minerals with original uranium, but with no original lead.	Uranium and lead by analyses and α-counts per unit of mass per unit of time.	10 million to 3–4 thousand million years. Not as accurate as with isotopic analysis.	There should be no leaching of uranium during life of mineral and no original lead in mineral at time of crystallization.
Uranium 238–Lead 206	Any igneous or metamorphic mineral or whole rock with original uranium 238 (especially uraninite, pitchblende, samarskite, cyrtolite, thorianite, zircon, and thorite.)	Quantities of uranium 238, lead 206, and lead 204 by mass spectrometer analysis. Lead 204 gives indication of amount of original lead, of all isotopes, that is present.	10 million to approximately 10,000 million years. More accurate than the chemical uranium–lead method.	Uranium 238 is more abundant than uranium 235 so less chance of analytical error. Radon 222 half-life is long enough (3.82 days) to suggest that there may be error through radon escape.
Uranium 235–Lead 207	Any igneous or metamorphic mineral or whole rock with original uranium 235. (Same minerals as above.)	Quantities of uranium 235, lead 207, and lead 204 by mass spectrometer analysis. Lead 204 gives indication of amount of original lead, of all isotopes, that is present.	10 million to several thousand million years. More accurate than the chemical uranium–lead method.	Radon 219 in series has such short half-life (3.92 sec.) that escape is much less likely than in uranium 238–lead 206 series.
Thorium 232–Lead 208	Accessory minerals in igneous rocks.	Thorium 232, lead 208 by mass spectrometer analysis.	10 million to several thousand million years. Accuracy 10 per cent.	Lead 208 may be removed selectively from thorium minerals resulting in ages which appear too low.
Lead–Lead	Galena or uranium minerals.	Lead 206, lead 207, lead 208, lead 204 by mass spectrometer analysis.	10 million to several thousand million years. 1 per cent over 100 million years. (Younger ages may be calculated with accuracy of 1 million years.)	Has the advantage that it involves an isotopic ratio instead of absolute measurement.
Lead 210	Fresh and marine water and glacial ice. Recently deposited sediments.	Count rate of bismuth 210 derived by radioactive decay of separated lead 210.	Up to about 100 years. Sequence of analyses gives fairly accurate estimate of rates of accumulation.	Part of chain of uranium 238–lead 206. Radon 222, to atmosphere from rocks, yields lead 210. Lead 210 washed from atmosphere by rain. This disintegrates to bismuth 210–polonium 210–lead 206. Lead 210 is rapidly removed from water by precipitation or absorption on sedimentary particles. Rate of removal from water and original amount of lead 210 in sediments are necessary for data for age determination.

Table C-1. Physical methods of age determination (cont.).

METHOD	KIND OF SAMPLE USED	PARAMETER MEASURED	RANGE AND ACCURACY	REMARKS
Isotopic or radioactive yield methods (continued)				
Potassium–Argon	Micas and feldspars.	Potassium 40, argon 40. Potassium measured by flame photometer, argon by mass spectrometer.	30,000 to several thousand million years. Error in micas less than 10 per cent.	Feldspars lose argon (up to 60 per cent). Metamorphism affects the results if rocks are heated.
Rubidium–Strontium	Mica, muscovite, biotite. Feldspar. Whole rock. Glauconite (in sediments).	Amount of strontium 86. Amount of strontium 87. Amount of rubidium 87. Ratios determined by mass spectrometry.	Not suited for ages below 50 million years due to long rubidium 87 half-life. Accuracy of method dependent on rubidium 87 half-life which is not accurately determined.	Decay constant of rubidium 87 uncertain due to low energy of β-particles emitted. Metamorphism may cause rehomogenation of rubidium and strontium isotopes so method may be used to date metamorphic events.
Thorium 230	Oceanic sediments, cores.	Count rate of thorium 230	Up to 300,000 years.	Part of chain of uranium 238–lead 206. Used to date marine sediments. Uranium 238 in sea water decays to thorium 230 which is precipitated and deposited on sea floor.
Thorium 230– Protactinium 231	Oceanic sediments, cores.	Count rate of thorium 230 and protactinium 231.	Up to 300,000 years.	Protactinium 231 is part of chain of uranium 235–lead 207. Accumulates similar to thorium 230 described above.
Helium	Uranium and thorium bearing minerals. (The helium is the α particles emitted.)	Abundance of helium. Uranium 238, uranium 235, and thorium 232 abundance by their α activity. Uranium/thorium ratio.	Several million to several thousand million years. Accuracy in dispute.	Method is not extensively used at present due to discovery of the ease of diffusion by helium.
Carbon 14	Charcoal, wood, shell.	β-activity of carbon in sample by β-counting techniques.	0 to 50,000 years. Accuracy good for last 20,000 years.	Assumes a steady state for the production of carbon 14 in the atmosphere by cosmic radiation. Assumes uniform distribution of carbon 14 throughout the earth's atmosphere.
Hydrogen 3 (Tritium)	Water samples, 2–6 liters.	β-activity of tritium.	Last 30 years.	Recent bomb testing has greatly affected the usefulness of this method.
Radiation damage methods				
Metamict	Zircon (best); euxenite; sphene; uraninite; davidite.	Amount of radiation damage (i.e. metamict condition) in mineral usually by X-ray diffraction. Amount of radioactive material by α-count.	Now considered a comparatively inaccurate method.	Heating will cause crystal to become crystalline again; therefore, method may be used to date metamorphic or igneous events.

Table C-1. Physical methods of age determination (cont.).

METHOD	KIND OF SAMPLE USED	PARAMETER MEASURED	RANGE AND ACCURACY	REMARKS
		Radiation damage methods (continued)		
Pleochroic halo (Radiation halo)	Biotite with an inclusion of zircon, apatite, or monazite.	Intensity of the coloration of the biotite. Width of the coloration. α-activity of the inclusion.	Late Precambrian to approximately Eocene. Useful for relative age determination.	Coloration is temperature sensitive and may reverse at a high level of radiation. Assumes identical irradiation sensitivity of all biotites and that the coloration, once formed, does not fade through time.
Thermoluminescence	Carbonate rocks, fluorite, quartz, feldspars.	Luminescence of sample when heated to temperatures up to 400°C. Glow-curve is measured by a recording microphotometer.	One thousand million year maximum. Accuracy ± 15 per cent.	May be used to determine periods of metamorphism. Has specific uses for paleoclimatological measurements.
Radiation pitting (Fission particle tracks)	Mica. Other minerals may also be usable.	Particle tracks counted by optical microscopy. Uranium determined by α activity.	One thousand million years. Accuracy has not yet been tested fully.	Very promising new method. May be sensitive to heating in metamorphic processes.
		Chemical methods		
Devitrification of glass	Glass, either natural obsidian or man made.	Crystallinity of the glass, especially on the surface.	Useful for relative age determination in past 5 million years for obsidian.	Rate of crystallization dependent upon chemical composition, temperature, and water present in proximity of the glass.
Degradation of amino acids	Shell or bone fragments.	Amino acids and peptide bonding.	Accuracy best in last 1 million years but amino acids are detectable in rocks over 500 million years old.	Involves a chemical process and rate is temperature dependent.
Protein-nitrogen	Shells of mature forms of mollusks.	Organic nitrogen content of water-soluble residues (by micro Kjeldahl method).	Probably limited to Pleistocene and Late Tertiary.	Only relative ages may be determined.

Appendix D

Rock Textures

Size in mm	This Book (all rocks)	Igneous and Metamorphic		Sedimentary		Pyroclastic	
				Sediment	Rock	Sediment	Rock
256		Very coarse *		Boulder	*Conglomerate* (fragment size is generally indicated as pebble con- glomerate, etc.)	*Block* **	*Breccia*
128		*Pegmatitic*		Cobble		*Bomb* **	
64		Phaneritic					
32							
16		Coarse *		Pebble			
8	Coarse- grained					Lapilli	Lapilli
4						Cinder	
2		Medium *		Granule			
1					*Coarse sandstone*	Coarse *Ash*	Coarse *Tuff*
1/2				Sand			
1/4		Fine *		*Sand*			
1/8					*Fine sandstone*		
1/16	Fine- grained	Aphanitic					
1/32				*Silt*	*Siltstone*	Fine *Ash*	Fine *Tuff*
1/64		Dense					
1/128							
1/256				*Clay*	*Shale* (most shales contain some silt)		

*These terms are generally used as modifiers such as medium-grained granite.
**Blocks are fragments broken from solidified rocks; bombs are molten when ejected.

Figure D-1. Summary of texture terms in common use and those used in this book. Terms used in this book are in italics.

Glossary

"When I use a word," Humpty Dumpty said, in rather a scornful tone, "it means just what I choose it to mean — neither more nor less."

"The question is," said Alice, "whether you can make words mean so many different things."

"The question is," said Humpty Dumpty, "which is to be master — that's all."

Through the Looking Glass
Chapter 6
Lewis Carroll

AA LAVA   Hawaiian term for blocky, rough surfaced lava.

ABRASION   Erosion by friction and impact — generally caused by rock particles carried by running water, ice, or wind.

ABYSSAL   Pertaining to deep ocean, over 3000 feet deep.

ADAMANTINE LUSTER   Diamond-like luster.

AEOLIAN   A term applied to wind erosion or deposition. Now obsolete, see *eolian*.

AFTERSHOCK   An earthquake following a larger earthquake, originating at or near the focus of the larger earthquake.

AGATE   Chalcedony with colors in bands, irregular clouds, or moss-like forms.

A HORIZON, SOIL   Uppermost layer of the soil. Topsoil. See Fig. 4-10.

ALKALI LAKE, FLAT   Lake or level area in desert region where water or soil contains soluble alkalis. The alkalis, potassium, sodium, and calcium salts, come from weathering and are concentrated by evaporation and poor drainage.

ALLUVIAL FAN   Low, cone-shaped deposit built by a river where it issues from a mountain into a lowland.

ALLUVIUM   Stream-deposited material.

ALPHA DECAY   Radioactive decay in which an alpha particle is emitted.

ALPHA PARTICLE   Helium nucleus. Consists of two protons and two neutrons, and is positively charged.

ALPINE GLACIER   Mountain glacier. A glacier in a mountain valley.

AMINO ACID   One of the nitrogenous organic compounds that are the building blocks of proteins.

AMPHIBOLE   A ferromagnesian mineral group. The most important rock-forming amphibole is hornblende.

AMPHIBOLITE   A metamorphic rock composed mainly of amphiobole, such as hornblende, and plagioclase feldspar.

ANDESITE   Fine-grained igneous rock. Intermediate in color and composition between rhyolite and basalt.

ANDESITE LINE   A line drawn on a map separating the oceanic areas of basaltic volcanism from the continental and island arc areas of andesitic volcanism.

ANGLE OF REPOSE   Maximum angle of slope at which loose material will come to rest.

ANGSTROM   A unit of length = 0.00000001 centimeters (approximately 0.000000004 inches).

ANGULAR UNCONFORMITY   An unconformity in which the older strata dip or slope at a different angle (generally steeper) from the younger strata.

ANORTHOSITE   A coarse-grained rock composed almost entirely of plagioclase.

ANTHRACITE   A type of coal formed by mild metamorphism of lower-grade coal.

ANTICLINE   A type of fold in which the strata slope downward in opposite directions away from a central axis, like the roof of a house. The oldest rocks are in the core.

APATITE   A mineral, Number 5 on the hardness scale. $Ca_5(PO_4)_3F$

APHANITIC   A rock texture in which the minerals are too small to see with the naked eye.

AQUIFER   A rock body that contains water and through which the water can move, a reservoir.

ARÊTE   A narrow, jagged ridge typical of glacially eroded mountains.

ARKOSE   A sandstone containing much feldspar.

ARTESIAN WELL   A well in which the water is under pressure so that it rises up the well bore. The water may or may not flow out onto the surface.

ASBESTOS   A name given to fibrous mineral material widely used as insulation. It may be composed of either an amphibole or a variety of serpentine.

ASH, VOLCANIC   Unconsolidated pyroclastic material less than 4 millimeters in size.

ASTEROID   One of a group of small planet-like bodies whose general orbit around the sun is between Mars and Jupiter.

ASTHENOSPHERE   The low velocity layer in the upper mantle.

ATOLL   A ring-like coral island or islands with a central lagoon.

ATOM   The smallest unit of an element. Composed of protons, electrons, and neutrons.

ATOMIC FISSION   A reaction in which an atom is split. The atomic bomb reaction is an example.

ATOMIC FUSION   A reaction in which two or more atoms are combined to form a different element. The hydrogen bomb reaction is an example.

ATOMIC MASS NUMBER   The total number of protons and neutrons in the nucleus of an atom.

ATOMIC MASS UNIT   The unit used by physicists to express the weight of a nucleus. 1.66×10^{-24} grams (based on oxygen).

ATOMIC NUMBER   The number of protons, or positive charges, in the nucleus of an atom.

AUGITE   The common rock-forming pyroxene mineral. Generally black with two cleavages at right angles. $(Ca,Na)(Mg, Fe^{II}, Fe^{III}, Al)(Si, Al)_2O_6$

AURORA   Luminous bands or areas of colored light seen in the night sky near the poles. Northern or southern lights.

AUTORADIOGRAPH   A "photograph" of radioactive material taken by placing film directly on a radioactive specimen. See Fig. 2-11.

AVALANCHE   The fall or slide of a snow mass. The term is sometimes used for *landslide*.

AVOGADRO'S NUMBER   The number of atoms in a molecular weight of a gas. 6.02×10^{23}

BARCHAN DUNE   A crescent-shaped dune in which the cusps face downwind.

BARRIER ISLAND   A low, long, sandy, wave-built island between the ocean and the continent.

BARRIER REEF   A reef between the main ocean and an island or continent.

BASALT   A fine-grained volcanic rock. It is dark in color and composed mainly of plagioclase feldspar and pyroxene. It is one of the most common rock types.

BASE LEVEL OF RIVER   The lowest level to which a river can erode. It is sea level for a river flowing into the ocean. The elevation of the mouth of a river.

BASEMENT ROCKS   The rocks that underlie the sedimentary cover. The basement is generally a complex of igneous and metamorphic rocks.

BATHOLITH   A very large discordant intrusive rock body, generally composed of granitic rocks. Some batholiths may be of metamorphic origin.

BAUXITE   Aluminum ore. Bauxite is the term given to a weathering product composed of several aluminum-rich minerals.

BAYMOUTH BAR   A bar extending either partly or completely across the mouth of a bay.

BEDDING   The layers or laminae of sedimentary rocks.

BED LOAD   The material not in solution or suspension that is moved along the bottom of a stream by the moving water.

BENCH, SURF-CUT   A flat surface cut or eroded by wave action.

BERYL   A mineral, an ore of beryllium (a metal). $Be_3Al_2Si_6O_{18}$

BETA DECAY   Radioactive decay in which a beta particle is released.

BETA PARTICLE   In radioactivity, an electron moving at high velocity.

B HORIZON, SOIL   Second or middle layer of the soil. See Fig. 4-10.

BIOTITE   A common mineral, a member of the mica family. Recognized by its shiny black or dark color and perfect cleavage. $K(Mg,Fe)_3(AlSi_3O_{10})(OH)_2$

BITUMINOUS COAL   Soft coal.

BLOCK FAULT   A fault that bounds an uplifted or down-dropped segment of the crust.

BLOCK, VOLCANIC   Pyroclastic fragment greater than 5 centimeters in size.

BLOCKY LAVA   Lava with a blocky surface. See *aa lava*.

BLOWOUT   Hollow caused by wind erosion.

BOMB, VOLCANIC   Pyroclastic material ejected from a volcano in a melted or plastic state. Its shape forms either in flight or upon impact.

BORNITE   A mineral, an ore of copper. Cu_5FeS_4

BOULDER   A rock fragment greater than 256 millimeters in size.

BOULDER TRAIN   A number of boulders from the same source deposited in a line by a glacier.

BOWEN'S REACTION SERIES   The sequence and reactions that take place in the crystallization of a basalt melt.

BRAIDED RIVER   A river with many channels that divide and join.

BRECCIA   A clastic sedimentary rock with angular fragments larger than 2 mm in diameter, a sharpstone conglomerate. A volcanic breccia is composed of pyroclastic fragments greater than 2 millimeters in size.

BROWN CLAY   A deep-sea deposit formed by fine material, mainly volcanic dust. (The color is probably caused by oxidation.) Also *red clay*.

CALCIC PLAGIOCLASE   Plagioclase feldspar with a high calcium content. Plagioclase is a continuous series between $CaAl_2Si_2O_8$ and $NaAlSi_3O_8$.

CALCITE   A common rock-forming mineral, composes limestone and marble. Recognized by three directions of cleavage not at right angles, and reaction with dilute acid. $CaCo_3$

CALDERA   A large volcanic depression containing volcanic vents.

CALICHE   A whitish accumulation of calcium carbonate in soil in dry areas.

CAPILLARY ACTION   The rising of water in narrow passages, caused by surface tension. See Fig. 12-2.

CARBON 14   Radioactive isotope of carbon having an atomic mass number of 14. Used in radioactive dating.

CARBONATE   A rock or mineral containing carbonate (CO_3). The main minerals are calcite and dolomite, and the main rock types are limestone and dolostone (dolomite).

CARBONATION   A reaction in chemical weathering in which carbon dioxide (CO_2) combines with the rock to form a carbonate (CO_3).

CARBONIC ACID   An acid important in weathering that is formed by the solution of carbon dioxide in water. H_2CO_3

CARBONIFEROUS PERIOD   A part of the geologic time scale. It is so called because many coal deposits were formed at that time. It lasted from about 345 million years ago to about 280 million years ago, the Mississippian and Pennsylvanian periods combined.

CARNOTITE   An ore of uranium and vanadium. $K(UO_2)_2(VO_4)_2 \cdot 3H_2O$

CASSITERITE   An ore of tin. SnO_2

CATACLASTIC   A metamorphic texture in which at least some minerals have been broken and flattened during dynamic metamorphism.

CATASTROPHISM   A doctrine that held that most geologic changes were caused by sudden violent events.

CEMENTATION   The process whereby binding material is precipitated or deposited between clastic grains, forming a rock.

CENOZOIC   An era of the geologic time scale. It lasted from about 65 million years ago to the present.

CHALCEDONY   An extremely fine-grained variety of quartz. Agate and chert are made of chalcedony.

CHALCOCITE   An ore of copper. Cu_2S

CHALCOPYRITE   An ore of copper. Fool's gold. $CuFeS_2$

CHALK   A soft, usually white variety of limestone. Composed of shells of microorganisms and very fine calcite.

CHEMICAL WEATHERING   The reactions that occur between a rock and its surroundings. The reactions tend to bring the minerals into equilibrium with the environment at or near the surface of the earth. Typically water, oxygen, and carbon dioxide are involved.

CHERT   An extremely fine-grained variety of quartz. See *chalcedony*.

CHLORITE   A compound containing chlorine. Also, a group of platy, micalike, usually green minerals.

C HORIZON, SOIL   The third layer of a soil, partially decomposed bedrock.

CHROMITE   An ore of chromium. $(Mg,Fe^{II})(Cr,Al,Fe^{III})_2O_4$

CINDER CONE   A cone-shaped small volcano composed of pyroclastic material.

CINDER, VOLCANIC   Pyroclastic material between 4 and 32 millimeters in size.

CINNABAR   An ore of mercury. HgS

CIRQUE   A glacial erosion feature. The steep amphitheater-like head of a glacial valley.

CLASTIC   A textural term for rocks or sediments composed of fragments broken from some other pre-existing rock.

CLAY   A mineral family. Most clays are very fine grained. Much of the material in soil and shales is clay.

CLAYSTONE   A rock composed largely of fine clay.

CLEAVAGE   The tendency to split along planes of weakness.

COAL   A sedimentary rock formed of altered plant material.

COBALTITE   An ore of cobalt. $CoAsS$

COBBLE   A rock fragment between 64 and 256 millimeters in size.

COL   A mountain pass, especially the high, sharp pass formed by the meeting of two headward-eroding cirques.

COLUMNAR JOINTING   Jointing that breaks the rock into typically rough five- or six-sided columns. Common in basalt, and probably caused by shrinking due to cooling.

COMPACTION   Reduction in volume. Generally caused by pressure in sedimentary rocks.

COMPOSITE VOLCANO   A volcano composed of both lava and pyroclastic material.

CONCHOIDAL FRACTURE   The rounded shell-like fracture typical of brittle materials such as glass.

CONCORDANT INTRUSIVE ROCK   An intrusive igneous body whose boundaries are parallel to the enclosing sedimentary rocks.

CONE OF DEPRESSION   The cone-shaped depression of the water table caused by withdrawing water from a well faster than it can move through the rocks to the well.

CONGLOMERATE   A coarse-grained clastic sedimentary rock. The fragments are larger than 2 millimeters.

CONTACT METAMORPHIC ROCKS   Thermally metamorphosed rocks near the boundary or contact of an intrusive body such as a batholith.

CONTINENTAL DRIFT   The movement of continental plates on the earth's surface.

CONTINENTAL GLACIER   A large, thick glacier caused by the buildup of an ice cap on a continent. Greenland has an example.

CONTINENTAL RISE   The transition between the continental slope and the deep ocean floor.

CONTINENTAL SHELF   The shallow ocean area between the shoreline and the continental slope that extends to about a 600-foot depth surrounding a continent. The underlying crust is generally continental.

CONTINENTAL SLOPE   The relatively steep slope between the continental shelf and the deep ocean basin.

CONTINUOUS MINERAL SERIES   A mineral group such as plagioclase feldspar in which complete or continuous change in composition between end members is possible.

CONTOUR LINE   A line on a map connecting points of equal elevation.

CONVECTION CURRENT   Movement in a fluid caused by differences in density. The density differences are generally caused by heating.

COQUINA   A type of limestone composed largely of shells and shell fragments.

CORAL   One of a group of marine animals that secrete an external skeleton of calcium carbonate.

CORE OF EARTH   The innermost layer of the earth. Consists of an inner core believed to be solid and an outer liquid core.

CORRELATION   The determination of the equivalence of two exposures of a rock unit.

CORUNDUM   A mineral. Number 9 on the hardness scale. Al_2O_3

COSMIC RAY   Very high-speed nuclei coming from space.

COVALENT BONDING   A bond formed by the sharing of electrons.

CRATER, IMPACT   A depression caused by impact of a falling object. Meteorites cause most large impact craters.

CRATER, VOLCANIC   A volcanic depression typically developed at the top of a volcano.

CREEP   Slow downslope movement of surface material under the influence of gravity.

CRETACEOUS   A period in the geologic time scale. It lasted from about 136 million years ago to about 65 million years ago.

CREVASSE   A deep crack in the surface of a glacier.

CROSS-BEDDING, CROSS-STRATA   Beds or laminations at an oblique angle to the main bedding.

CRUST OF THE EARTH   The outermost layer of the earth. Above the Moho.

CRUSTAL PLATE   A relatively thin plate of the outer part of the earth. The depth is much less than the other two dimensions.

CRYSTAL   A geometrical solid bounded by smooth planes that result from an orderly internal atomic arrangement.

CRYSTALLINE   A material with a crystal structure; that is, with an orderly internal atomic arrangement.

CRYSTALLIZATION   The processes of crystal formation, of becoming crystalline.

CRYSTALLOGRAPHY   The study of crystals.

CRYSTAL SETTLING   The sinking of early-formed crystals in a melt.

CRYSTAL STRUCTURE   An orderly internal (atomic) arrangement in a crystal.

CURIE POINT   The temperature above which a substance loses its magnetism.

CYCLE OF EROSION   The sequence of changes that occur during erosion of a landscape.

DACITE   A fine-grained extrusive igneous rock corresponding in composition to quartz diorite or tonalite.

DARCY'S LAW   A formula relating the flow of fluids to the pressure gradient and the properties of the rocks.

DAUGHTER ELEMENT   An element formed from another element in radioactive decay.

DEBRIS SLIDE   Landslide involving mainly unconsolidated material.

DECAY, RADIOACTIVE   The spontaneous change of one element into another.

DEEP-FOCUS EARTHQUAKE   Earthquake with a focus below 180 miles.

DEFLATION   Sorting and removal of material by wind erosion.

DELTA   The material deposited by a river at its mouth.

DENDRITIC DRAINAGE   Branching stream pattern. See Fig. 8-41.

DENSITY   Denseness or compactness. The mass of a material in grams per cubic centimeter.

DEUTERIUM   An isotope of hydrogen.

DEVONIAN   A period in the geologic time scale. It lasted from about 395 to 345 million years ago.

DIAMOND   A mineral composed of carbon. Number 10 on the hardness scale.

DIATOMITE   A rock composed largely of the siliceous shells of diatoms.

DIFFERENTIATION   The process by which different rock types are derived from a single magma.

DIKE   A tabular discordant intrusive body.

DIKE SWARM   A number of dikes, either more or less parallel, or radial from a single source.

DIORITE   A coarse-grained igneous rock intermediate in composition between granite and gabbro.

DIPOLE MAGNETIC FIELD   The type of magnetic field produced by a bar magnet. See Fig. 16-4.

DISCHARGE   The quantity of water flowing past a point in a river; volume of water per unit of time.

DISCONTINUITY, MOHOROVIČIC (M-DISCONTINUITY, MOHO)   The level at which a distinct change in seismic velocity occurs. It separates the crust from the mantle.

DISCONTINUOUS MINERAL SERIES   That part of the Bowen reaction series in which abrupt changes in mineral occur.

DISCORDANT INTRUSIVE   An intrusive rock body that cuts across the bedding of the enclosing sedimentary rocks.

DISSEMINATED MAGMATIC MINERAL DEPOSIT   A mineral deposit in which the ore is scattered throughout an igneous body.

DISSOLVED LOAD   The material carried in solution by a river.

DOLOMITE   A mineral $CaMg(Co_3)_2$, or a rock composed largely of dolomite (dolostone).

DORMANT VOLCANO   An inactive volcano. A volcano that is not presently erupting, but may have erupted in historic time, and which is considered likely to erupt in the future.

DREIKANTER   A pebble faceted by sandblasting by wind-blown sand. Same as *ventifact*. Also, specifically having three wind-cut facets.

DRIFT   (see *continental drift*)

DRIFT, GLACIAL   A general term for glacier-transported and deposited material.

DRUMLIN   A smooth hill of glacial drift with its long axis parallel to the direction of glacial movement.

DUNE   An accumulation of wind-blown sand.

DUNITE   A coarse igneous rock composed almost entirely of olivine.

DYNAMIC METAMORPHISM   Metamorphism caused by deformation of the rock.

EARTHFLOW   A slow flow of wet overburden.

EARTHQUAKE   Vibrations of the earth caused by sudden internal movements in the earth.

ECLOGITE   A rock that forms under very high pressure, composed of garnet and pyroxene. Its chemical composition is similar to that of basalt.

ELECTRON   A fundamental sub-atomic particle. It has a very small mass and a negative charge.

ELEMENT   A material that cannot be changed into another element by ordinary chemical means. Just over 100 elements are known. All atoms of an element have the same atomic number (number of protons).

EMERGENT COASTLINE   A coastline formed by either uplift of the continent or lowering of sea level in which areas formerly under water are now dry land.

ENARGITE   An ore of copper. Cu_3AsS_4

EOCENE   A subdivision of the most recent era in the geologic time scale. It lasted from about 54 to 36 million years ago.

EOLIAN   A term applied to wind erosion or deposition. (*aeolian*, obsolete)

EPICENTER   The point on the earth's surface directly above the focus of an earthquake.

EPIDOTE   A metamorphic mineral. $Ca_2(Al,Fe^{III})_3(SiO_4)_3OH$

EPOCH   A subdivision of a period in the geologic time scale.

ERA   A major subdivision of the geologic time scale.

EROSION   The wearing away and removal of material on the earth's surface.

ERRATIC   A boulder transported, commonly far, from its source and deposited on a different bedrock. Also, such a boulder, specifically transported by a glacier.

ESKER   Long curving ridges of glacially deposited material.

ESTUARY   Tidal inlet along a sea coast. Many form at river mouths.

EVAPORITE   A sedimentary rock formed by the evaporation of water, leaving the dissolved material behind.

EXFOLIATION   The breaking or peeling off of layers of a large rock mass. See Fig. 4-4.

EXTRUSIVE IGNEOUS ROCK   An igneous rock erupted on the earth's surface. A volcanic rock.

FACETED SPUR   A ridge that has been cut off by glacial, stream, or wave action, or by faulting.

FAULT   A break in the earth's crust along which movement has occurred.

FELDSPAR   The most abundant mineral family. Consists of plagioclase $CaAl_2Si_2O_8$ and $NaAlSi_3O_8$, and potassium feldspar $KAlSi_3O_8$ (orthoclase and sanidine). Mixtures of potassium and sodium feldspar are termed perthite.

FELSITE   A light-colored, fine-grained igneous rock. Most have chemical compositions similar to rhyolite.

FERROMAGNESIAN MINERAL   A mineral containing iron and magnesium. Most are dark colored.

FISSILITY   The ability of a rock to split easily along parallel planes.

FISSION, ATOMIC   A reaction in which an atom is split. The atomic bomb reaction is an example.

FLINT   A variety of chalcedony generally dark in color. See *chalcedony*.

FLOOD BASALT   Basalt that forms thick extensive flows that appear to have flowed rapidly (flooded).

FLOOD PLAIN   The flat part of a river valley that is subject to floods at times.

FLUORESCENCE   The emission of light when exposed to ultraviolet light.

FLUORITE   A mineral. Number 4 on the hardness scale. CaF_2

FLUVIAL   Pertaining to rivers.

FOCUS, EARTHQUAKE   The point where an earthquake originates.

FOLDING   The bending of strata.

FOLIATION   A directional property of metamorphic rocks caused by the layering of minerals. Generally caused by crystallization under pressure. Also the flat arrangement of features in any rock type.

FORMATION   A rock unit that can be mapped.

FOSSIL   Any evidence of a once-living organism.

FRACTURE ZONE   Long linear areas of apparent faulting and irregular topography on the deep-sea floor.

FROST HEAVING   The lifting of the surface by the expansion of freezing water.

FROST POLYGONS   Polygonal patterns of stones caused by repeated freezing.

FROST WEDGING   Mechanical weathering caused by expansion of freezing water.

FUSION, ATOMIC   A reaction in which two or more atoms are combined to form a different element. The hydrogen bomb reaction is an example.

GABBRO   A coarse-grained igneous rock composed mainly of plagioclase feldspar and augite.

GALENA   An ore of lead. PbS

GARNET   A mineral family.

GEODE   A hollow space in a rock, generally with a lining of another mineral such as chalcedony and an inner lining of crystals.

GEOSYNCLINE   A large elongate downsinking area in which many thousand feet of sedimentary rocks are deposited. It is later deformed to form a mountain range.

GEYSER   An intermittent hot spring or fountain from which hot water and/or steam issues.

GLACIER   A mass of moving ice.

GLASS   A supercooled liquid. A glass does not have an orderly internal structure like a crystal.

Glossopteris FLORA   A fossil assemblage characterized by the distinctive *Glossopteris* leaves. The flora occurs in rocks in the southern hemisphere and is an important line of evidence for continental drift.

GNEISS   A coarse-grained foliated metamorphic rock. This rock contains feldspar and is generally banded.

GOETHITE   A mineral. FeO(OH)

GOUGE, FAULT   The fine-grained ground rock in a fault zone.

GRADE, METAMORPHIC   A measure of the intensity of metamorphism.

GRADED BEDDING   Bedding that has a gradation from coarse at the bottom to fine nearer the top.

GRADED RIVER   An equilibrium river that has just the slope and velocity to carry all the debris carried into it.

GRADIENT OF A RIVER   The slope of a river.

GRANITE   A coarse-grained rock containing quartz and feldspar. See Fig. 3-42.

GRANITIZATION   The formation of granite or granite-like rocks by metamorphism.

GRANODIORITE   A coarse-grained rock containing quartz and feldspar. See Fig. 3-42.

GRANULE   A fragment between 2 and 4 millimeters in size.

GRAPHITE   A soft mineral composed of carbon.

GRAVEL   Unconsolidated fragments larger than 2 millimeters.

GRAVITY ANOMALY   An area where the attraction of gravity is larger or smaller than expected.

GRAVITY FOLDING   Folding caused by beds sliding down a slope.

GRAVITY METER   An instrument for measuring the attraction of gravity.

GRAYWACKE   A type of sandstone, generally poorly sorted with a clay or chloritic matrix.

GREENSCHIST   A schist with much chlorite and other green minerals. Generally formed by the metamorphism of basaltic rocks.

GUYOT   A flat-topped seamount.

GYPSUM   A mineral. Number 2 on the hardness scale. $CaSO_4 \cdot 2\ H_2O$

HALF-LIFE   The time required for one-half of the atoms of a radioactive element to decay.

HALITE   A mineral. Common salt. NaCl

HANGING VALLEY   A tributary valley whose floor is higher than that of the main stream valley at this junction. Commonly caused by glaciation.

HARDNESS   A mineral's resistance to scratching.

HEMATITE   An ore of iron. Fe_2O_3

HOLOCENE   The youngest subdivision in the geologic time scale. The *Recent*.

HORN   A sharp, pyramidal peak shaped by headward erosion of glacial cirques.

HORNBLENDE   A common rock-forming mineral. An amphibole. $Ca_2Na(Mg,Fe^{II})_4$ $(Al, Fe^{III},Ti)(Al,Si)_8O_{22}(O,OH)_2$

HORNFELS   A metamorphic rock formed by thermal or contact metamorphism of any rock type.

HUMUS   Dark decomposed organic material in soil.

HYDRATION   A process in chemical weathering that involves the addition of water.

HYDROTHERMAL DEPOSIT   A mineral deposit emplaced by hot water solutions.

ICECAP GLACIER   (See *continental glacier*.)

IGNEOUS ROCK   A rock that formed by the solidification of a magma. A rock that was once melted or partially melted.

ILMENITE   An ore of titanium. $FeTiO_3$

INDEX FOSSIL   A fossil that can be used to date the enclosing rocks.

INFILTRATION   The seeping of water into rocks or soil.

INTENSITY, EARTHQUAKE   A measure of the surface effects of an earthquake. See Table 13-1.

INTERLOCKING TEXTURE   A texture in igneous and metamorphic rocks in which the mineral grains grew together, filling all of the available space.

INTRUSIVE IGNEOUS ROCK   A rock that solidified below the surface.

INVERTEBRATE   An animal without a backbone.

ION   An atom that has gained or lost electrons and so has an electrical charge.

IONIC BONDING   The atoms are attracted to each other by electrical charges caused by concurrent gain and loss of electrons.

ISLAND ARC   A curving group of volcanic islands associated with deep-focus earthquakes, deep ocean trenches, and negative gravity anomalies.

ISOSTASY   The theory that the crust floats in equilibrium on a substratum in the mantle.

ISOTOPE   A variety of an element having a different number of neutrons in its nucleus.

JOINTING   A fracture in a rock along which no appreciable movement has occurred.

JURASSIC   A period in the geologic time scale. It lasted from about 195 to 136 million years ago.

KETTLE   A depression in glacial deposits probably caused by the melting of a buried block of ice.

LACCOLITH   A concordant intrusive igneous rock body that domed up the overlying sedimentary rocks to some extent.

LACUSTRINE   Pertaining to lakes.

LAGOON   A shallow body of water with restricted connection to the ocean.

LAMINAR FLOW   Flow without turbulence. See left part of Fig. 8-1.

LANDSLIDE   Rapid downslope movement of bedrock and/or overburden. Also used as a general term for all downslope movements.

LAPILLI   Pyroclastic fragments in the size range of 1 to 64 millimeters.

LAPILLI TUFF   A pyroclastic rock composed of lapilli in a fine tuff matrix.

LATERAL FAULT   A fault with horizontal movement. Also called strike-slip fault, and wrench fault.

LATERAL MORAINE   The material deposited by a glacier along its sides.

LATERITE   The red-brown soil formed by chemical weathering in the tropics.

LAVA   Molten volcanic flows on the earth's surface; also, the rock solidified from it.

LEACHING   The dissolving of minerals by water passing through rocks.

LEVEE   The raised bank along a river.

LIGNITE   A brown-black coal that forms from the alteration of peat.

LIMESTONE   A sedimentary rock composed largely of calcite.

LIMONITE   A general term for brown hydrous iron oxides.

LITHIFICATION   The process whereby a sediment is turned into a sedimentary rock.

LOAM   A soil composed of clay, silt, sand, and organic material.

LOESS   Wind-deposited silt.

LONGITUDINAL DUNE   An elongate sand dune parallel to the wind direction. *Seif dune* is a type of longitudinal dune.

MAGMA   A natural hot melt composed of a mutual solution of rock-forming materials (mainly silicates) and some volatiles (mainly steam) that are held in solution by pressure. It may or may not contain suspended solids.

MAGNETITE   A mineral. An ore of iron. Fe_3O_4

MAGNITUDE, EARTHQUAKE   A measure of the amount of energy released by an earthquake.

MANTLE   The middle layer of the earth, between the crust and the core.

MARBLE   A metamorphic rock most commonly composed largely of calcite. Formed by the metamorphism of limestone. There is also dolomite marble.

MARE   A large, low, dark, smooth area on the moon. (Plural: *maria*)

MASCON   From "mass concentration." Areas on the moon with higher gravity, believed to be caused by concentrations of mass.

MASS NUMBER, MASS UNIT   (See *atomic mass number, atomic mass unit.*)

MATRIX   Fine-grained interstitial material.

MATURITY   The stage between youth and old age in the cycle of erosion.

MEANDERS   A series of more or less regular loop-like bends on a river.

MECHANICAL WEATHERING   The disintegration of a rock by physical processes.

MESOZOIC   An era in the geologic time scale. It lasted from about 225 to 65 million years ago.

METALLIC BONDING   The atoms are closely packed and the electrons move freely among the atoms.

METALLIC LUSTER   The luster characteristic of metals.

METAMORPHIC GRADE   A measure of the intensity of metamorphism.

METAMORPHIC ROCK   A rock that has undergone change in texture, mineralogy or composition as a result of conditions below the depth affected by weathering processes.

METAMORPHIC ZONE   Essentially equivalent to metamorphic grade. Zones are generally thought of as related to depth at which metamorphism occurs. Generally recognized by index minerals.

METAQUARTZITE   Quartzite.

METEOR   A shooting star. A meteorite burning due to friction in the atmosphere.

METEORITE   Matter that has fallen on the earth from outer space.

METHANE   Marsh gas. CH_4

MICA   A mineral family. Biotite and muscovite are common rock-forming micas.

MICROMETEORITE   A tiny meteorite that must be viewed under the microscope. (See *meteorite.*)

MICROTEKTITE   A glassy, usually spherical object, less than a millimeter in diameter, which may be related to tektites.

MINERAL   A naturally occurring, crystalline, inorganic substance with a definite small range in chemical compositon and physical properties.

MIOCENE   A subdivision of the Cenozoic Era of the geologic time scale. It lasted from about 26 to 7 million years ago.

MISSISSIPPIAN   A period in the geologic time scale. It lasted from about 345 to 310 million years ago.

MOHOROVIČIĆ DISCONTINUITY (M-DISCONTINUITY, MOHO)   The level at which a distinct change in seismic velocity occurs. It separates the crust from the mantle.

MOLYBDENITE   An ore of molybdenum. MoS_2

MONOCLINE   A flexure changing the level of any given bed. See Fig. 14-12.

MORAINE   Unsorted material deposited by a glacier. Also, the distinctive landforms that characterize these deposits.

MOUNTAIN   An area elevated above the surrounding area.

MOUNTAIN GLACIER   *Alpine glacier.* A glacier in a mountain valley.

MOUNTAIN ROOT   The low density rocks below a mountain range. The thickening of the crust below mountain ranges.

MUD CRACKS   The irregular polygonal cracks that form due to the shrinkage of mud when it dries.

MUDFLOW   The downslope movement of water-"lubricated" debris.

MUDSTONE   A fine-grained sedimentary rock, without fissility, that formed from the consolidation of mud.

MUSCOVITE White mica. Recognized from its color and perfect cleavage. An important rock-forming mineral. $KAl_2(AlSi_3O_{10})(OH)_2$

MYLONITE A fine-grained metamorphic rock formed by the milling of rocks on fault surfaces.

NAPPE A large plate of rocks moved from its place of origin by thrust faulting or recumbent folding.

NEBULA A cloud of gas or dust found in space.

NEUTRON A fundamental particle with no electrical charge found in the nuclei of atoms.

NON-CLASTIC SEDIMENTARY ROCK A chemically or biologically precipitated sedimentary rock.

NORMAL FAULT A fault caused by tension or uplift in which the hanging wall is depressed. See Fig. 14-16.

NUÉES ARDENTES A French term meaning "glowing clouds." A volcanic pyroclastic eruption of hot ash and gas.

OBSIDIAN Volcanic glass.

OCEANIC CRUST That part of the earth above the Moho and under the oceans. Composed largely of basaltic rocks.

OLD AGE A stage in the cycle of erosion near its end.

OLIGOCENE A subdivision of the geologic time scale. It lasted from about 36 to 26 million years ago.

OLIVINE A common rock-forming mineral. $(Mg,Fe)_2SiO_4$

OMPHACITE A dense, green pyroxene found in the rock eclogite.

OOZE A fine-grained sediment found on deep ocean floors composed of more than 30 per cent organic material.

OPAL An amorphous material with composition $SiO_2 \cdot nH_2O$. Loosely considered a mineral by most people.

ORDOVICIAN A period in the geologic time scale. It lasted from about 500 to 435 million years ago.

ORE Material that can be mined at a profit.

OROGENY The process of making the internal structures of mountains, especially folding, faulting, igneous and metamorphic processes.

ORTHOCLASE A common rock-forming mineral. Potassium feldspar. $KAlSi_3O_8$

OVERBURDEN The surface material overlying bedrock.

OVERTURNED FOLD A fold at least one limb of which has been rotated more than 90 degrees so that the beds are overturned.

OXBOW LAKE A crescent-shaped lake formed in an abandoned river curve.

OXIDATION The process of combining with oxygen.

OXIDE A compound containing oxygen.

PAHOEHOE LAVA Hawaiian term for lava with a smooth ropy surface.

PALEOCENE A subdivision of the geologic time scale. It lasted from about 65 to 53 million years ago.

PARABOLIC DUNE A sand dune with the general shape of a parabola. The cusps point in the opposite direction to the movement of the wind.

PEAT Dark brown material formed by the partial decomposition of plant material.

PEBBLE A fragment between 4 and 64 millimeters in size.

PEDALFER A type of soil found extensively in eastern United States. It is enriched in iron and aluminum.

PEDIMENT An erosion surface cut in bedrock and superficially resembling an alluvial fan.

PEDOCAL A type of soil found extensively in western United States. It is enriched in calcium.

PEGMATITE A very-coarse-grained igneous rock. The crystals are larger than 1 centimeter.

PENEPLAIN The smooth, rolling erosion surface that develops very late in the cycle of erosion.

PENNSYLVANIAN A period in the geologic time scale. It lasted from about 310 to 280 million years ago.

PERCOLATION Movement of water through rocks.

PERIDOTITE   A coarse-grained igneous rock composed mainly of olivine and pyroxene with little or no feldspar.

PERIOD, GEOLOGIC TIME   A subdivision of the geologic time scale.

PERIOD, OCEAN WAVE   The time it takes a wave to travel one wavelength.

PERMAFROST   Permanently frozen subsoil.

PERMEABILITY   The interconnection of pore space allowing fluids to move through a rock.

PERMIAN   A period in the geologic time scale. It lasted from about 280 to 225 million years ago.

PERTHITE   A type of feldspar formed by the intergrowth of potassium feldspar and sodic plagioclase.

PHANERITIC   A textural term meaning that the crystals or grains can be seen with the naked eye.

PHENOCRYST   One of the larger crystals in a porphyritic rock.

PHYLLITE   A rock intermediate in texture between slate and schist.

PILLOW LAVA   Lava extruded underwater or in wet mud develops a structure that resembles a pile of pillows.

PIRACY, STREAM   The diversion of the headwaters of a stream by the headward erosion of another stream. Also *stream capture*.

PITCHBLENDE   An ore of uranium.

PLACER DEPOSIT   Accumulation of valuable heavy, durable minerals. Made by agents of transportation and deposition such as running water, wind, or waves.

PLAGIOCLASE   One of the feldspar minerals. A continuous series from $CaAl_2Si_2O_8$ to $NaAlSi_3O_8$. Recognized by its hardness, two cleavages, and striations on one of the cleavages.

PLASTIC   Can change shape permanently without rupture.

PLATEAU   An elevated fairly flat surface.

PLEISTOCENE   A subdivision of the geologic time scale. The last glacial age.

PLEOCHROIC HALO   A colored zone or halo surrounding a tiny radioactive mineral grain. The color of the halo changes when the grain is rotated under a polarizing microscope. Also *radiation halo*.

PLIOCENE   A subdivision of the geologic time scale.

PLUCKING   Glacial erosion in which blocks of rock are torn out by the action of freezing water.

PLUNGING FOLD   A fold that is not horizontal. See Fig. 14-10.

PLUTON   Any body of igneous rock that formed below the surface.

POLAR WANDERING   Movement of the geographic poles (as opposed to continental drift).

POLYMORPH   Two or more minerals with the same composition but different crystal form, such as graphite and diamond.

POROSITY   The open space in a rock.

PORPHYRITIC TEXTURE   An igneous rock texture in which some of the crystals are much larger than the matrix.

PORPHYRY   A rock with a porphyritic texture.

POTHOLE   A hole worn into bedrock by moving water.

PRECAMBRIAN   The oldest subdivision of the geologic time scale. It includes all the time from the origin of the earth to about 570 million years ago.

PRECESSION OF THE EQUINOXES   A slow change from year to year of the position of the stars at the time of equinox.

PRECIPITATE   Separate from a solution.

PRECIPITATION   Rain or snowfall, *OR* being separated from a solution.

PROGRESSIVE METAMORPHISM   Metamorphism of increasing grade.

PROTON   A fundamental sub-atomic particle with a positive charge. Found in the nucleus of an atom.

PUMICE   Very light, cellular, glassy lava.

P WAVE (PUSH WAVE, PRIMARY WAVE)   One of the waves caused by earthquakes. The fastest traveling wave. A compression wave.

PYRITE   A common mineral. Fool's gold. FeS_2

PYROCLASTIC ROCK   A rock formed by volcanic ejecta.

PYROXENE   A mineral family. Augite is the common rock-forming member.

PYROXENITE   A rock composed essentially of pyroxene.

QUADRANGLE   Topographic map produced by the United States Geological Survey.

QUARTZ   A common rock-forming mineral. SiO_2 Recognized by its hardness, 7, and lack of cleavage.

QUARTZ DIORITE   A coarse-grained igneous rock. Composed largely of quartz, plagioclase, and dark minerals.

QUARTZ LATITE   A fine-grained igneous rock corresponding to granodiorite or quartz monzonite in composition.

QUARTZ MONZONITE   A coarse-grained igneous rock composed largely of quartz, potassium feldspar, and plagioclase.

QUARTZ (QUARTZOSE) SANDSTONE   A sandstone composed largely of quartz.

QUARTZITE   A metamorphic rock composed mainly of quartz. It forms from the metamorphism of sandstone or chert.

QUICK CLAY   A rock type that becomes fluid if vibrated or disturbed.

RADIAL STREAM PATTERN   A stream pattern like the spokes of a wheel.

RADIATION HALO   A colored zone or halo surrounding a tiny radioactive mineral grain. The color of the halo changes when the grain is rotated under a polarizing microscope. Also *pleochroic halo*.

RADIOACTIVITY   The property of some elements to change spontaneously into other elements by the emission of particles from the nucleus.

RAIN SHADOW   Area of diminished rainfall on the lee side of a mountain range.

REACTION SERIES   See *Bowen's reaction series*.

RECENT   Most recent subdivision of the geologic time scale. The *Holocene*.

RECESSIONAL MORAINE   A glacial moraine deposited by a retreating glacier, either during a temporary pause in the retreat, or during a minor readvance in a general retreat.

RED BEDS   Red-colored sedimentary rocks due to the presence of iron oxide.

RED CLAY   A deep-sea deposit formed by fine material, mainly volcanic dust. (The color is probably caused by oxidation.) Also *brown clay*.

REEF   A rock built of organic remains especially of corals.

REGIONAL METAMORPHIC ROCKS   Rocks such as schist and gneiss formed by metamorphism affecting a large region, rather than a restricted location.

REMANANT MAGNETISM   Permanent magnetism induced in a rock by a magnetic field.

RETROGRESSIVE METAMORPHISM   Metamorphism at conditions of lower grade than those under which the original rock formed.

REVERSED MAGNETISM   A magnetic field with polarity reversed from that of the present field.

REVERSE FAULT   A fault caused by compression. The hanging wall has been raised. See Fig. 14-15. See also *thrust fault*.

RHYODACITE   Quartz latite and dacite. A fine-grained igneous rock similar in composition to granodiorite.

RHYOLITE   A fine-grained igneous rock similar in composition to granite.

RIFT VALLEY   The valley formed along a rift fault or lateral fault.

RILLE   A valley-like feature found on the moon.

RIP CURRENT   A short-lived, rapid current caused by the return to the ocean of water carried landward by waves.

RIPPLE-MARK   An undulating surface on a sediment caused by water movement.

ROCHES MOUTONÉES   Rounded knobs of bedrock that have been overrun by a glacier. French for *sheep's backs* which they resemble in the distance.

ROCK   A natural aggregate of one or more minerals *OR* any essential and appreciable part of the solid portion of the earth (or any other part of the solar system).

ROCK CYCLE   The sequence through which rocks may pass when subjected to geological processes.

ROCK FLOUR   Very finely ground material from glacial abrasion.

ROCKSLIDE   Rapid downslope movement of bedrock.

ROOT OF MOUNTAIN   The low density rocks below a mountain range. The thickening of the crust below mountain ranges.

ROPY LAVA   Lava with a ropy surface. (See *pahoehoe lava*.)

ROSCOELITE   An ore of vanadium. $K(V,Al)_2(OH)_2AlSi_3O_{10}$

RUNOFF   The water that flows on the earth's surface.

RUTILE   An ore of titanium. TiO_2

SALT DOME   A structure caused by the upward movement of rock salt through overlying sedimentary rocks.

SAND   Fragments that range in size from 1/16 to 2 millimeters.

SAND BAR   An accumulation of sand built up by waves or running water.

SANDBLASTING   Erosion by impact of wind-blown sand.

SANDSTONE   A clastic sedimentary rock with fragments between 1/16 and 2 millimeters in size.

SANDSTORM   Strong winds carrying sand. Most of the sand moves close to ground level.

SANIDINE   A potassium feldspar found in volcanic rocks. $KAlSi_3O_8$

SCARP   A steep slope, cliff.

SCHEELITE   An ore of tungsten (wolfram). $CaWO_4$

SCHIST   A foliated metamorphic rock generally containing conspicuous mica.

SCORIA   Volcanic rock with coarse bubble holes.

SEAMOUNT   A submarine mountain. Generally basalt volcanoes.

SECONDARY WAVE (SHAKE WAVE, S WAVE)   One of the waves generated by an earthquake. This transverse wave is second to arrive.

SEDIMENTARY ROCK   Rocks formed near the earth's surface, generally in layers.

SEIF DUNE   An elongate sand dune parallel to the wind direction. A type of *longitudinal dune.*

SEISMIC SEA WAVE (TSUNAMI)   A commonly destructive wave generated by submarine earthquakes. Commonly mistermed *tidal wave.*

SEISMOGRAPH   A device that records earthquakes.

SERPENTINE   A mineral family. $Mg_3Si_2O_5(OH)_4$

SERPENTINITE   A rock composed largely of serpentine.

SHAKE WAVE, SHEAR WAVE, SECONDARY WAVE   (See *S wave.*)

SHALE   A fine-grained sedimentary rock composed largely of clay and silt and characterized by its fissility.

SHARPSTONE CONGLOMERATE   A conglomerate composed of angular pebbles. (See also *breccia.*)

SHEAR   To force two adjacent parts of a body to slide past one another.

SHELF, CONTINENTAL   (See *continental shelf.*)

SHIELD   A continental area that has been relatively stable for a long time. Shields are mainly composed of ancient Precambrian rocks.

SHIELD VOLCANO   A broad, gently sloping volcano built by flows of very fluid lava.

SILICATE MINERAL   A mineral containing SiO_4 tetrahedra.

SILICON-OXYGEN TETRAHEDRON   (SiO_4 tetrahedron) The basic unit of the silicate minerals. Composed of four oxygen atoms in the shape of a tetrahedron surrounding a silicon atom.

SILL   A concordant igneous intrusive body.

SILLIMANITE   A mineral found in metamorphic rocks. Al_2SiO_5

SILT   A clastic sediment ranging in size between 1/256 and 1/16 millimeter.

SILTSTONE   A clastic sedimentary rock composed largely of silt.

SILURIAN   A period in the geologic time scale that lasted from about 435 to 395 million years ago.

SINK   A depression with no outlet.

SLATE   A fine-grained metamorphic rock characterized by well-developed foliation.

SLOPE, CONTINENTAL   (See *continental slope.*)

SLUMP   A landslide that develops where strong, resistant rocks overlie weak rocks. A block slips down a curving plane.

SMALTITE   An ore of cobalt. $(Co,Ni)As_{2-3}$

SOAPSTONE   A metamorphic rock composed of talc.

SODIC PLAGIOCLASE   The sodium-rich end member of the plagioclase feldspar series. $NaAlSi_3O_8$

SOIL   The mixture of altered mineral and organic material at the earth's surface that supports plant life.

SOLIFLUCTION   Downslope movement of overburden saturated with water.

SPECIFIC GRAVITY   Ratio of the mass of a material to the mass of an equal volume of water.

SPHALERITE   An ore of zinc. $(Zn,Fe)S$

SPHEROIDAL WEATHERING   Formation of rounded boulders by expansion of surface layers. See Fig. 4-5.

SPIT   A sand bar connected on one end to the shore.

SPODUMENE   An ore of lithium. $LiAlSi_2O_6$

SPRING   A place where water issues naturally from the ground.

SPUR, TRUNCATED   A minor ridge that has been cut off by a moving glacier.

STALACTITE   A deposit formed on the roof of a cave.

STALAGMITE   A deposit formed on the floor of a cave.

STAUROLITE   A metamorphic mineral. $FeAl_4Si_2O_{10}(OH)_2$

STIBNITE   An antimony ore. Sb_2S_3

STOCK   An igneous intrusive body smaller than a batholith.

STOPING   The sinking of blocks of heavier pre-existing rock through molten magma.

STRAIN   Deformation caused by an applied force.

STRATIGRAPHIC CORRELATION   (See *correlation*.)

STREAK   The color of a powdered mineral.

STRESS   Force per unit area. Force that causes strain.

STRIATION, GLACIAL   Scratch on bedrock made by rocks carried by a glacier.

STRIATIONS, MINERAL   Parallel lines that result from intergrown crystals (twins). Commonly seen on one cleavage of plagioclase feldspar.

SUBDUCTION   The process by which a crustal plate descends below another plate.

SUBMERGENT COASTLINE   A coastline along which either the land has sunk or the ocean level has risen.

SUBSOIL   The middle layer or B horizon of a soil profile.

SUPERNOVA   A very luminous exploding star.

SURF   Nearshore wave activity.

SURFACE WAVE   One of the waves caused by an earthquake. They move slowly with large amplitude and cause much damage.

SUSPENDED LOAD   The material actually carried by wind or water, as opposed to bedload which is rolled along.

S WAVE (SECONDARY, SHAKE, or SHEAR WAVE)   One of the waves generated by an earthquake. This transverse wave is second to arrive.

SYENITE   A coarse-grained igneous rock with very little or no quartz.

SYNCLINE   A fold in which the beds slope inward toward a common axis. The youngest rocks are in the core.

SYSTEM, GEOLOGIC   The rocks deposited during a geologic period.

TACHYLITE   Volcanic glass of basaltic composition.

TALC   A metamorphic mineral. Number 1 on the hardness scale. $Mg_3Si_4O_{10}(OH)_2$

TALUS   The slope at the foot of a cliff composed of fragments weathered from the cliff.

TARN   A small lake in a glacially gouged basin in a cirque. Also, any landlocked small lake.

TEKTITE   A piece of glass commonly of unusual shape. Such pieces are believed to result from either meteorite impact or to have come from space.

TELLURIDE   A mineral containing tellurium.

TERMINAL MORAINE   The moraine formed by a glacier at its farthest advance down-valley.

TERRACE   A flat-topped surface.

TERTIARY   A period in the geologic time scale.

TETRAHEDRON   A solid body with four faces, each of which is an equilateral triangle.

TEXTURE   The size and arrangement of grains in a rock.

THERMAL METAMORPHIC ROCKS   Rocks whose metamorphism was controlled primarily by temperature near the boundary or contact of an intrusive body such as a batholith. Also *contact metamorphic rocks*.

THRUST FAULT   A reverse fault with a fairly flat fault plane.

TIDAL CURRENT   A current caused by tidal movement of water.

TIDAL WAVE   (See *seismic sea wave*.)

TIDE   The periodic rise and fall of the ocean, caused by the gravitational attraction of the moon and to a lesser extent the sun.

TILL   Material deposited directly by glacial ice.

TILLITE   A sedimentary rock made of till (or a till-like sediment).

TONALITE   Quartz diorite. A coarse-grained igneous rock composed of quartz, plagioclase feldspar, and dark minerals.

TOPAZ   A mineral. Number 8 on the hardness scale. $Al_2(SiO_4)(F,OH)_2$

TOPSOIL   The upper or A horizon of a soil profile.

TOURMALINE   A mineral group.

TRACHYTE   A fine-grained igneous rock with composition similar to that of syenite.

TRANSFORM FAULT   A fault associated with mid-ocean ridges in which the actual motion is opposite to the apparent offset of the mid-ocean ridge. See Fig. 16-22.

TRANSVERSE SAND DUNE   An elongate dune with its long axis at a right angle to the wind direction.

TRENCH   A very deep, elongate area of the ocean, associated with volcanic island arcs.

TRIASSIC   A period of the geologic time scale. It lasted from about 225 to 190 million years ago.

TRITIUM   An isotope of hydrogen with two neutrons and a proton in its nucleus.

TROILITE   A mineral. FeS

TRUNCATED SPUR   A minor ridge that has been cut off by a moving glacier.

TSUNAMI   (See *seismic sea wave.*)

TUFF   A pyroclastic volcanic rock with fragments smaller than 4 millimeters.

TURBIDITY CURRENT   A current or flow caused by the movement downslope of water heavy because of its suspended material (muddy water).

TURBULENCE   A type of flow in which the fluid moves in a disorderly, irregular way.

TYPE AREA   The place at which a formation or system is well exposed and is defined.

UNCONFORMITY   A surface of erosion and/or non-deposition. A gap in the geologic record.

UNDERTOW   The bottom, nearshore water returning to the ocean after waves moved the water onshore.

UNIFORMITARIANISM   The concept that ancient rocks can be understood in terms of the processes presently occurring on the earth.

URANINITE   An ore of uranium.

VAN ALLEN RADIATION BELT   A more or less doughnut-shaped zone of radiation in space high above the equator.

VARVED CLAY   Fine, thinly bedded glacial sediments deposited in still water. Each bed is assumed to be a yearly deposit.

VENTIFACT   A pebble faceted by sandblasting by wind-blown sand. See *dreikanter.*

VERTEBRATE   An animal with a backbone.

VISCOSITY   The resistance to flowage of a liquid. Caused by internal friction.

VITREOUS LUSTER   Glassy luster. The shiny luster of broken glass.

VOLATILE   A material easily vaporized.

VOLCANIC SEDIMENTARY ROCKS (VOLCANIC SANDSTONE and VOLCANIC CONGLOMERATE)   Sandstone and conglomerate composed of volcanic fragments.

WATER TABLE   The surface below which the pores are saturated with water.

WEATHERING   The alteration that occurs to a rock at the earth's surface.

WEIGHT   The mutual attractive force between a mass and the earth.

WELDED TUFF   A tuff that was erupted when hot and whose fragments have fused together as a result of the action of heat and the gases present.

WOLFRAMITE   An ore of tungsten (wolfram). $(Fe,Mn)WO_4$

YOUTH   The first stage in the cycle of erosion.

ZIRCON   A mineral. $ZrSiO_4$

Index